Domesticating Drones

The public debate over civilian use of drones is intensifying. Variously called "unmanned aircraft systems," "unmanned aerial vehicles," "remotely piloted aircraft," or simply "drones," they are available for purchase by anyone for a few hundred to a few thousand dollars. They have strikingly useful capabilities. They can carry high-definition video cameras, infrared imaging equipment, sensors for aerial surveying and mapping. They can stream their video in real time. They have GPS, inertial guidance, magnetic compasses, altimeters, and sonic ground sensors that permit them to fly a preprogrammed flightplan, take off and land autonomously, hover and orbit autonomously with the flick of a switch on the DRone OPerator's (DROP) console. The benefits they can confer on law enforcement, journalism, land-use planning, real estate sales, critical infrastructure protection and environmental preservation activities are obvious.

However, their proliferation in response to these demands will present substantial risks to aviation safety. How to ensure the safety of drone operations perplexes aviation regulators around the world. They are inexpensive consumer products, unsuited for traditional requirements for manned aircraft costing hundreds of thousands or millions of dollars and flown only by licensed pilots who have dedicated significant parts of their lives and their wealth to obtaining licenses. Regulatory agencies in Europe and Asia are ahead of US regulators in creating spaces for commercial use.

Over the next several years, legal requirements must be crystallized, existing operators of helicopters and airplanes must refine their policy positions and their business plans to take the new technologies into account, and all businesses from the smallest entrepreneur to large conglomerates must decide whether and how to use them. *Domesticating Drones* offers rigorous engineering, economic, legal, and policy theory and doctrine on this important and far-reaching development within aviation.

Henry H. Perritt, Jr. is a Professor of Law and former Dean, Chicago-Kent College of Law, the law school of Illinois Institute of Technology, which he served as Vice President and Chief Academic Officer for the Downtown Campus from 1997 to 2002.

Eliot O. Sprague is a Professional ENG helicopter pilot, Director of Market Development and Commercial Pilot, AM Air Service, and helicopter flight instructor, Midwestern Helicopter.

Domesticating Drones

The technology, law, and economics of unmanned aircraft

Henry H. Perritt, Jr. and
Eliot O. Sprague

LONDON AND NEW YORK

First published 2017
by Routledge
2 Park Square, Milton Park, Abingdon, Oxon OX14 4RN

and by Routledge
711 Third Avenue, New York, NY 10017

Routledge is an imprint of the Taylor & Francis Group, an informa business

British Library Cataloguing in Publication Data
A catalogue record for this book is available from the British Library

Library of Congress Cataloguing in Publication Data
Names: Perritt, Henry H., author. | Sprague, Eliot O., author.
Title: Domesticating drones : the technology, law and economics of unmanned aircraft / Henry H. Perritt Jr. and Eliot O. Sprague.
Description: Abingdon, Oxon; New York, NY : Routledge, 2016.
Identifiers: LCCN 2016005695 | ISBN 9781472458629 (hardback) | ISBN 9781315577999 (ebook)
Subjects: LCSH: Drone aircraft. | Aeronautics, Commercial. | Aeronautics, Commercial–Law and legislation.
Classification: LCC TL718.P47 2016 | DDC 338.4/7629133–dc23
LC record available at http://lccn.loc.gov/2016005695

ISBN: 978-1-472-45862-9 (hbk)
ISBN: 978-1-315-57799-9 (ebk)

Typeset in Times New Roman
by Out of House Publishing

Contents

Figures

Tables

Preface

This book began with a conversation we had driving back to the Chicago area from a helicopter flying lesson in Wisconsin. A year earlier, Eliot had taught Hank, an airplane pilot for more than forty years, how to fly helicopters. Hank had earned his private helicopter rating and now, under Eliot's continued tutelage, was working on his commercial rating.

Both of us had seen the 60 Minutes interview with Jeff Bezos, in which Bezos unveiled a secret plan to use small drones to deliver packages to customers. DJI had begun advertising the thousand-dollar Phantom microdrone. "What do you think about drones?" Hank asked. "I think the technology is awesome," Eliot said," but it's going to be a long time before they deliver packages."

The conversation then shifted to the Phantom. Ever since we had known each other, we had regularly speculated and argued about various aspects of aviation technology, new and old, and had begun to get immersed in the political battle over two competing proposals for commercial heliports in downtown Chicago.

For the remainder of the trip, we wondered about the details of the Phantom technology and argued about how it represented advantages or disadvantages over manned helicopter designs. By the time we arrived at our destination, we said, "Let's buy one!"

We did, and flew it for the first time on a subzero night in a nearby athletic field covered with about a foot of snow. We got cold before the Phantom's battery ran down, but we were hooked.

During the next several months we flew the Phantom to test the limits of its capabilities, argued some more about its potential to compete with helicopters, and began to write what became more than a dozen magazine and law review articles about drone technology, politics, regulation, and economics. We talked to lots of aviation technology enthusiasts, pilots, mechanics, and other members of the traditional aviation community, and members of the general public.

As our discourse broadened, we realized that few people saw the whole picture. Interest was growing, but helicopter and airplane pilots saw drones mainly as irrelevant toys, or as a threat to safety or their jobs. Model aircraft hobbyists getting the drone bug knew how to wire electric motor controls and how to set up radio control systems, but they didn't know much about the Federal Aviation Regulations. Lawyers were mostly just beginning to think about what might be a suitable regulatory regime, as the FAA struggled and failed to meet a statutory deadline for releasing a set of rules for drones. The general public mostly talked about drones delivering their Amazon packages.

Our articles addressed the issues that were beginning to crystallize in our minds. The first few talked about the competition between drones and helicopters. The longer ones offered a broader foundation to help people develop their positions and plans. Before long we had submitted a formal proposal to the FAA for a rule that would put the already-available automated safety features of the Phantom and its competitors at the center of drone regulation.

Our inventory of new article ideas grew. By the fall of 2014 we said, "Maybe we should write a book." We were well aware that good books come out of tough-minded answers to two questions. First, who is the audience? How will the book help them do their jobs, entertain them, or broaden their horizons? Second, why are we qualified to write it?

We were convinced by then that drones would be far more than a passing fancy for the latest high-tech consumer toy. The interest from players in a variety of industries was serious. Engineers were excited about the demonstrated capacity of miniaturized radio transmitters and receivers, integrated-circuit navigation boards, and powerful small electronic propulsion systems, but they had thought little about safety regulation. Lawyers were beginning to wrap themselves around regulatory issues and participate in litigation over the FAA's efforts to enforce a complete ban on commercial use of small drones while stepping aside when hobbyists flew exactly the same vehicles for fun. Lawyers, however, struggled to understand the technology.

Our more entrepreneurial friends were intrigued about the money-making possibilities in specific applications—newsgathering, event photography, real estate marketing, motion picture and television production, and law enforcement. Most of them were clueless about both the technology and the regulation.

Politics was developing beyond its infancy as the more powerful, experienced, and well-funded aviation lobbies began to take pro and anti-drone positions. Their early pronouncements relied more on myth than reality.

Even as their involvement grew, most participants saw only part of the picture—whatever their own specialties and professional perspectives revealed to them. We could help fill these gaps.

As we talked about the book idea, we began to realize that our partnership offered unique advantages. Hank, in addition to being a lawyer, was trained as an aeronautical engineer at MIT, and is an Extra Class radio amateur. Eliot flies news helicopters every day for a living, and understands the practical realities of aviation support as well as aerial videography.

Eliot, having grown up in a rural farming community and worked on farms, understood agriculture. Hank has written and produced short movies and thus understands the practical basics of moviemaking. Eliot started a successful business when he was in high school and served as the base manager for a helicopter operator trying to develop new business in a medium-sized city. Hank had worked developing new civilian airplane concepts at Lockheed Aircraft Corporation and selling them to potential customers. Both of us had investigated and written about organizational frameworks to improve the accessibility of helicopter support for law enforcement. Both of us know a good bit about small computers and radio and television technology.

We know not only about the product side, including technology development and successful marketing, but also about the labor market. All forms of aircraft are automated to a considerable degree, but human beings will always be involved. Hank's early work as a labor lawyer and author of many books and articles on labor and employment law and his work with the Labor Department make him attentive to how labor markets function in general and in specific industries. Eliot is an active participant in the aviation labor market, and as a helicopter flight instructor, has guided many young people in charting their aviation careers.

Hank's work on the White House Staff and as Deputy Under Secretary of Labor gives him a sophisticated understanding of how policy is made in the federal government. Eliot's experience in several different kinds of traditional aviation organizations gives him a practical view on how actual business leaders will embrace or avoid the drone phenomenon.

We have flown small drones, often testing their potential to perform specific tasks important in various industrial applications. We know what they can do and what they cannot do. Consulting for a variety of small businesses enthusiastic about earning money from commercial drone operations and creation of our own small business have confronted us with the realities of establishing viable startup enterprises.

Hank's legal work and Eliot's assistance for a dozen clients who applied for specific FAA authorization to fly their drones commercially provide practical lawyer perspectives on legal compliance. Our writing and the public exposure our petition for FAA rulemaking engendered have made us active participants in the political arena. We were actively involved in crafting a new drone ordinance for the city of Chicago.

We already knew from working on the articles and setting up a small company that we would argue a lot. Hank instinctively takes the position that the new technologies have exciting potential, and that regulators must cast aside long-standing approaches to aviation regulation that make no sense when applied to small drones. Eliot, though equally excited about the technology, instinctively takes the position that helicopters and good helicopter pilots will always be able to provide aerial support better than little drones, and that most traditional aviation rules are there for a reason. We know from experience that our arguments produce better insight than either of us could come up with on our own.

We have discovered that our substantial age difference is an advantage. Hank's decades of experience illuminate how markets and legal institutions are likely to accommodate the excitement about drones. Eliot's early achievements in the early stages of a decades-long future in aviation tees up questions about the future in a more concrete and personal way than traditional analytical projections that are a mainstay of Hank's academic work.

We agreed at the outset on two philosophical principles that would guide the book. First, it would not simply present abstract theoretical ideas; it would, in all cases, give concrete examples. When these were not yet available, it would work out concrete and possible scenarios. This principle means that we examine every specific industry in which drone use seems promising rather than lumping everything together and ignoring important differences in economics and culture.

The second principal mirrors the first. We would consistently build a theoretical foundation for everything we talk about, from the basic constitutional and statutory authority for aviation regulations to the precepts of aerodynamics and radio necessary to understand how drones fly and why they are designed like they are. We do not simply pass on marketing hype from vendors.

This book, as a result, is aimed at people who are interested in civilian drones—the more politically correct name for which varies among small unmanned aircraft systems (sUAS), unmanned air vehicles (UAV), remotely piloted vehicles (RPV) and other terminology.

We begin, in chapter 1, by portraying drones already at work, in a variety of industries. This chapter provides detailed mission profiles, showing how drones can enhance economic activity in news gathering, law enforcement and other public safety endeavors, utility infrastructure patrols, agricultural oversight, surveying, real estate marketing, and eventually, logistics such as package delivery. In each area of application, it explains what microdrones are good at and which tasks may require macrodrones.

Chapters 2 and 3 probe the key technologies in depth, in explaining how they enable or inhibit effective and safe drone flight, identifying further advancements needed and assessing how likely they are. They provide considerable detail on underlying theories of aerodynamics, photographic imaging, electronics, and radio. They reflect the authors' conviction that one can make better decisions when one understands the technology, and that one can understand the practical behavior of complex machines better if one understands the foundational technological principles.

Chapter 4 explores human factors. It shifts focus from the workshop to the cockpit and explores the unique capabilities that human pilots and operators bring to safe and effective mission performance. Training appropriate for manned aircraft pilots is not suited at all for microdrone DROPs, and it must be rethought and reworked for macrodrone DROPs. Assessment of human factors requires specifying the tasks that must be performed by DROPs, the safety issues they must deal with, recruiting large numbers of them in light of desirable skills and personality traits, training, and supervision.

Chapter 5 considers drone emergencies and the interaction of the human operator and vehicle automation.

Then there is politics, which shapes the law. Chapter 6 explores the political dynamics of wider drone use, which will ultimately determine the content of drone regulation.

Chapter 7 lays a foundation for understanding drone regulation by explaining the structure of the modern administrative state, and briefly summarizes aviation regulation as it applies to drones in the nation states most interested in the technology. Chapter 8 provides much greater detail on the regulatory framework in the United States.

Chapter 9 returns to technology and explains the role that high levels of automation play in the interface between man and machine. It argues that microdrones should be regulated like consumer products by requiring built-in safety features, rather than by trying to enforce detailed pilot licensing, aircraft design, and operating rules.

Economics, the subject of chapter 10, will ultimately determine how drones are used. How effectively will they be able to compete with existing ground-based resources and with other aerial platforms such as airplanes and helicopters? How will the cost structures and business models of drone operators differ from those of airplane and helicopter operators? Will development costs for macrodrones swamp any operating-cost advantages for the foreseeable future? This chapter uses basic microeconomics to evaluate the determinants of profitability for a drone operator and the considerations that influence their potential customers.

Economics must be put into practice. To be successful, a drone operator must assess the market accurately, persuade potential customers to consider what his enterprise offers, acquire and equip a drone fleet and appropriate ground facilities, hire the right people, and manage them effectively. Chapter 11 integrates and applies the ideas developed in the preceding chapters to identify considerations and decisions involved in starting a drone business.

Only one thing is certain: the future will be different from the present. No one can predict the details, but anyone with a stake in the future of drone flight should be able to perceive the major trends, identify the most important open issues, and have a strategy for dealing with them. Chapter 12 undertakes those tasks.

All of the chapters recognize that the audience brings with it diverse backgrounds and knowledge. It deals with this reality by starting with basic principles and concepts and then moving more deeply in the subject matter, so that its content is accessible and useful, regardless of whether one is a novice or an expert.

We have been willing to take risks. We have always tried to be concrete and to take a position, rather than to offer essentially unsatisfying waffles that something might go this way or it might go that way; no one can know. We have been determined to offer our own ideas; not simply to describe others'.

When we think it would simplify discussion, we have coined our own terms and acronyms.

We recognize that even among the most expert and experienced professionals in any industry there can be—and often are—vehement disagreements about best practices and future plans. We expect that industry experts are unlikely to agree with everything we have said about their industries. We respect differences of opinion but have been clear about our own, in order to get the professionals thinking.

Acknowledgments

Albert J. Plawinski, a law student at Chicago-Kent College of Law, read every chapter in draft and made material contributions on video technology and state regulation.

Scott Vanderlin, Research Librarian in the law library at Chicago-Kent College of Law, did a great job in updating foreign drone law sources in chapter 7. Thanks, Scott!

Aran Quinn, a law student at Chicago-Kent College of Law and certified public accountant, reviewed chapter 11 and helped improve its explanation of how to start a new drone business.

Steve Idler, audio and video engineer, reviewed chapter 3 and helped us improve the analysis and description of radio control and video processing technologies.

Henry Strickler did a great job as a freelance editor in helping us identify opportunities for improvements in organization and style.

The authors appreciate the dedicated work by Patricia O'Neal in checking over the manuscript and assembling the final package to the publisher.

The authors thank them for their invaluable help. They are not responsible, of course, for the views expressed; they are our own.

Eliot says: I am truly lucky to be surrounded by family and friends who always shed a positive light. I would like to thank my parents for their support and encouragement on whatever crazy life decisions I've made. A special thanks to Uncle Kevin: you subconsciously sparked my interest in aviation so many years ago and continue to provide career advice, and I'm sure glad we have the opportunity to share our flying adventures. To my first chief pilot, Chris Laskey, thank you for never doubting me as a young flight instructor and providing opportunity for an aspiring pilot. Last but not least I would like to thank Hank; I certainly wouldn't have tackled co-authoring a book without your perseverance.

Hank says: Being an American has provided me with a phenomenal opportunity to do a rich variety of things of my own choosing, motivated always by the conviction that each of us has the capacity to make the world a better place. I thank Chris Laskey for getting me hooked on helicopters. Above all, in the context of this book, I thank Eliot for our friendship and our hundreds of amusing and enlightening explorations and arguments as we refined our thinking about drones—and so many other things.

As soon as the final FAA regulations are published, the authors will provide the text of the regulation and an analysis on www.domesticatingdrones.com, with no additional charge.

1 Drones at work and play

1.1 Introduction

UAVs, UASs, RPVs—unmanned air vehicles, unmanned aircraft systems, remotely piloted vehicles—are invading the skies. Everyone calls them *drones*, ignoring the best efforts of political-correctness enforcers to call them something else. Despite the intensity of political debate over them, often focused on trivialities, they are the wave of the future in global aviation. They will not displace manned airplanes and helicopters, because they cannot do the same jobs as well. But they will expand aviation's capabilities to support human endeavor.

Commercial drone use was already widespread in some parts of the world before the United States Federal Aviation Administration (FAA) began to relax its ban on commercial flights. In Japan, for example, a medium-size drone, the Yamaha RMAX, is a routine tool for applying chemicals to agricultural crops.[1]

In the United States, the FAA responded to a statutory mandate to integrate drones into the National Airspace System by releasing a notice of proposed rulemaking (NPRM) in February 2015, allowing commercial activities by microdrones and proposing new rules designed around their capabilities and the risks they present. To relieve pressure for immediate action while the proposed rule was being finalized, the agency granted more than five thousand section 333 exemptions in response to applications by particular operators. Section 333 exemptions—so called because of the statutory section authorizing them—are the regulatory mechanism the FAA uses to permit commercial drone operations pending final adoption of a general rule.

Drones, particularly rotary-wing, multi-copter microdrones, are useful in many applications now. They excel at low-level aerial photography and have demonstrated their ability to collect high-quality aerial news imagery for TV stations and movie production. They operate well in hazardous environments that would present unacceptable risk to aircrews of helicopters and airplanes. Even at the low end of the price range, drones can be programmed to follow search or survey grids precisely. They are regularly used to augment real estate marketing by capturing overhead views of the property and to monitor agricultural crops. Amazon and Google are investing substantial capital in proving drones' utility for delivering small packages.

The barriers to their wider use are almost entirely political and regulatory, not technological. The FAA cannot quite figure out how to regulate them.

Macrodrones, predominantly fixed wing, with a few helicopter configurations in the early prototype phases, are being introduced into commercial markets more slowly than microdrones, primarily because their size and flight profiles create risks similar to those of airplanes and helicopters, warranting a more rigorous regulatory regime. Their acceptance depends more on their cost competitiveness with helicopters and airplanes than on their ability to perform completely new types of missions.

This chapter explores uses of drones. It distinguishes between microdrones, weighing less than 55 pounds, from macrodrones, those weighing more than 55 pounds. Microdrones have flooded the market already. They are being used for activities where manned helicopters are for the most part infeasible because of cost or risk to aircrews, providing aerial support to economic activities that have not enjoyed it before. In a sense, their missions are being invented, not from past practice with aircraft, but crafted around the capabilities of the new technology—relatively short flights—substantially less than one hour—proximate to the operator—with light payloads, usually a camera. This chapter evaluates the types of missions and the kind of support being provided by hundreds of microdrone flights occurring now.

It also explores possible missions for macrodrones, although this analysis necessarily is more speculative, because few commercial macrodrones are on the market, and the FAA faces much larger challenges in crafting rules for them; their greater weights and more expansive flight envelopes present greater risks.

The chapter begins by reviewing the pattern of section 333 exemptions granted by the FAA. Then it works its way through the types of commercial activities identified by the pattern of the exemptions. After addressing specific sector activities, it explains why the subjects addressed in other chapters of the book enable—and constrain—expanded drone applications. It refers to specific regulatory limitations explained more fully in chapter 8.

1.2 Section 333 exemptions: patterns of usage

While it was finalizing a general rule for commercial microdrone operation, the FAA granted more than five thousand exemptions from its rules for manned aircraft, allowing commercial microdrone operations proposed by specific individuals and enterprises. Figure 1.1 shows the number of exemptions granted by type of application.

The chart shows that a few types of applications are most common: survey, real-estate marketing, motion picture and television production, mapping, and agricultural support. Many exemption petitioners are keeping their options open by filing for generic aerial imagery or inspection. Others

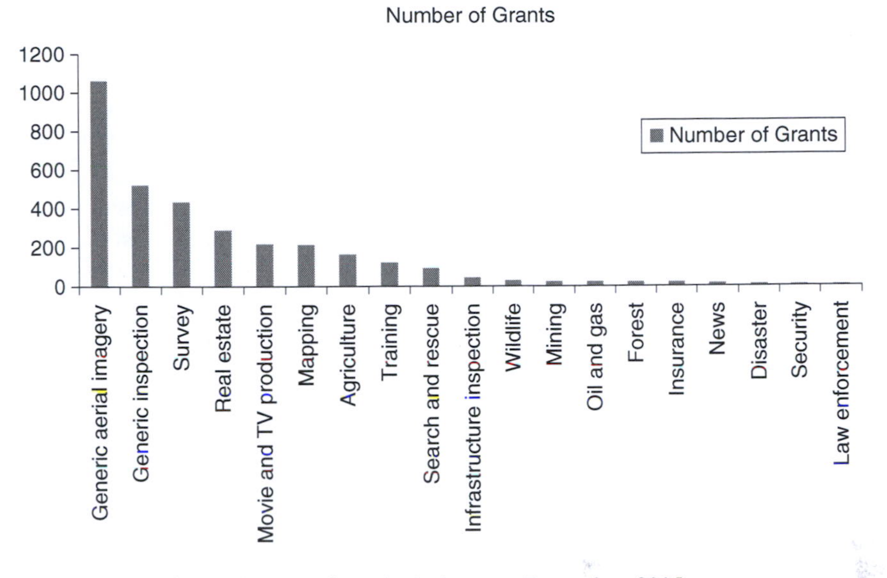

Figure 1.1 Section 333 exemptions, by industry—December, 2015

proposed more narrowly defined missions. In many cases, potential users are taking a wait-and-see attitude, hoping to learn from the experiences of the early adopters. Moreover, many of those with exemptions are making only limited use of the new technology, expecting to refine their plans based on experience.

The content of the final rules will remain uncertain for another couple of years, the criteria for granting section 333 exemptions undoubtedly will evolve, and the technology is changing rapidly. The universe of what is feasible now will expand considerably. Still, basic patterns of adoption revealed by the early section 333 exemptions are unlikely to change much for several years because of the relationship between microdrone capabilities and user needs.

Over the longer term, however, designs for airspace management systems allowing intensive use of drones for a broader range of applications—such as delivery of packages—will prove themselves and be deployed, and basic commercial macrodrone configurations will crystallize and be priced. Then, the patterns of activity may shift dramatically.

1.3 Aerial surveying

Surveying is a broad term. Agricultural surveying is considered in § 1.6. This section considers surveying of construction sites.

Before construction begins, a surveyor must prepare a contour map, which shows the ground level at various points on the site, and then superimpose a grid. She then calculates the amount of material that must be removed (for a *cut*) or added (for a *fill*) to produce the desired surface levels.

Traditionally, surveyors used instruments like theodolites (optical tools for measuring angles), levels, rods, and tapes to prepare the contour map and then calculated the cubic yards of fill or cut from the resulting contour map. Now, airborne sensors, usually employing lasers, precisely measure the distance between an airborne platform and the surface at finely defined GPS points on the ground. Such laser sensing systems are known as LIDAR. Mapping software takes the resulting data and automatically computes what must be done in each of several arbitrarily defined grid squares. The same sensor hardware and mapping software is used, regardless of the type of aerial platform that collects the data—airplane, helicopter, or drone.[2] Aerial surveying relies on the ability to fly a precise pattern over the area to be surveyed, so that elevations can be matched with GPS coordinates.

Microdrones provide two advantages over manned aircraft: lower acquisition and operating costs and autonomous navigation systems that fly precise patterns to collect the data.

More sophisticated microdrones can be equipped with the necessary sensors and precise navigation subsystems. Their principal limitation is their short endurance. Fifteen to twenty minutes can be adequate to survey a small field or construction site, with reasonable resolution. Larger scale surveying, however, is inefficient if the drone must land every fifteen minutes to get a battery swap.

Macrodrones powered by internal combustion engines have much longer endurance and thus greater utility. Their ability to fly at higher altitudes may also be advantageous to survey extensive areas. Macrodrones' greater payloads, however, are of no particular benefit.

It is no surprise that aerial surveying uses have taken a prominent position in the inventory of section 333 exemption applications and grants, because these operations typically occur over specific construction sites, which are under control by customers.

As operators fly their drones under these exemptions, the resulting experiential data will help them refine their strategies for the most cost-effective drone deployment.

1.4 Real estate marketing

As soon as microdrones with high-quality cameras were on the market, real estate agents began to use them to capture new photographic perspectives to help sell properties—never mind whether it was legal. As the section 333 process developed, the FAA had little difficulty in approving these operations, because they occur over property controlled by the customer. Of course some realtors also want bird's-eye views of the neighborhood.

High-quality photographs have long been part of marketing packages for real estate, in print flyers and on webpages. Only the largest and highest priced sites, however, have warranted the cost of aerial photography by helicopter. The aesthetic values of any promotional photography must be high: crisp images, good color, and compelling points of view that emphasize the most attractive features of the property and show off its neighborhood environment.

The necessary flight profiles fit comfortably within the performance capabilities of microdrones like the Phantom, and the Inspire. A skilled DROP and photographer can get all the imagery needed in a 15- or 20-minute flight, and cameras at the GoPro level have adequate resolution and sensitivity. No particular technological development is necessary to make this a cost-effective marketing tool.

There is no particular need for automated flight plan capability, because the best results are obtained when the DROP and the photographer watch the downlinked image as it is being captured and make adjustments in drone position and camera angle to get exactly the shots they want.

Nor are the regulatory challenges particularly great. As with agricultural uses, the drones fly over property controlled by the operator or the person contracting for the service. There is also no reason to fly over people on the ground while the imagery is being captured, although some promotional strategies might want to feature residents or models around an outdoor pool or volleyball net. In those cases, it is relatively easy to obtain advance approval by whoever is going to be in the frame.

The relatively short duration of the necessary flights, and modest size and weight of camera payloads suggest that microdrones can do whatever is necessary. There is no particular need for macrodrones, unless the business model of the photographer requires taking multiple shots at widely separated locations on the same day. Then the short transit times of macrodrones might be an advantage over drones that must be transported on the ground from one location to another. When this is the case, vertical takeoff and landing (VTOL) configurations, such as tilt-wing or tilt-rotor designs, would be advantageous, because they can fly like an airplane to transit from one location to another, while retaining the capability of slow flight, low-level operations, and hovering when they are on scene.

1.5 Movie and television production

Motion picture and television production on closed sets was the first activity to receive FAA approval via section 333 exemptions. Risk management and regulation were straightforward because the activity takes place on closed sets where the customer exercises detailed control. Risk reduction benefits were obvious because of the dangers of typical helicopter operations involving low level flight and aggressive maneuvers.

Moviemaking, including television program production, involves some of the most demanding types of aerial photography. Compelling story telling is at the center of an acceptable product, and good video storytelling requires a variety of camera angles and a mixture of close shots and long shots. Creative lens manipulation is also necessary to enable objects or people in the foreground to be in sharp focus while the background is blurred, or vice versa. Lighting intensity and direction are important; they must be consistent from shot to shot.

Ultimately, the movie is assembled in an editing process separate from, and later in time than, principal photography. The editor needs shots with multiple camera angles and framing to afford creative choices in the subsequent editing process. When a director discovers during the editing process that he doesn't have what he needs, one or more scenes must be reshot, and costs soar.

Locations for movie or television shoots involve property controlled by the producer (as on a studio set) or it may involve a public setting for which local government permits have been obtained and security established. Long established practices for principal photography assure the safety of everyone on the location. The incremental risk associated with drone photography is minimal. That is why the FAA was comfortable granting the first set of section 333 exemptions for closed-set movie production.

Cinematographers already use multiple tools so cameras can capture the imagery the director wants. Some cameras are operated from fixed tripods for shots not involving much subject movement. Others are installed on dollies or tracks to permit the camera to move along with the action being photographed. Cherry pickers carry cameras for overhead shots. Cameras mounted on helicopters get overhead imagery of complex action like car chases and other stunts. Drones represent an additional tool.

When the budget permits, almost all principal photography involves more than one camera running at the same time to get different perspectives of the same scene. Otherwise, multiple takes are necessary, moving the camera around to get the necessary range of choices for the editing process.

Helicopter cinematography places great demands on both the pilot and the cinematographer. The cinematographer must position the camera to capture exactly what he and the director envision; while a pilot flies the helicopter to facilitate what the cinematographer wants to do. Often, effective helicopter shooting requires a helicopter to be operated at low levels and to make abrupt maneuvers, in close proximity to actors and production personnel.

Published interviews with production personnel reveal considerable enthusiasm for microdrones because of their flexibility, reduced risk, and no compromise in quality.[3] They can fly low and maneuver nimbly. If they crash, they do not jeopardize an onboard crew, and the danger posed to actors and production personnel is lower because of their lighter weights and the absence of flammable fuel.

Existing microdrone products have the requisite performance capabilities for most movie shoots. Often, low altitude is an advantage rather than a disadvantage; there is no need for speeds in excess of 25 or 30 knots—except for chase scenes—and it would be an unusual scene in which both director and cinematographer do not want to keep their camera in visual sight. So line-of-sight (LOS) operation is acceptable for most of what they want to do. For other scenes, however, such as some high-altitude long shots, and certainly for chase scenes, helicopters can do things that microdrones cannot.

Similarly, limited endurance, on the order of 15 or 20 minutes, is not a significant disadvantage for most shots. Typically, a director wants a series of relatively short clips of each part of the scene. It is hard to imagine a scene that would require continuous operation of a drone-mounted camera for longer than 30 minutes. Multiple takes of the same scene segments are the norm, as the director and cinematographer work to obtain just the right nonverbal interaction between the actors and just the right expression of dialogue. Using conventional techniques, a director lets the actors perform a part of a scene for 3 to 5 minutes, then stops video and audio recording, and does another take of the same scene or repositions the actors for a different part of the scene. In the editing process, it is better to have separate short clips for different takes and different scenes, rather than having to cut up longer video sequences.

Drones are useful tools for moviemaking when they carry high-quality cameras with gimbals that eliminate vibration and enable precise camera control. Most of the necessary technology now exists, as evidenced by the increasing use of microdrones to make high-quality, big-budget movies and television programs. Movie makers use small drones carrying inexpensive GoPro-level cameras with modest-quality lenses and sensors to feed monitors for directors and cinematographers. They use larger drones carrying various models of more sophisticated RED or Black Magic cameras to capture imagery for incorporation into the actual movie.

Helpful technology improvements relate to camera capability and more sophisticated gimbals, especially those that can maintain the framing as the microdrone changes orientation.

When multiple airborne cameras—therefore multiple drones—are used, DROPs must avoid collisions. This would be extremely difficult to program in advance; the best way to assure drone separation is by attention and coordination between skilled DROPs.

Accommodating regulation is made easier by the fact that movie budgets are large, and can easily cover the relatively modest expense for top-quality drone capability. Similarly, substantial personnel budgets make it relatively easy-to-satisfy licensed pilot and separate observer requirements. Regardless of the size of the budget, rational economic decisionmaking compares the cost of introducing a new camera angle via drone versus doing it from the ground. For example, a producer will compare the costs of a drone with a motorized dolly.

For lower budget movies, the imagery that can be obtained by a 2 1/2 pound Phantom or Phantom competitor with a GoPro camera is almost as good as what can be captured by a digital single-lens reflex (DSLR) camera typically used for low-budget movie making.

It is unlikely that camera and gimbal payloads require macrodrones rather than microdrones. On the other hand, many movies require long shots of activities such as movements of armies or the progress of floods. The visual context of pursuit and chase sequences may benefit aesthetically by photography from higher altitudes, say 2,000 ft. Helicopters are necessary for such shots until macrodrones at suitable price levels are commercialized.

Remote locations require transporting actors, production personnel, equipment, and sets from one point to another. When the budget permits, helicopters can do this more efficiently than ground vehicles—and sometimes ground vehicles cannot do it at all when a location is inaccessible, as on a mountaintop. It will be many years before people will consent to be transported by unmanned aircraft, so helicopters are safe from replacement by drones for that reason as well.

The FAA granted the first section 333 exemptions to motion picture and television producers and their contractors for closed-set drone operations. The video entertainment industry continues to keep the pipeline of applications and grants full. Already, drone operators and their customers are demonstrating the utility of drones for moviemaking. Their safety and cost advantages make them attractive as a supplement to, and sometimes as a replacement for, helicopters.

1.6 Agricultural applications

Agricultural applications received early approval via the section 333 process because most of them occur in rural areas, over property controlled by the customer, and because of the political power of agricultural interests.

Data-driven farming and automation of all kinds of farm machinery are significantly improving agricultural productivity. Microdrones reinforce these trends by providing additional sources of data to feed analysis of photographic and infrared imagery of crops.[4] Aerial imagery of fields can identify anomalies in irrigation and soil quality and early signs of insect or fungal infestations. Often, an aerial image reveals problems not apparent during conventional inspection with the naked eye from ground level. Infrared imagery can be combined with conventional photography to reveal patterns of growth not otherwise evident. Software can stitch together a series of photographs or video clips into a mosaic for an entire field. Time-lapse video can reveal trends in crop growth in a particular field.

The availability of data such as this enables farmers to concentrate supplemental irrigation where it is needed instead of wasting water where it is not needed, and to concentrate pesticide and fungicide applications on spots where they are needed, thus reducing cost and limiting environmental damage.

These capabilities mainly involve advances in software, sensors, and image processing, independent of vehicle type. But the basic collection of sensor data can be accomplished by microdrones much more cheaply and precisely than from manned aircraft. Necessary sensors, whether cameras or infrared sensors, are well within the payload and power capabilities of microdrones now on the market. The vehicles also need precise GPS-based navigation systems that permit them to fly a preprogrammed rectangular grid to capture strips of imagery covering an entire field. Adequate flight-plan navigation is available now in popular drone models such as the DJI Phantom 3, and Inspire. Full exploitation of automatic flight plan execution is limited, however, by line of sight requirements.

For large-scale crop surveys, live video streaming may not be important, because the ultimate product will be assembled by software after the flights are complete. Additionally, the flight path must be precisely controlled by navigation software rather than by DROP intervention based on what he sees, so LOS DROP control is relatively unimportant.

The principal shortcoming of microdrones results from short battery life and endurance. Surveying a large field requires several hours of image collection. It would take nine hours cover a one-section (640-acre) field, flying at 30 miles per hour, at a height of 25 ft above ground level (AGL), with a camera angle of view of 127 degrees (the GoPro angle at a resolution of 1080p). This tactic would result in images of a 20-ft wide strip on each pass. Doubling the height doubles the strip width and halves the time required to survey the entire field. So, flying at a height of 200 ft would produce image strip widths of 160 ft, and a dwell time of a little over an hour. The tradeoff is the greater level of detail available from images of 20-ft strips compared with the lesser detail available from images of 160-ft strips.

If the microdrone must return to the DROP every 15 or 20 minutes for a battery swap, productivity is far less than it would be with longer flight times. Multiple microdrones could be operated over different parts of the same field simultaneously, however. Their autonomous navigation systems would permit one DROP safely to handle several of them. Fixed-wing drones might be more efficient because of their higher speed and longer endurance, assuming that camera sensors are fast enough to capture the requisite detail at higher speeds.

Desirable technology development involves improvements in battery technology and drone endurance, improving the precision of GPS-based navigation systems, and improving the power and flexibility of image processing systems.

As noted at the beginning of this section, regulatory issues are not particularly difficult. The necessary flights occur over land controlled by the operator or by a farmer contracting with the operator, over relatively flat terrain with few obstacles, and mostly in rural areas. So the need for limiting proximity to people and structures on the ground, for keeping out of controlled airspace, and for systems that handle emergencies while avoiding collisions is minimal.

Aerial application of chemicals has been a tool of farming since the early days of aviation. Cropdusting, increasingly referred to as *ag applications*, was one of the first commercial aviation activities.

Cropdusting involves flying overlapping rows over a field at relatively low levels so that the chemical being applied—fertilizer or pesticide—goes where it is meant to as opposed to drifting to other parts of the field. The operator must be precise in positioning each pass, equally precise in triggering the release of the chemical on and off at the beginning and the end of each pass, and lose little time in making the 180° turns from the end of one pass to the start of the next.

Macrodrones can apply insecticides, fungicides, and fertilizers with great precision when they have appropriate automated navigation systems and application release control systems. The principal limitation on their utility is useful load. Agricultural chemicals weigh at least as much as water—8.3 pounds per gallon, and many chemical agents must be applied in quantities of 5 to 10 gallons per acre.

Microdrones certainly do not have the useful load capacity to carry any significant quantity of agricultural chemicals. Even the largest microdrones lack the capability to do agricultural applications at acceptable productivity levels. The DJI S1000 has a maximum payload of about 15 pounds. That would accommodate about two gallons of pesticide, which would, under DuPont's recommendation of 5-10 gallons per acre, cover only one-quarter to one-half acre.

Even the largest macrodrones now on the market have significantly less useful load than high-end manned cropdusters. Typical cropduster aircraft have maximum payloads of 6640 pounds, or 800 gallons, enabling them to cover 80-160 acres without refilling their tanks. A Reaper-sized macrodrone has a payload capability of 3800 pounds, or slightly more than that of a manned cropduster.

The Yamaha RMax, widely used in Japan and elsewhere to support agriculture, has a maximum payload of 62 pounds or about 8 gallons. That permits it to apply pesticide to about one acre before reloading.

In late 2015, DJI introduced DJI-Agras MG-1, an octocopter capable of carrying 22 pounds of liquid payload. It can cover 7-10 acres per hour, using radar to ensure accuracy. The vehicle was initially offered for sale only in China and Korea.[5]

There is no reason, of course, that a macrodrone developer cannot design an aircraft from the bottom up to carry any payload required. An engineer could start with the characteristics of a manned cropduster, take the onboard pilot-oriented systems out, and replace them with the requisite automated radio-control navigation subsystems. The economic question then would be whether the market is large enough to cover the development costs and produce an adequate rate of return for the developer. As widespread commercial drone flight becomes legal, the agricultural community will learn more about routine use of macrodrones in Japan and elsewhere, and preliminary design and business model development will accelerate.

The regulatory issues associated with macrodrone delivery of agricultural chemicals are not particularly daunting, except that the design of new vehicles for the heavier payload will invite regulatory scrutiny of design, flight testing, and manufacturing processes. Regulators are likely to revert to traditional airworthiness and type certification, with its attendant costs, delays, and inflexibility to accommodate innovation.

Scores of agricultural firms have acted on their belief in the potential of drone support for precision agriculture by dominating the field of section 333 exemptions and grants. As grantees begin commercial operations, appreciation of what does and does not work economically will advance rapidly.

1.7 Event photography

Microdrone imagery of events such as weddings occurs over property controlled by the customer, but it involves drone flight in proximity to large numbers of people. The FAA has generally granted section 333 exemptions, while routinely imposing a limitation that drone flight cannot occur over uninvolved people.

Professional photography of major life events like weddings, bar mitzvahs, graduation parties, and major birthdays, have long been a mainstay of the professional photography industry. In many cases, customers prefer still photographs over full-motion video, but all of the microdrone cameras on the market have the capability to take still photographs or video at the option of the photographer. Helicopter-based aerial photography has not been a common technique; helicopters are more often used to permit the celebrant, such as a recently married couple or someone celebrating a birthday to make a grand entrance arriving by helicopter or to give attendees rides while they are at the event.

Inexpensive microdrones with standard cameras widen the options available to event photographers and are already in wide use for this industry segment.[6] As with real estate promotion, the photography mission is well within the capability of microdrones. Relatively short flights suffice, and the mission equipment does not require heavy payload capability.

No particular technological breakthroughs are required. The best results will be obtained by equipping the DROP's console (DROPCON) and the photographer's console (PHOTOCON) with video displays that reveal a fair amount of detail from video downlinked in real time.

The regulatory environment is more challenging than for real estate promotion because many people are present at a party. Because they all have prior relationships with whoever is contracting for the microdrone photography, it should not be difficult to get their advance permission for the drone flights. The subject matter is on property under the control of the person or entity contracting for the microdrone services, and this will simplify regulatory compliance.

Macrodrones are not necessary. Their greater range, endurance, and payload capability offer no particular advantages for this type of photography; microdrones can do the job.

Section 333 exemptions for "aerial photography" dominate the inventory of granted exemptions. Aerial photography is a generic category, including many of the specific industry categories addressed in this chapter. Little doubt exists, however, that much of the exemption activity reflects the desire by professional event photographers to use drone technology to expand their offerings.

1.8 Search and rescue

Aerial search and rescue in remote areas or at sea has been an important helicopter mission as long as helicopters have flown.

Microdrones can be useful adjuncts in the search process, but not for rescues necessitating the extraction of human beings. Even their short range and endurance permit them to be deployed strategically to supplement other search techniques.[7] One or more microdrones can be programmed to fly a grid over part of a larger search area. Even if the drone finds nothing, it permits searchers to exclude that area from the larger-scale search, allowing resources to be concentrated where they are likely to be productive.

Desirable equipment includes high-quality video cameras, infrared imagery, and precise and reliable capability to fly a preprogrammed search grid. While the drone will fly the grid autonomously most of the time, the DROP and a separate systems operator should have available live, high resolution video imagery so they can monitor the search and take control of the drone to direct it to objects or activities of particular interest.

Current technologies are sufficient to permit microdrones to perform these missions. Improvements in photo processing and object identification will vastly improve search capability, but they are several years away.

No particular regulatory challenges are associated with these missions, because they will be typically conducted at relatively low levels over remote areas.

Macrodrones might be necessary for some searches, especially those covering large areas, because of their greater range and endurance. For many searches, however, the flexibility of being able to take a microdrone with the search party to the command post, launch it, and program it on an ad-hoc basis would give microdrones advantages over macrodrones.

Similarly, tethered drones can be useful to monitor trends flood water movement and in logistics support.

The technology and regulatory challenges are essentially the same for macrodrones as for microdrones in the context of search missions, except for the challenge of airworthiness and type certification of new types of aircraft configuration—far more likely for macrodrones than for microdrones.

The same technologies that can be used to repair bridges and towers can be used to rescue people. Equivalent technological and regulatory challenges exist. One inherent limitation of using an unmanned aircraft for rescue is that no crewmember is available to assist an unsophisticated or injured victim to secure herself to the hoist apparatus. Such human assistance is a centerpiece of helicopter rescues.

1.9 Pipeline and powerline inspection

Many of the early section 333 exemptions involved pipeline and powerline inspection. In these applications, risks are low because most pipeline inspection occurs in remote areas. The issues relate more to economics than risk.

Aerial inspection of natural gas and petroleum pipeline is common, using fixed-wing and rotary-wing aircraft. Fixed-wing aircraft offer the advantages of greater range and speed and generally lower costs. Rotary-wing inspections permit inspection personnel to fly low and slow and land on site.

Typical aerial patrols look for secondary signs of leaks and right-of-way encroachments that might disrupt the integrity of the pipeline. More sophisticated airborne systems can detect leaks based on plumes of gas. For example, the ALPIS (Airborne Lidar Pipeline Inspection Service) detects natural gas leaks by means of a specialized airborne laser system. It avoids the limitations of ground-based sniffer teams, which can cover only 8-10 miles a day, scan the swath of only one vehicle width, are ineffective over rough terrain, and unwelcome over private land.[8]

Microdrones can perform the same basic tasks as airplanes or helicopters, carrying cameras with the necessary resolution, recording, and downlink capabilities, and integrating imagery with GPS position. GPS-based flight plan systems available now on mass-market microdrones permit them to follow a pipeline. Live downlinking of imagery to the DROP and inspectors permit them to stop and hover the drone to capture more detailed imagery of potential problem areas.

Their limited range is a problem, however. Pipelines run through remote areas, often with rugged terrain which limits access by ground crews, including DROPs. Hundreds of miles of pipeline must be inspected on a typical mission. A microdrone flying at 30 miles per hour with an endurance of 30 minutes, can cover only 15 miles before it must return to the DROP for battery replenishment. If terrain permits the DROP to follow it, it could cover only 30 miles before it would have to land.

Improvements in battery technology allowing greater endurance and range would improve drones' utility for inspecting pipelines.

Regulatory issues are not especially problematic. Missions would be flown over land owned or controlled by the pipeline company, and most above-the-ground pipelines are in rural areas, where flight over people is not a hazard.

Macrodrones offer greater range, especially if they have gasoline or diesel engines rather than electric propulsion and have fixed-wing rather than multirotor configurations. They also carry heavier payloads. One state-of-the-art laser system to detect gas leaks weighs 250 pounds.

A significant number of section 333 applications and grants for this industry segment shows that drone contractors and their customers are actively exploring the benefits and limitations of drones.

Powerline inspections were cited by the FAA in the preamble to its February 2015 notice of proposed rulemaking (NPRM), because performing the activity by personnel climbing structures or being lowered by helicopter poses obvious risks. It was willing approve section 333 exemptions for these reasons, and because most of the activity occurs in rural areas.

As with pipeline inspections, aerial inspection of powerlines is common. Helicopters rather than airplanes are the norm because of their ability to fly low and slow to focus on possible defects.

High Definition photography provides an overview of the powerline corridor and helps with right-of-way (ROW) and vegetation management. It detects missing bolts and corrosion, damaged, burned, or rotting structures, broken or damaged cross arms, missing aerial marker balls, and unreadable structure numbers. GPS-determined position information is typically overlaid on the video stream or still photos.

Infrared (IR) sensors can identify faults based on heat anomalies. Faulty components in powerlines often increase resistance in that part of the line. Higher temperatures result as the electricity forces it way through the greater resistance. IR sensors thus can identify faults due to poor connections and current-related problems like corroded joints, problems with bolted clamp jumper connections, bad splices, dead end shoes, switch components, and jumpers.

Ultraviolet (UV) sensors make corona discharges visible through flashes in the video. Coronas indicate high electrical fields due to inadequate equipment design, the effects of pollution, in-service damage, or simple wear and tear. Broken or chipped insulators show high electrical discharges and often indicate insulation failures. Broken wires or those with loose strands are visible with a UV camera and can be detected reliably.

Typical microdrone cameras have adequate resolution, and widely available gimbals have sufficient flexibility to point the camera or other sensors at areas of concern. GPS-based autonomous navigation on existing microdrones permit macro-level surveying, until the DROP and engineers on his team spot a potential problem, stop the drone, and hover it. Live, downlinked video is necessary to permit the crew to interrupt autonomous navigation and move in closer under manual control to check out potential problems.

Limited range and endurance are problems for the same reasons they are for pipeline inspection. The weight of sophisticated IR and UV imaging systems drive up payload requirements.

Technology development needs and regulatory approach are the same as for pipeline inspection. Miniaturization of sensor components will enhance the capability of smaller microdrones.

Macrodrones offer increased range, which is a big advantage, considering the length of powerlines and pipelines. Their inability to hover, however, limits their utility for investigating suspected faults.

A significant number of section 333 applications and grants for this industry segment shows that drone contractors and their customers are actively exploring the benefits and limitations of drones.

1.10 Tower inspection and repair

Owners and operators of towers that support antennas, wind turbines, and infrastructure such as electric power lines must perform maintenance. Hazard lighting can go out, antennas or turbine blades can be damaged by weather or bird strikes, connectors can come loose. Owners and operators of the structures must inspect them periodically to detect faults or troubleshoot known faults. Once they have identified what repairs need to be made, they must get the necessary hardware in place and install it. Until the advent of drones, there were four basic ways to do this:

- Inspection personnel on the ground can use binoculars and other magnification devices and cameras to troubleshoot
- For smaller structures, they can use ground-based cranes or cherry pickers to perform the inspection and to accomplish the repair, by lifting personnel to where the fault exists
- A structure can be designed so that it has ladders or steps, permitting maintenance and inspection personnel to climb it
- Helicopters can be used for close-up inspection with the unaided eye, with high definition cameras with zoom lenses, or a combination of the two. Helicopters also can be used to lift personnel to perform repairs.

All of these approaches, except for standing on the ground and looking with binoculars or pointing a camera, are dangerous. Suitably equipped microdrones can perform the inspection functions with no significant risk to personnel. To do so, they need to be equipped with high-resolution cameras that have good close-up capability, unlike aerial cameras for other kinds of missions that may need to be optimized for longer shots. Some degree of gimbaling may be helpful, although most basic close-up inspection jobs can be done simply by maneuvering the drone to aim the camera at the right place. Camera stability is extremely important because of the need for clear detail in the imagery. Full motion video is less important than a series of good still shots. The images must be imprinted with camera location and orientation so that there is no confusion when they are analyzed as to what the camera was pointing at.

Additionally, depending on ambient lighting conditions, relatively low intensity light sources proximate to the camera may be helpful. Ordinary consumer flash capability would suffice. Streaming video back to the DROP or photog is necessary so that the camera can be positioned appropriately for the full range of images needed. Live streaming beyond the microdrone crew is not necessary. While a separate photog might provide benefits, this is a mission where the DROP can fly the drone safely and also operate the camera, especially if camera angle is determined mainly by orientation of the drone.

Current microdrone products can easily perform this kind of mission.[9] Necessary onboard equipment is well within the payload capability of microdrones on the market. There is no reason that the mission cannot be performed within 10 or 15 minutes aloft, and therefore battery capacity limitations in existing products are not particularly constraining. Sophisticated collision avoidance technologies are not necessary either, because the microdrone will stay well within the line of sight of the crew, and there is no need to operate it except in close proximity to the tower being inspected, where manned aircraft cannot fly safely in any event.

Existing and probable regulations for general commercial microdrone operations are not problematic, except that some towers are higher than 400 ft above the ground, and that constraint must be relaxed for these operations.

The inspection function can be performed adequately by microdrones, and macrodrones would provide no significant advantage. On the other hand, microdrones cannot perform repairs once problems have been diagnosed.

Macrodrones could hoist repair personnel proximate to the fault or actually perform repairs by means of specialized robotics. Such operations pose significant regulatory and technological challenges. With respect to human-hoist operations, existing requirements for operations by helicopters would need to be adapted to macrodrone operators, and the FAA would need to be assured of equivalent levels of safety. Unlike the typical drone operation where the risk to personnel is reduced because no people are onboard, human hoist operations would place the person hoisted in jeopardy.

Precise control of the position of the vehicle is necessary to compensate for the tendency of any load—human or inanimate—to swing, thereby changing center of gravity and making it more difficult to remain in position. Moreover, the control systems to do this must have extremely high reliability. The statistical probability of failure of the automatic control systems must be no higher than the statistical probability of system failure or pilot error on a helicopter performing the same function.

Moreover, specialized robotics to perform repair functions would add considerable weight and involve significant engineering design.

1.11 Bridge inspection and repair

The FAA's NPRM cites bridge inspection as an activity where risk can be reduced by using drones instead of ground personnel who must climb or be hoisted by cranes or helicopters. The agency was therefore receptive to section 333 petitions proposing bridge inspection.

The equipment and flight profiles necessary to perform bridge inspection are similar, for the most part, to those required for tower inspection. Relatively short flights proximate to DROP suffice to get the job done, and cameras now available on mass-market microdrones are adequate.

There are differences between bridge inspection and tower inspection, however. The objects and features to be inspected on bridges are more obscured than the features on towers, and a view from below is essential in many cases. Close-up high resolution photography with good lighting is even more necessary for bridge inspection than for tower inspection. An initial inspection may reveal the need to look at the same part of a bridge structure from multiple angles. Both DROPs and engineers on the team must have high-quality video imagery in real time so they can position the drone precisely to get the imagery they need for post-flight materials-science and engineering-structures analysis.

Automated navigation is not particularly necessary, because the drone is close to the subject, with the DROP nearby to position it precisely for the best perspective on the bridge element of interest.

Supplementary high-intensity lighting will be necessary for many missions, which will add to the payload, although the photography that needs to be lit is close-up. Smaller lighting packages are adequate; a searchlight with significant range is unnecessary. A consumer-level small LED flashlight mounted so that its beam is aligned with the camera lens might be sufficient. The light should toggle on and off as the camera is turned on and off remotely.

Actual repair of bridges cannot be accomplished by microdrones, because they cannot lift personnel or any but the smallest parts. Some relatively minor bridge repair could be accomplished by macrodrones suitably equipped for human-load operations or with sophisticated robotics designed for the particular repair task, as with tower repair. But the role that even the largest macrodrones can play is likely to be limited, because the objects necessary to repair a bridge, in many cases, are heavy and dimensionally awkward. It is likely to be a long time before a macrodrone would be a cost-effective substitute for heavy-lift helicopters, ground-based cranes, and apparatus slung from the bridge itself.

Regulatory issues would be the same as for drone-based tower repair, except for bridges in densely populated urban areas, many of which carry large volumes of vehicles. Any general regulatory restrictions on operating close to structures or to people on the ground would have to be relaxed. The regulatory regime for this microdrone activity must allow microdrone operations close to the structures being inspected, and that may bring them close to people and vehicles operating over the bridges or underneath them.

Appropriate safety measures for bridge repair, such as blocking traffic and erecting barricades, have been in use for a long time, and they are adequate for drone-based bridge repair as well as for repairs accomplished through more traditional means.[10]

1.12 Railroad track inspection

A few section 333 exemption petitions for railroad inspection have been filed. Granting the first, from BNSF, was straightforward because the vast majority of track inspections occur in rural areas, over property controlled by the customer.

About a third of railroad accidents are caused by track structure defects. To reduce the accident rate, railroads regularly inspect track, usually from hi-rail vehicles—pickup trucks equipped to operate on the rails.

Microdrones can supplement these conventional track inspection activities. Indeed, BNSF railroad applied for and obtained a section 333 exemption to use an AirRobot AR180, a quadcopter that weighs about 11 pounds; an AirRobot AR200, a hexacopter that weighs about 19 pounds; and a

3Drobotics Spektre Industrial Multi-Rotor Aerial Vehicle, a quadcopter of about 10.5 pounds, to operate over land owned or controlled by the railroad for track inspection.[11]

Some, but not all defects can be identified by aerial surveillance. Complete breaks in rails, seriously misaligned switches, or collapse of the track bed, are gross defects that can be spotted by ordinary aerial photography. More sophisticated laser-based aerial measurement equipment can also detect gage anomalies. Gage is the distance between the two rails comprising a track. Close-up, high resolution photography also can detect common rail defects such as spalling (fragments separating from the steel surface), burned rail, flaking, shells, flat spots, and head checking. Encroachment of foliage or trackbed flooding also easily can be at detected from the air.

The necessary photographic and measurement equipment is not particularly heavy, and so microdrones are capable of performing the work.

Other defects, however, require test equipment to be actually in contact with the rails, allowing sonic and magnetic techniques to determine internal metallurgical faults not easily visible from the outside.

The principal technological limitation is endurance. If DROPs have to remain within 15 or 20 minutes of each track inspection drone, the ground crew may as well perform track inspection from a hi-rail vehicle, and use additional diagnostic equipment in contact with the rails.

Regulatory limitations on height and requirements for consent by the property owner are no problem. The drone flights occur only over right-of-way owned by the railroad operator, and flights, in order to be useful, must be conducted at low levels—10 to 200 ft over the track. Limitations on operation from moving vehicles and beyond line of sight are obstacles to effective use, however. Operation of a drone from a hi-rail vehicle is desirable, for example.

There is no particular need for macrodrones, except for achieving greater endurance. Speed is not particularly helpful; nor is high-altitude flight.

1.13 Insurance claims inspection and risk assessment

The property and casualty insurance industry is enthusiastic about using microdrones to support claims processing and risk assessments and to get overviews of the scope of natural disasters.[12]

Insurance adjusters are using microdrones to facilitate inspection of claims for damage to specific property. Microdrone portability, ease of set up, and simple operation are assets in facilitating close-up inspection of roofs, chimneys, and other structural components. They help detect sources of water leaks.

By using them, adjusters avoid the need to rent cherry pickers and scissor lifts, to erect scaffolding, and to climb ladders and slippery surfaces in adverse weather with objects in their hands. Microdrones also make risk assessment more productive by enabling nonspecialists to collect video imagery, which then can be processed by risk specialists at remote locations. For the same reason that drones aid precision agriculture, they aid claims adjusters evaluating agricultural claims.

The limitations contained in section 333 exemptions and proposed in the NPRM present few difficulties for insurance claims adjustment. The necessary imagery must be acquired close up and at relatively low heights above the ground, well below any 400 ft limitation (except for very tall buildings). Line of sight is not a problem; it is desirable. Insureds presenting claims readily grant permission for drone flights over their property. It does not make much difference whether the property suffering a loss is located in a remote or a congested area.

Macrodrones are not particularly necessary, except for surveys of large-scale disasters to aid in getting a macroscopic view, similar to their utility for disaster relief.

1.14 Electronic news gathering

Microdrones capturing video of newsworthy events are attractive to journalists.[13] Microdrones enable ground-based teams of reporters and photographers to supplement their coverage with overhead imagery. The problem is that news occurs almost anywhere and involves people not under the control of the customer. The FAA has granted section 333 exemptions to freelance photojournalists, and television networks, and stations have been using microdrone contractors. More widespread exploitation of the potential awaits relaxation of regulatory constraints, particularly those prohibiting flight over people, prohibiting nighttime operations, and requiring overflight approval by property owners.

The typical customer for electronic newsgathering is a television station, although print media outlets and Internet broadcasters increasingly use still photographs and video imagery for stories, putting them on their websites.

For breaking news, TV stations prefer live imagery that they can include in their over-the-air broadcasts, on their live Internet streams, or on their websites immediately as the story develops. For other longer form feature stories, imagery delivered after it is shot is satisfactory. Typically, editors and producers take a series of still photographs or video clips of a minute or two and edit them down to scenes of 10-30 seconds for inclusion in a "package"—a story to be broadcast.

A package can include drone coverage to enhance ground-based reporting and photography. A field team will launch its microdrone only when an aerial perspective adds value to what can be done from the ground, as in the case of a fire, a large-scale law enforcement operation, a natural disaster, or a story subject whose scale makes it difficult to represent it photographically with ground-based imagery.

In most of these cases, microdrone flights of 10 to 15 minutes are adequate, unless the scene to be shot is a considerable distance away from where the reporter and photographer are located, as might be the case in a large-scale fire or large-scale law enforcement incident, when the authorities set up barricades some distance from the incident. Then, more flight time is required simply to get the microdrone in a desirable position to shoot its imagery, and line of sight restrictions may be problematic as well.

Onboard equipment to gather news imagery includes a camera of at least the GoPro class, capable of capturing HD video, recording it, and downlinking it to a photog console on the ground. The camera must be mounted to the microdrone with a three-axis stabilized gimbal that can be operated separately from the microdrone itself. To achieve good results, a photog separate from the DROP operates the gimbal and camera from a console separate from the DROPCON. This PHOTOCON has a transmitter to transmit photog commands to the camera and gimbal, a receiver to receive downlinked imagery, and a video display. The aircraft must have a receiver to receive photog commands, which may be multiplexed on the DROP's control link and thus handled by the basic receiver on the aircraft, and a transmitter to send the video stream to the PHOTOCON. The system for transmitting live video to the customer typically would be a bonded cell modem, such as those sold by LiveU and Dejero. That equipment would be connected to the PHOTOCON; it need not be on board.

Microdrone packages now on the market, priced in the low-to mid-thousands of dollars have these capabilities. Payloads of 8 pounds or less are more than adequate.

Experience with GoPro-level cameras on microdrones will provide data from which broadcasters can decide whether this level of imagery is good enough or higher grade cameras and video subsystems are necessary to push the quality closer to what is available from ENG helicopters. The product desired by ENG customers is the same, whether the imagery is captured by microdrone, a manned helicopter, or macrodrone.

A hallmark of helicopter-captured footage is artful use of zoom lenses by the photogs. A typical traffic shot or breaking news shot involves a close shot of the scene, immediately followed by a gradual zoom out to a long shot.

And, of course, zoom capability means that the air vehicle can stand off further and still capture details of the scene. The ability to operate effectively at a distance is particularly desirable for breaking news such as fires and critical law enforcement incidents, where drone operations can interfere with public safety personnel.

The standard for acceptable drone imagery is not static, however. Helicopter ENG quality will not necessarily remain the norm. Increasingly, television programming uses stringer and amateur imagery captured by GoPro and iPhone cameras.

Nevertheless, consumer preferences are more likely to shift toward even higher resolution imagery, and camera technology can be expected to evolve naturally beyond the current HD norm. Technological innovation is desirable mainly with respect to image capture and remote zooming. Image capture allows the photog to designate an object within a frame, causing the camera and gimbal system to lock on to it and keep the camera pointed at it even as the microdrone's attitude changes. Remote zooming is necessary because zooming in and out to highlight different parts of an image are routine parts of good news photography and videography.

Industry trends, mainly the integration of multiple channels of distribution for the same programming, will increase the demand for good imagery of all kinds, and this will enlarge the opportunities for individuals and small enterprises who want to contract to do ENG work either on a continuing or an ad-hoc basis.

Macrodrones provide capabilities beyond the reach of microdrones, principally longer-range and endurance, the capability of getting to a news site quickly, and moving from one site to another. Their capabilities will be closer to those offered by manned helicopters. Instead of being operated by field teams of reporters and photogs, they would be operated from a central facility for an entire metropolitan area.

Newsgathering requires a persistent presence. A break for refueling or crew change may cause a news helicopter to miss the most newsworthy event at a scene. The problem is greater with drones because of their short endurance. The solution is to have multiple drones on the scene so that one can be launched to cover the scene while another returns for battery change.

But one reason that helicopters are attractive newsgathering tools is their mobility. An assignment desk with only a vague report of the location or character of breaking news can dispatch its helicopter. The helicopter can quickly search intersections and backyards to pinpoint the location of the news scene. As it does so, the single camera mounted on the helicopter allows close inspection of potential sites. A slow-flying and low-flying drone would be incapable of performing such a search effectively. Additionally, the helicopter's altitude permits it to receive police and fire radio signals not receivable on the ground, also allowing air crews to identify breaking news.

Multiple drones controlled from one news van or helicopter also allow simultaneous coverage from different angles. Switching among their live images is as simple as switching among studio cameras, long a mainstay of television production.

The important technological innovations relate mainly to reducing weight and power requirements for the onboard equipment, and to meeting sense-and-avoid regulatory requirements. Collision avoidance technological capabilities and related regulatory requirements are the most daunting, as discussed in chapters 3 and 8.

The regulatory environment for such operations is far more complex than that appropriate for microdrones. ENG macrodrones would require reliable systems to handle equipment failures such as loss of propulsive power, interruptions in the control link, or loss of GPS signal and other navigational inputs. Because the aircraft would be operating at higher altitudes in unpredictable locations, emergency capabilities must extend far beyond simple return-to-home or land-immediately responses appropriate for ENG microdrone operations.

Tethered drones, evaluated in § 1.19, are attractive for covering traffic congestion at major interchanges and for covering special events and breaking news expected to last several hours.

Given the obvious utility of microdrones now on the market to supplement ENG helicopters, it is surprising that few television networks and stations and their major ENG helicopter contractors have applied for section 333 exemptions. CNN is the exception, and it has taken the lead on cooperative R&D ventures with the FAA.

Public presentations by broadcasters and private conversations with them suggest that the industry is taking a cautious approach, unwilling to take the risk of violating the law by operating without section 333 exemptions before a final rule is promulgated. They have abstained from seeking section 333 exemptions because of their concern that the FAA's limitations in the exemptions requiring that DROPs be rated pilots and that drones not be operated over people circumscribe the utility of drones for newsgathering. They are taking a wait-and-see attitude, hoping that the limitations will be eased in a final rule.

Discussions with people in the industry reveal the following specific concerns. The restrictions in the section 333 exemptions, reinforced by the terms of the NPRM, interfere with ENG drone deployment under usual photojournalism practices. The fluidity of station broadcast schedules means news directors and assignment editors assign photojournalists on a last-minute basis, daily or hourly. This is obviously true for breaking news, and it is also true for a significant part of video supplementation for long-form packages. A TV station does not plan most photo shoots a week in advance. That means that the 24 hour minimum notice for NOTAMs is a significant obstacle.

Some of the other restrictions are obstacles, and some are not. The prohibition against flying at night is problematic because much news—and some of the most evocative photographs of developing news—occur at night. The requirement to have a licensed pilot—even a sport pilot—serve as pilot in command is a significant problem. Most of news photographers are not pilots, and most pilots are not particularly good photographers. So that adds additional staffing and complexity in allocating responsibility for flying the drone and manipulating the camera to get the best imagery.

Conversely, the height and speed restrictions do not pose problems. Given the short focal length of most microdrone camera lenses, 200 ft is high enough to get a good long view, and a close shot requires flying the drone much lower. Similarly, restricting speed to 80 knots is not a problem. Most microdrones cannot fly even half that fast, and their short endurance means that they are not useful for quick transit to another, more distant, location.

The availability of news helicopters in larger markets also blunts any immediate need for ENG microdrones. The large-market stations are the ones with larger news and technology budgets, and they generally are at the forefront of news technology innovation. They are the ones most likely to be out in front with microdrone deployment, but they do not need it except as a supplement for helicopter coverage. Smaller market stations are less likely to be innovators for budgetary or other reasons, even though they could use drones to offer aerial imagery—something they have not had before.

At the same time, a number of *would-be* ENG drone contractors have applied for—and have been granted—section 333 exemptions, and it is clear that some networks and stations are more than willing to gain experience by contracting with such operators.

1.15 Law enforcement support

Microdrones are beginning to be used by law enforcement agencies to supplement helicopter support and to provide aerial imagery when helicopters are not available.

Aerial imagery is useful in seven distinct law enforcement situations:

- Initial incident investigation
- Patrol
- Pursuit
- Perimeter enforcement
- Surveillance
- Accident reconstruction
- Transport and coordination of tactical response teams

Initial incident investigation puts the first responding unit in the position of checking dwellings and businesses for physical security, detecting possible perpetrators of crimes, and finding complainants. Certain relevant information easily can be obtained by units or officers on the ground. In other cases, aerial imagery can facilitate checking the immediate vicinity, including roofs and back and side yards. The first responding unit calls in helicopter support or launches a microdrone carried in the trunk of a patrol vehicle.

Patrol keeps an aircraft in the air functioning as an airborne patrol car. It flies over areas to detect crimes in progress, to identify suspicious behavior, to monitor ground units for safety as they make traffic stops, and to be readily available to be dispatched to assist ground units.

Aerial pursuits may involve vehicles seeking to escape from crime scenes or suspects on foot. The speed and visual perspective of aircraft exceeds that of ground units, which may easily lose sight of the suspect or be outrun.

Perimeter enforcement involves more or less stationary aerial surveillance to detect someone crossing a perimeter set up at a crime scene or tactical situation in which suspects or hostages are barricaded.

Surveillance involves keeping a geographic area in view to detect possible criminal activity or monitoring the actions of one or more individuals suspected of criminal activity in order to obtain evidence useful in prosecuting them.

Accident reconstruction involves obtaining a perspective on a scene that facilitates understanding the physics of what happened through visible features such as skid marks.

Aerial insertion of tactical teams quickens the response to active shooter-and-hostage situations.

Helicopters are regularly used to perform these functions, depending on whether the law enforcement agencies involved have helicopter assets readily available. The Los Angeles Police Department is the paradigm of an agency that has assets that regularly patrol and thus are always in the air to handle specific incidents. Most police agencies, however, do not have their own helicopters, and regional mechanisms for sharing helicopter assets are in their infancy.[14]

An important difference exists between ENG and law enforcement use of microdrones. In the ENG case, the aerial imagery vehicle has to get to the scene of breaking news; in the law enforcement case, ground units are already there.

Microdrones can contribute to law enforcement effectiveness in some, but not all of these missions.[15] For initial incident investigation, a microdrone permits the first unit on the scene quickly to check out parts of structures and their environs for movement of people or signs for forced entry. The drone must be equipped with a basic video camera and systems for streaming live video to the DROP and an accompanying tactical flight officer (TFO); equipment beyond basic video capture and downlink and display to the DROP and TFO is not necessary.

Payload requirements thus are well within the capability of mass-market microdrones. Limited endurance, on the order of 15 or 20 minutes, is not a problem, because the vast majority of scenes can be checked out in that time period. Returning the drone to the launching point and refreshing the batteries are not particularly limiting, either.

If ground units carry microdrones in their trunks and if patrol officers are trained to fly them, they are readily available in case the responding officers believe they would be useful.

Patrol with microdrones is not useful because of their limited endurance and range.

Depending on the nature of a pursuit, microdrones can be useful, especially if it is a foot pursuit, involving modest speeds and distances. For vehicle pursuits, the suspect vehicle is likely to outrun microdrones or quickly to exceed their range. Despite their limited endurance and range, microdrones nevertheless can be essential in answering questions such as, "Did he turn left or right? Did he enter the freeway?," permitting resources with greater capability to take up an effective pursuit.

Microdrones obviously lack the capability to insert law enforcement personnel into a tactical situation. They can, however, be useful in the early stages of perimeter enforcement. Initially, when large numbers of ground personnel have not yet been assembled, and the initial unit or units are struggling to maintain visual coverage of all sides of a perimeter; a microdrone can fly for 15 or 20 minutes to fill gaps in observation until other assets arrive.

Accident reconstruction is perfectly suited to microdrone capability. They can be a regular tool for investigators. Microdrones' limited endurance and range are not problematic because the typical accident investigation involves a relatively small area, and all the imagery that can be helpful can be captured in a few minutes.

All of these missions can be accomplished with sensors and downlink systems that require a modest payload and are well within the capabilities of existing mass-market technology.

Regulatory requirements for public-use aircraft are more flexible than those for civil aircraft. Height restrictions are unlikely to be a problem, because the missions benefit from low-level flight. Line of sight may be a problem, however, because these missions may require flying the drone behind obstacles such as buildings or foliage.

Aerial patrol, surveillance, and some pursuits require greater endurance, range, and speeds than are available from microdrones. Macrodrones with performance similar to that of light helicopters are necessary to perform those missions effectively.

Autonomous execution of a pre-programmed flight plan is of marginal utility, because even the most regular patterns flown by patrol helicopters depend on pilot and TFO judgment as to where criminal activity is most likely to occur. Instead, the kind of live imagery available to DROPS and SYSOPS for military drones is more useful, and technology developments regarding image capture and presentation to DROPs and SYSOPs will shape utility.

Legal uncertainty discourages early adoption. The process for getting FAA approval is even more confusing than for other civilian applications. The FAA has a special process for public safety agencies to obtain COWAs, but has also opened up the section 333 process to state and local government agencies.

Many law enforcement agencies and their lawyers erroneously believe that the Fourth Amendment's restrictions on warrantless searches and warrant requirements must be rethought in the context of drones for violence. In addition, many states, as chapter 8 explains, have enacted statutes specifically restricting the law enforcement use of drones. As more legal analysis becomes available, law enforcement drone use will proliferate.

Multi-agency co-operation will emerge to make it easier for small agencies to evaluate specific products and technologies, to understand legal ramifications, and to train their personnel.

1.16 Other public safety activities

§ 1.15 considers drones as support for law enforcement activities. They are also useful for other public safety activities, such as fire suppression and disaster relief. A few fire departments are using microdrones to gain situational awareness on a fireground.[16] The first firefighters to arrive on the scene can launch a microdrone and fly it over and around the fire to determine whether the roof is involved, whether the fire is spreading to nearby structures, and to gain a first impression of hotspots. Such early intelligence improves the positioning of firefighting equipment.

Microdrones also can be useful in certain aspects of disaster relief. One of the first and most important tasks in disaster relief is collecting intelligence. To deploy relief resources effectively, the emergency response teams must know the spatial extent of the disaster. For example, they may need to discern the path traversed by a tornado. They must have location information about the victims, such as those stranded by a flood. They must understand continuing risks to rescuers, such as whether a river is surging or receding.

After such an initial assessment is made, keeping track of resources is a challenge. Aerial inspection can show where personnel are located, likely to be unfamiliar with the geography and therefore finding it difficult to navigate on their own. It can show where excess relief supplies are positioned, which might otherwise be forgotten.

Drones, airplanes, and helicopters all can provide such aerial perspectives. Microdrones, because of their lower costs, can be deployed in greater numbers than manned aircraft, thus providing more eyes in the sky. Microdrones can safely inspect structures more closely than personnel can. They can also provide a macro view of relatively small areas of afflicted neighborhoods.

For other aspects of disaster relief, macrodrones are necessary: surveying large areas, delivering meaningful quantities of relief supplies, and eventually, perhaps, rescuing victims in circumstances in which a helicopter rescue would present unacceptable risks to the flight crew.

Macrodrones, moreover, add considerably to the capability of microdrones for fire suppression. Microdrones cannot carry enough water to be useful for quenching the fire. It is worth considering whether macrodrones could be designed to resemble helicopters and airplanes that deliver large quantities of water or other chemicals to wildfires. If they are feasible, they could be attractive because they offer similar capability without the high risks to aircrews. Kaman Corporation announced the restart of its production line for the K-Max heavy lift helicopter in 2015, and along with it, development and testing of unmanned versions for firefighting and humanitarian missions.

Use of drones (macrodrones would be necessary) to effect medical evacuation are farther into the future, because of the instinctive opposition to carrying people aboard a pilotless aircraft. This remote possibility is considered in chapter 12.

1.17 Border patrol

Macrodrones are interesting adjuncts for patrolling US borders for many of the same reasons that they are interesting for military and intelligence operations abroad. The Bureau of Customs and Border Protection within the

US Department of Homeland Security operates a fleet of General Atomics Predator B and Guardian drones to conduct missions that are inaccessible by ground because of remoteness or rough terrain.

The drones typically fly at about 20,000 ft and use photographic and infrared sensors to detect ground movement. While the imagery they capture is useful, a 2014 Inspector General report questioned their effectiveness and cost.[17]

It found a high incidence of malfunctions and accidents, low daily utilization, and costs exceeding those of helicopters performing similar missions.

The experience illustrates the challenges associated with adapting military macrodrone technology to civilian applications.

As civilian macrodrone technology matures and becomes more affordable, border patrol operations will broaden their deployment of the technology. They also will use tethered drones to monitor critical border areas.

1.18 Logistics and package delivery

Much of the growing public interest in commercial applications of microdrones began with a December 2013 interview of Jeff Bezos, the CEO of Amazon, on the television news program *60 Minutes*. In that interview, Bezos theatrically showed off a microdrone that Amazon was designing to deliver packages.

Delivery of Amazon consumer packages is only one of many different logistical applications involving delivery of objects by drone. These missions are distinguished from almost all the others because they involve external payloads, unlike most missions, in which the payload comprises the mission equipment and stays attached to the aircraft.

Microdrones can deliver small packages over relatively short distances, say 5 to 10 miles, at a cruise speed of 30 mph and an endurance of 20 to 30 minutes. Reliable GPS-based navigation systems are necessary, because it is unlikely that the DROP could always keep the vehicle within his line of sight; the routes are not chosen by the DROP, but dictated by the delivery destination. Specialized machinery must be installed on the drone to hold the package securely and to release it at the destination.

More general missions, like those envisioned by Amazon, require aircraft with ranges of 10 miles or so and with payload capabilities on the order of 10 to 20 pounds. As with other missions benefiting from fast transit time and rotary wing performance at the destination, VTOL configurations are interesting. In late 2015, Amazon previewed a revised configuration that employs eight horizontal rotors to generate lift in conjunction with wings, and a single vertically mounted propeller for forward propulsion. It featured sense-and-avoid technology to allow BLOS operation for more than 10 miles, to deliver packages up to five pounds in 30 minutes or less, to a consumer's welcome mat.[18]

In addition to routine use to deliver packages ordered through e-commerce, macrodrones with adequate endurance and range can deliver specialty items to persons in remote areas. Examples include delivery of pharmaceuticals in remote areas of Denmark and delivery of relief supplies to persons stranded by natural disasters. The ultimate macrodrone design will be influenced by the Marine Corps' Kmax, which flew in Afghanistan to deliver millions of pounds of logistics packages in a contested military environment.

The technology development required and the regulatory regime must be much more sophisticated than anything available now. Automatic systems for landing immediately, returning to home, and keeping out of certain airspace must be sophisticated, flexible, and completely reliable, because the microdrones will operate in densely populated residential and commercial areas, with many people abroad. Obstacles abound, such as utility poles and wires, trees, and buildings of various sizes and shapes. Chapter 3 describes NAMID, the kind of low-level airspace management system that would be appropriate. Such a regulatory regime must be built from the ground up around the technologies that will permit mission accomplishment while protecting safety in some of the most difficult environments that can be flown by drones.

The limited range and payload capability of microdrones will not allow the implementation of large-scale package delivery systems, although they might fly ad-hoc missions. Lifeguard microdrones are regularly deployed in Chile to drop life preservers to distressed swimmers.

The technology development and regulatory creativity required for macrodrones in these applications include all of the challenges associated with microdrones, plus those associated with higher altitude flight vehicles intermingled in airspace regularly used by manned aircraft.

1.19 Weather

Microdrones can be flown in weather conditions that are below safe minimums for helicopters and airplanes. Even a half-mile visibility and ceilings of 600 ft or so permit line of sight operation within the range of the control and video links. Precipitation is a potential limitation, however, because both the drone and its mission subsystems must be waterproof and snow-resistant. But if motors and onboard electronics are waterproof, their mission profiles can extend into bad weather.

While few microdrone missions discussed in the preceding sections can be accomplished successfully in a driving rain or heavy snowstorm, the possibility remains for operating in weather that would frustrate manned aircraft operations. Of course the impingement of raindrops or snowflakes on lenses can impair utility.

1.20 Tethers

Ordinary microdrones can be attached to tethers, either to secure them against substantial winds, to feed them electrical power, or both. The two major considerations are the weight of the tether—a kite string might be appropriate for small drones, but wires for electrical power would add to the weight of the tether considerably. In addition, the tether must be attached to the drone in such a way that the drone's center of gravity is not disrupted as the tether comes under tension. Tethered drones can stay in the air indefinitely and thus are attractive as stationary eyes in the sky for monitoring traffic congestion and for law enforcement situations, fires, or natural disasters likely to persist for hours or days.

Their tethers ensure that if they fall to the ground, they will do so within a limited radius that can be made secure.

CNN is actively exploring tethered drones for newsgathering, in a cooperative R&D venture with the FAA and tethered drone vendors, under its requested section 333 exemption.

1.21 Consumer drones

As useful as drones are commercially, experience shows that thousands of people want to fly microdrones for fun, or in conjunction with personal leisure activities.[19] This phenomenon is distinct from the long-standing model aircraft community's activities. RC modelers almost always fly their airplanes and helicopters as part of a group activity. They are tinkerers at heart and have developed a safety culture that US law has long embraced.

The appeal of consumer drones spans the range of leisure activities. How consumers use digital cameras is a good starting point for understanding how their use of microdrones will grow. When they go to an outdoor event, they will take their drones just like they take their cameras or plan on capturing video with their cellphones. Of course it is more trouble to take the drone along; their phones are always in their pockets. The exploding market for GoPro cameras and their more recently introduced competitors, however, shows that consumers are more than willing to use specialized cameras to capture recreational activities, rather than sticking with their cellphones.

The leading drone manufacturers are taking advantage of this. The market is beginning to reflect a distinction between the consumer submarket and the commercial submarket. Both DJI and 3Drobotics have introduced new models: The Phantom3 and Inspire 1, in the case of DJI, and the Solo, in the case of 3Drobotics, are aimed at consumers, although they also can be entry points for professionals who want to use them commercially. GoPro itself may introduce a consumer-drone product line.

The CEO of 3Drobotics, Chris Anderson, has been particularly articulate in public statements about what he thinks consumers want: significant autonomy, not only for flying the drone, but also for using the camera to orbit while pointing at an object of interest, to take selfies, and to follow the DROP as he moves. Consumers want what Anderson calls "no pilot systems."

Consumer drones represent a particular problem for crafting an appropriate regulatory regime. They are neither commercial uses that fit comfortably within the NPRM or the section 333 exemption process; nor are they model aircraft, mostly flown in conjunction with RC model club activities. A bill introduced by California Senator Dianne Feinstein (analyzed more fully in chapter 7) appropriately focuses regulation on point-of-sale rather than on deploying enough inspectors and police officers to detect operating rule violations. It thus embraces the approach advocated by chapter 9, to make consumer drones law-abiding right out of the box.

But that approach, as chapter 9 recognizes, challenges technological development. Most of the autonomous safety features now being offered as an option or a default by major drone manufacturers depend upon the drone's having GPS signals—being able to get a "GPS lock," something not available indoors, and sometimes unreliable outdoors. Designers, manufacturers, and the FAA have more engineering work to do to make sure that GPS-based autonomy is available; to ensure that DROPs cannot override it, thus negating its risk mitigation; and to explore possibilities for supplementing GPS-based safety features with inertial navigation when the GPS lock is lost.

1.22 Different purposes, demand, regulation, and politics

Microdrones and macrodrones serve different purposes. Macrodrones perform the same missions as helicopters. Microdrones perform missions never flown before.

The civilian demand for macrodrones arises from the same sources that give rise to the demand for airplanes and helicopters: a desire to transport people and goods from one point to another and, secondarily, to have eyes in the sky thousands to tens of thousands of feet overhead. Their competitive advantages depend upon reducing costs and reducing crew risk. The commercial demand for microdrones arises from a new ability to do something that has never been possible before: to have overhead imagery from new perspectives closely connected to activities on the ground. Demand evolves from entrepreneurial imagination of new benefits from flight.

The risks associated with macrodrones and microdrones are different. Macrodrones, because of their weight and speed, have kinetic energy similar to that of airplanes and helicopters. They present mid-air collision risks comparable to manned aircraft, and the consequences of an accident are profound for flight crews and passengers and for persons and property on the ground. Microdrones, because of their smaller weight and lower speeds, possess much less kinetic energy. While they pose some risks to people and property on the ground, and, depending on their flight profiles, some collision risk to other aircraft, by and large they do not present any kind of catastrophic risk.

Regulatory regimes suitable for macrodrones are unsuitable for microdrones, and vice versa. Macrodrones fly among manned aircraft. In doing so, they will be safest if they fly in accordance with rules developed for aircrew members, flight vehicles, and operations designed for manned aircraft. The content of traditional regulation for manned aircraft needs improvement. But the reforms needed to enhance safety, to take advantage of new technologies, and to reduce costs are the same for macrodrones and manned aircraft.

Traditional aviation regulation is manifestly ill-suited for microdrones. Their low heights above the ground, their proximity to ground-based operators and observers, their flight among buildings and people, and their high degree of autonomy all beg for a different regulatory regime. That regime should limit height, speed, weight, and range; require certain autonomous behaviors; and recognize the infeasibility of using large numbers of coercive enforcement agents to detect and punish rule violations.

Conversely, regulatory regimes designed for microdrones are unsuitable for macrodrones. Macrodrones cost much more and have much more expansive flight profiles. Their greater reliance on point-to-point transportation rather than close flight and return mean that height and range restrictions, autonomous features such as return to home and built-in avoidance of controlled airspace suitable for microdrones are ill-suited for macrodrones. As entirely new systems to channel low-level drone flight for delivery of packages are developed, they will be irrelevant to macrodrone operations far above them.

The economics of macrodrones and microdrones are quite different. The main question for a potential purchaser or user of macrodrones is straightforward, head-to-head competition between a particular model of macrodrone and airplanes or helicopters that closely resemble it. A potential user decides which can fly an existing mission best: a macrodrone or a comparable model of airplane or helicopter.

The central criterion is relative cost for similar capability. The competing vehicles might offer different levels of capability, but if in a particular competition, the manned competitor has deficient capabilities compared to the unmanned competitor, the features of the macrodrone can relatively easily be built into airplanes or helicopters. Similarly, if the manned competitors have capabilities that the macrodrone lacks, that capability can be relatively easily be built in to the macrodrone. The similar payloads, range, speed, and fuel consumption make it possible, in many cases, simply to remove a system from one and install it in the other.

For microdrones, the question is not whether they can replace manned aircraft; they cannot. Nor is the question whether potential users of microdrone services should instead use an airplane or a helicopter; they cannot, because airplanes and helicopters cannot do what microdrones can do. The question is whether some new activity that benefits from flight should be performed by microdrones, even though the activity has never occurred before. The potential user decides whether performing the activity is worth the relatively modest cost and risk of adverse public response and possible increases in insurance premiums.

The politics of microdrone and macrodrone operations are manifestly different, as well. Macrodrones will mostly take off and land from the same places that airplanes and helicopters use for that purpose. Usually, the general public will not even know whether a particular aircraft operation is performed by a manned or unmanned aircraft. Opposition to macrodrones will come more from the aviation community concerned about the risks of collisions with manned aircraft operations and about the threats posed to pilot jobs.

Microdrones stir up a different kind of opposition. If they are successful in stimulating new uses and developing new markets, they will be buzzing around everywhere. The general public can hardly fail to notice them. Their reaction will depend upon concerns about noise, trespass to land, potential invasions of personal privacy, and the risk of a microdrone falling on a person, on an automobile, or on a house, even though it is unlikely to cause substantial damage or to inflict serious injury.

These differences permeate every chapter and section of this book and influence almost every facet of its analysis.

1.23 Technologies

Specific technological developments have made drones possible. Those are reviewed in chapter 2. Wider use of drones depends on further technology development, the subject of chapter 3. Electric battery development is necessary to improve endurance and extend the range of microdrones. Further miniaturization is necessary to reduce costs and improve the already quite good payload-to-gross weight capabilities. Additional wireless bandwidth and more efficient use of available bandwidth is necessary to accommodate control links and download of imagery.

Improvements in automatic control system technologies are necessary to improve response times to vehicle perturbations and the envelope within which autonomous subsystems can operate reliably. That, in turn, depends in large part on greater computing power in small packages.

Further improvements in structures and materials will enable advances in strength and permit still lower weight for the same level of performance.

The game changers are battery technology, automatic control system refinement, and better management of the wireless spectrum, especially for macrodrones, which require longer range control links.

1.24 Human factors

Drones, although pilotless, typically require human inputs for flight control, mission planning and overall enterprise management. Chapter 4 compares the essential functions and tasks performed by airplane and helicopter pilots with the corresponding functions and tasks to be performed by DROPs. As the market for drones moves into adolescence, drone operators and regulators must crystallize appropriate screening criteria for recruiting DROPs, determine what level of training they need, build the institutions to deliver it, and discover the right organizational structures, leadership, and motivational forces.

At the same time, sustainable diffusion of drone technology requires not only good talent to sit behind a DROPCON and fly the drone, but also requires human resources to market the services effectively and to build and manage the enterprises that deliver drone services. Chapter 4 analyzes these issues, as well.

1.25 Emergencies

Regulatory requirements for drones and restrictions on their use depend largely on the perception of regulators. That perception is informed in turn by the views of experts and the general public on how drones and DROPs will handle emergencies that, if not properly managed, threaten life and property. One risk of flight—jeopardy to the flight crew—is completely eliminated by drones, but others remain. Drones may lose power, harming persons or property on the ground. They may collide with other aircraft. Manned aircraft primarily rely on pilots in the cockpit to deal with emergencies, assisted at the margins by ground-based air traffic controllers. Drones depend more on onboard automation, supplemented by control inputs from DROPs. Careful and realistic risk assessment and design of emergency responses is a prerequisite to the development of sufficiently accommodative regulatory approaches, especially for macrodrones, where the risks are much higher because of their size, more expansive flight profiles, and fuel loads.

Chapter 5 considers the ability of DROPs and automated systems to handle drone emergencies and compares that capability to emergency responses in manned aircraft.

1.26 Politics and law

All societies regulate technology. Aviation technology is subject to detailed regulation of the design of aircraft, the selection, training, and licensing of aircraft operators, and the manner in which aircraft are operated. Politics determines the content of the regulation, which is expressed in law. Widely publicized aviation accidents spawn public pressure to adopt new regulations to prevent similar accidents. Fear of flight activities that may jeopardize physical, economic, and legal interests shape the regulatory tone, even when no invasion of those interests yet has occurred. New aviation technologies threaten job skills embedded in old technologies, and individuals who hold those skills lobby to delay introduction of technologies that threaten them.

Basic legal principles circumscribe political responses. Constitutional protections of personal liberty and property may encourage or discourage particular regulatory initiatives. Constitutional and statutory protections of due process channel how the political will can be turned into law, while affording opportunities for conflicting interests to determine the content.

Chapter 6 analyzes the political response to drones, probing the determinants of political passions and positions and exploring the dynamics of interest group politics. Chapter 7 takes a broad view of regulation, considering how different social values shape the response in different states. Chapter 8 digs more deeply into the content and evolution of drone regulation in the United States.

Autonomous technologies allow drones to internalize regulatory limitations: to limit height and range, to exclude drones from protected airspace, to cause them to return to the launching point if the control link is lost. Chapter 9 explores requirements for such law-abiding drones.

1.27 Economics and entrepreneurship

Because resources are limited everywhere, economics determines what is possible. Drones offer no escape from this reality. Aeronautical and electrical engineering constantly reshape how underlying scientific principles determine the microeconomics of aviation operations. Sound decisions whether to use microdrones, macrodrones, airplanes, or helicopters depend on rigorous economic analysis, informed by actual experience. Chapter 10 provides analytical templates for those considering drone acquisition and operation and for those that purchase drone services.

Good economics, however, is not enough; viable enterprises must be launched, managed, and adapted to changing technologies and market forces. Chapter 11 brings to bear the extensive intellectual capital developed in universities and in marketplaces to identify strategic, managerial, marketing, and operational-supervision factors that are the hallmarks of a successful business.

1.28 Is it a fad?

The industry-specific sections of this chapter make it clear that practical people in many parts of the economy are interested in what drones can do, and are willing to try them out. That does not mean, however, that drones will prove their mettle. Drone trade shows are attended by the tens of thousands, but more attendees are vendors and operators than customers who will buy their products and services. Many of them will be disappointed and move on to other things.

The FAA, other policymakers, the general public, and too many drone proponents are preoccupied with the question of whether an aircraft without a pilot can fly safely. Too often, they engage this question without regard to the size of the pilotless aircraft. Many pilots and commercial operators of manned aircraft mainly regard drones as a threat.

Those concerns obscure the ultimate question: how much will drones be flown at all? Whether and how often macrodrones will be flown civilly and commercially depends on an assessment by potential users of whether their flight activities require a pilot. Macrodrones will be flown when they can do a better job than a piloted aircraft at or below equivalent cost. Microdrones will be flown because there are many activities that benefit from low-level aerial photography that were not possible before

Potential users of macrodrones will decide whether they are better off with a pilot in their aircraft or without. That decision drives their decision on what kind of vehicle to select. That question is not asked by microdrone users, because having a pilot is not a possibility.

The criteria for macrodrone use are

- How well do the automated flight profiles in the available macrodrones work?
- What parts of the mission, if any, require the flexibility and adaptability of an onboard decision maker? How much advantage does a human perspective from a cockpit confer, in terms of mission performance or safe operation?
- What is the level of risk for the aircrew?
- What is the likelihood of aircrew boredom?

The analysis focuses on what pilots do best, what they do badly or unhappily, and what puts their lives at risk.

Much of the competition pits automated flight against manned flight. Some flight activities are already automated in airplanes. Most airplanes above the basic trainer level have autopilots, and most pilots use them for routine cruise and many instrument approach procedures. The same technology is easily adaptable to flying search and rescue grids, surveying, and maybe utility infrastructure patrol. Both macrodrones and airplanes use it under the watchful eye of either a DROP or a pilot in the cockpit. Experimentation and operational experience will reveal who can do a better job. For both, boredom is a problem. Usually, nothing goes wrong. When it does, it is a surprise, and the DROP or the pilot must act quickly and correctly. DROPs and pilots have similar instrumentation. Pilots have an advantage in that they can feel and hear a deviation from the expected behavior of the aircraft. The competition depends on the advantages of the additional perceptual cues available from inside a cockpit.

Automation has limits. Many autopilots do just fine in cruise flight, can fly departure and arrival procedures, but only the most sophisticated—and therefore most expensive—can land or take off the aircraft. Otherwise, takeoffs and landings must be done manually.

Pilots have advantage over DROPS in maneuvers for which the autopilot does not provide autoland capability. Taking off and landing are complex tasks the performance of which benefits from the widest possible range of perceptual cues, intellectual, kinesthetic, aural, as well as two-dimensional visual and three-dimensional binocular. The competition between automation and pilot control resolves into relative precision and the capacity to handle conditions that far exceed the normal.

Dealing with conditions that stray significantly past the ordinary is the crux of the competition. Pilots can improvise; automated robots cannot. The logic of flight automation depends on programming responses to customary events. Not everything is equally likely to happen, and some flight conditions are so remote that the automation is unlikely to be programmed to deal with it: an angle of attack greater than 45 degrees; an inverted spin; complete loss of elevator movement; runaway trim. When conditions exceeding their knowledge occur, most autopilots simply shut off and hand things back to the pilot. Debate grows over the readiness of average pilots to handle the sudden responsibility, but anecdotal evidence abounds supporting the superior performance of exceptional pilots. Captain Bryce McCormick's landing of American Airlines Flight 96 in 1972 after collapse of the floor in the aircraft severed control cables is a dramatic example.

The choice between macrodrone and airplane is not only a competition between pilot and DROP on handling emergencies when automation runs out of options. It also is a choice between levels of automation. Unmanned flight requires more automation. Manned flight does not require any. The cost of sophisticated automation is substantial, and that affects price.

Whether piloted or pilotless is best depends on whether commercially available macrodrones can perform the requisite flight activity with sufficient precision at affordable cost. No user is going to choose a macrodrone over an airplane or helicopter merely because it can do the job *almost* as well as a manned aircraft at the *same* cost. There is no incentive to replace a manned aircraft with a macrodrone. If a pilot can fly a mission with a less automated airplane than a DROP can fly the same mission with a more automated macrodrone, the total cost for the manned airplane will likely be lower, and an economically rational operator will choose the airplane instead of the macrodrone.

Other types of civil missions, however, are candidates for macrodrones because they present high levels of risk to onboard aircrews. These missions resemble those that caused armed forces and intelligence agencies to begin using drones in the first place. Cropdusting and aerial firefighting involving dropping fire retardant on fires are examples. Other high-risk activities, such as medical evacuation are not candidates for macrodrones, because they carry passengers and the complexity of the tasks that the pilots perform are very difficult to automate.

Missions that can be automated well because they involve highly predictable flight profiles are early candidates for macrodrones. As control systems and automation continue to improve, and the price of sophisticated automation systems drops, the range of missions that can be performed better by a largely automated aircraft under the control of a DROP will expand.

Part of the pilot versus pilotless question, and therefore the choice between airplanes and macrodrones relates to the quality of the workforce that can be attracted for each type of vehicle. Chapter 4 considers criteria for DROPs and the determinants of motivation for both pilots and DROPs. Boredom, romance, and pride are important parts of the equation.

The fate of microdrones depends on different considerations. Potential users of microdrones want to do something that can only be done by aircraft too small to accommodate a pilot. Microdrones can do many things that an airplane or helicopter cannot do, simply because of their size. Conversely, there are also many things that airplanes and helicopters can do that microdrones cannot do because of their limited ceilings, endurance, speeds, and the need to keep them within the DROP's line of sight and within the range of their control links. There is, some overlap, however. The overlap area comprises missions requiring relatively slow flight at 500–1,000 ft above ground level. A helicopter with a good camera and a telephoto lens can get essentially the same imagery as a microdrone flying lower. For those overlap missions, microdrones' order of magnitude lower acquisition and operating costs are likely to give them an advantage, once user reluctance to be the first mover fades and regulatory regimes crystallize.

Drones are unlikely to be a fad that will fade altogether. Their use may bloom and then recede somewhat, however, as early enthusiasts are unable to gain an economically viable foothold.

Notes

1 Akira Sato, *Civil UAV Applications in Japan and Related Safety and Certification*, (Sep. 2, 2003), www.uvs-international.org/phocadownload/03_5ac_Relevant_Information/Applications_Civil-UAV-Applications-in-Japan.pdf (unmanned vs manned aircraft aerial chemical application).
2 PIX4D website, www.pix4d.com (describing software that automatically processes terrestrial and aerial imagery acquired by lightweight drone).
3 Angela Watercutter, *Drones Are About to Change How Directors Make Movies*, Wired (Mar. 6, 2015), www.wired.com/2015/03/drone-filmmaking/.
4 DRONELIFE News, *Drones Set to Take Off over Farm Fields*, (Sep. 24, 2015), www.dronelife.com/2015/09/24/drones-set-to-take-off-over-farm-fields/, (example of aerial crop imagery collection).
5 *DJI Introduces Company's First Agriculture Drone*, (Nov. 27, 2015), www.dji.com/newsroom/news/dji-introduces-company-s-first-agriculture-drone (vendor news release providing details).
6 Matt McFarland, *Drones: The next big thing in wedding photography, or a tacky intrusion?*, Washington Post (Feb. 24, 2015), www.washingtonpost.com/news/innovations/wp/2015/02/24/drones-the-next-big-thing-in-wedding-photography-or-a-tacky-intrusion/ (reporting on growing use of drones to get unique perspectives on weddings).
7 Matt McFarland, *Drone operators assist search and rescue efforts after devastating floods in Texas*, Washington Post (May 29, 2015), www.washingtonpost.com/news/innovations/wp/2015/05/29/drone-operators-assist-search-and-rescue-efforts-after-devastating-floods-in-texas/ (reporting on use of drone to facilitate search-and-rescue operation at much lower cost than helicopters).
8 Lasen, Airborne Lidar Pipeline Inspection Service (ALPIS™), www3.epa.gov/gasstar/documents/workshops/houston-2007/LaSen_April_2007.pdf (slide briefing explaining laser-based pipeline inspection system for helicopter use).
9 Cloud9Drones, *Tower Inspections*, www.cloud9drones.com/tower-inspection/ (explaining how inspection of cellphone and electricity distribution towers by drone improves safety).
10 Barritt Lovelace, *Unmanned Aerial Vehicle Bridge Inspection Demonstration Project*, (Jul. 2015) www.dot.state.mn.us/research/TS/2015/201540.pdf, (UAS bridge inspection study).

11 David Z. Morris, *Why BNSF Railway is using drones to inspect thousands of miles of rail lines*, Fortune (May 29, 2015), www.fortune.com/2015/05/29/bnsf-drone-program/ (reporting on BNSF use of drones for more frequent track inspection and the challenges posed by short range and appropriate software).

12 Cognizant, *Drones: The Insurance Industry's Next Game-Changer?*, www.insurance-canada.ca/ebusiness/canada/2015/Cognizant-drones-next-insurance-game-changer-1501.pdf. Cognizant is a major insurance industry consulting firm.

13 *"Drones will have an enormous impact on news production," Keith Laing, Feds approve CNN for drone flights*, The Hill (Dec. 14, 2015), www.thehill.com/policy/transportation/263103-feds-approve-cnn-for-drone-flights (reporting on grant of section 333 exemption to CNN; quoting CNN petition).

14 Henry H. Perritt, Jr., Eliot O. Sprague, and Christopher L. Cue, *Sharing Public Safety Helicopters*, 79 *J. Air L. & Comm.* 501 (2014).

15 Matt Alderton, *To the Rescue! Why Drones in Police Work Are the Future of Crime Fighting*, (Apr. 30, 2015), lineshapespace.com/drones-in-police-work-future-crime-fighting/ (quoting police sources on how drones can revolutionize law enforcement); Susan Greene, *Mesa County, Colo. A National Leader In Domestic Drone Use*, Huffington Post (Jun. 6, 2015), www.huffingtonpost.com/2013/06/06/mesa-county-colo-a-nation_n_3399876.html (reporting on police agency use of drones for general law enforcement and accident reconstruction).

16 Fire Engineering, *Illinois Firefighters Use Drone At House Fire*, (Aug. 30, 2015, www.fireengineering.com/articles/2015/07/il-firefighters-drone-house-fire.html (firefighters use drones to help direct fire suppression).

17 US Dept. of Homeland Security, Inspector General, *US Customs and Border Protection's Unmanned Aircraft System Program Does Not Achieve Intended Results or Recognize All Costs of Operations*, (Dec. 24, 2014), www.oig.dhs.gov/assets/Mgmt/2015/OIG_15-17_Dec14.pdf (reporting on use of drones for border patrol).

18 Amazon Prime Air, www.amazon.com/b?node=8037720011 (Amazon promotion featuring VTOL drone configuration).

19 HobbyTown, www.hobbytown.com/search.aspx?searchtext=Drones&refine=, (consumer site offering drones for sale).

2 Genesis

2.1 Introduction

In part, the development of civilian drones was inspired by the widely publicized use of military drones in the conflicts in Kosovo, Iraq, and Afghanistan. Entrepreneurs and engineers asked themselves about the possibilities if they took the weapons off and adapted the sensors. Some of the engineers who had developed the military technology turned their attention to adapting it to civilian uses.

The evidence suggests, however, that another discovery process was at least as important. Enhancements in technology and the affordability of greater capabilities for model aircraft inspired hobbyists to consider whether they could make money off of the model aircraft they were playing with. They began to talk about flying camera-equipped model airplanes to do wedding and real-estate photography and showed off their efforts on YouTube. Entrepreneurs thought about extending the applications to electronic news gathering, land surveying, law enforcement, and powerline and pipeline patrol.

Microdrones are largely an outgrowth of model airplane development, while macrodrones are emerging from military developments.

An understanding of the distinction between microdrones and macrodrones—the differences between small vehicles and larger ones—is essential to have a sound understanding of everything else. Microdrones weigh less—usually much less—than 55 pounds. They typically are multicopters, air vehicles propelled by 4–8 electrically driven rotors. Macrodrones weigh more than 55 pounds. They are usually powered by internal combustion engines driving conventional propellers or fan jets.

Both types of drone became possible only because of advances in electronics, computer, and manufacturing technologies—far more than advances in aerodynamics, but powered flight was the starting point. The chapter starts there. The middle part of the chapter focuses on the crucial technology developments, some of which benefited aviation in general, some of which made drones feasible, and some of which was particular to the two types of drone. Basic aerodynamic and powerplant technology improved all kinds of air vehicles. Advances in control systems and automation likewise have benefited aircraft at large. Radio technology advances are pertinent to both types of drones, but had less influence on manned aircraft. Internal-combustion propulsion system development mainly relates to macrodrones, while electric propulsion system development mainly relates to microdrones. Miniaturization has benefited all types of aviation, but was a critical pathway to making microdrones feasible.

The capabilities of drones depend on the same principles of aerodynamics, propulsion, and structures as those applicable to manned aircraft. Accordingly, the history of drones is intertwined with the history of airplanes and helicopters. But the underlying principles of flight have been known for a century or more. They are directly derived from Newtonian mechanics. Far more significant than breakthroughs in these areas were breakthroughs in material science, mechanical engineering, and electronics. Accordingly, this chapter focuses mainly on innovations in those supporting technologies.

This chapter addresses historical developments of technologies that made drones possible. Chapter 3 explains how the technologies coalesce into practical microdrones and macrodrones and considers what remains to be done to extend these technologies and thereby enhance the utility of drones.

2.2 Different origins

Lumping micro- and macrodrones together leads to several important errors in thinking about each type of vehicle. Their origins are different. Macrodrones evolved from an effort by governments to develop intelligence-gathering and weapons-delivery vehicles that would reduce the risk to aircrews already flying airplanes and helicopters in armed conflicts. They were designed for intelligence collection and combat in environments posing unacceptable risks to human pilots. Microdrones evolved from miniaturization of imaging, flight, and electrical propulsion systems that, for the first time, enabled air vehicles to fly in close quarters, down low, and in circumstances where manned aircraft would simply be too big. The purpose was not to get rid of the human pilot, but the result was a vehicle too small to carry one. Macrodrones first entered wide use in the military context and are now spreading to the civilian world; microdrones first entered wide use in the model aircraft hobbyist community and are now spreading to the commercial world.

2.3 What made drones possible?

Drones emerged in the twilight years of the twentieth century from developments in military and naval aviation and from parallel developments in the model aircraft hobbyist community. In each case, technology advances relaxed constraints that limited flight possibilities. The following sections focus on the particular technology developments that made drone use attractive. As suggested in the introductory sections, the analysis distinguishes between microdrone/model aircraft technologies and macrodrone/military technologies. Many of the crucial technological developments, of course, benefited both the model-aircraft and military realms.

Before that could occur, however, longstanding fantasies of flying above the ground had to be turned into reality. So the story starts with the technologies of flight itself.

2.3.1 Making flight possible

The major milestones in the history of human flight are well known. The Greeks, the Romans, and the Chinese dreamed of it, drew sketches, and built models of machines that might make it possible. Leonardo da Vinci worked out a sophisticated design for a rotary wing aircraft.[1] But the technologies did not yet exist to turn the aspirations into reality. Practicable heat engines were far in the future. No one yet knew how to build the necessary light-weight structures. Electricity was yet to be discovered, and so it did not occur to anyone that electrical power, electronic controls, and wireless links for navigation might aid human operators.

As scientific theory deepened during the Enlightenment and turned into engineering advances during the Industrial Revolution, interest in flight awakened more concrete concepts for how flight might be achieved. Best known is the Wright Brothers' 1903–1905 demonstration of manned flight.[2] Orville and Wilbur Wright's early interest in aviation was fueled by a model French helicopter driven by a rubber band, and their design efforts were informed by Otto Lilienthal's experiments with gliders in Germany and Samuel Pierpont Langley's 1896 flight of an unmanned powered aircraft over the Potomac River.

The problems that frustrated the aeronautical pioneers around the turn of the twentieth century were not how to generate lift or how to modify an internal combustion engine to provide thrust; they related instead to controllability. Controllability fascinated the Wright Brothers. The unsuccessful tests, most notably those by Louis Bleriot in France and Samuel Pierpont Langley in the United States, ended in failure because the aircraft pitched up and stalled or went into an uncontrollable roll and caught a wing tip on the ground. Langley, secretary of the Smithsonian institution, spent $500,000 of government money producing a couple of spectacular early accidents that were ridiculed by the press.

Langley successfully flew an unmanned steam-powered aircraft for 90 minutes in 1896. With support from Assistant Secretary of the Navy Theodore Roosevelt, he scaled the model up, named it the "aerodrome," and crashed into the Potomac River twice in 1903.[3] Langely's first crash resulted from improper calculations of structural stress, causing the aircraft to break up right after its catapult launch from a houseboat. The second time the "aerodrome" crashed because it was uncontrollable. The pilot, Langley's assistant, Charles M. Manly, nearly drowned.

Bleriot began his aviation experiments with ornithopters—aircraft that fly by flapping mechanical wings, like birds. These experiments were unsuccessful, leading Bleriot to concentrate on fixed-wing configurations that flew, but had control problems that nearly killed him several times. Eventually, he managed to fly across the English Channel—but it was after the Wright Brothers' first flight. Competition among the Wrights, Bleriot, and Langley spurred all of them. As World War I approached, interest in aviation's possibilities grew in the public consciousness.

The Wright Brothers were mechanics, but they also, despite lack of formal education in the subject (MIT offered the first course in aeronautical engineering in 1914), were technically astute. The brothers spent many hours discussing and experimenting with ways to achieve equilibrium in flight and to permit the pilot to correct for perturbations and to command different flight regimes. The opening language of their 1906 patent says that the purposes of their invention was "to provide means for maintaining or restoring the equilibrium or lateral balance of the apparatus [and] to provide a means for guiding the machine both vertically and horizontally …"[4]

The result was a biplane, with an elevator mounted in front of the wings to control pitch, and a rudder mounted aft to control yaw. The most important contribution was a means for the pilot to change the twist of the wings differentially, causing one to generate more lift, and the other to generate less—a process they called *wing warping*. The pilot lay prone in order to lower the center of gravity.

Working with relatively meager funds of their own, they selected Kitty Hawk, North Carolina, as their test site because of its nearly constant winds. They tested the airframe first as a glider and then with a custom-built aluminum engine designed and fabricated by Charlie Taylor, a talented mechanic they hired for their bicycle shop.

Scores of mishaps occurred, but after each one, they had the patience to figure out what was wrong, to determine how to prevent it happening in the future, and to rebuild the aircraft to fly again. Initially, they received little publicity, because the press, many distinguished scientists, and most members of the public believed that powered flight was impossible and that the Wright brothers must be con artists. But eventually, the brothers started to gain credibility, and they performed highly publicized demonstrations, attended by thousands of people, including senior government officials, in France and then in Washington. After these successful demonstrations, the race was on to build airplanes. Many would-be pioneers followed the Wright Brothers' design exactly; some others, like Glenn Curtis, experimented with new ideas, such as ailerons to control roll.

After their successful flights in Kitty Hawk, the Wright Brothers were widely acclaimed—more enthusiastically in Europe than in the United Sates. They were as good at self-promotion and entrepreneurship as at engineering. Their Wright Company dominated the aviation field under the protection of a number of aggressively drawn patents. They brought patent enforcement actions against anyone who used an airplane with ailerons. Their aggressiveness stalled further airplane development by hamstringing competitors and by diverting their attention from engineering into litigation.

Finally, the US government, desperate for airplanes to support its involvement in World War I, forced the establishment of a patent pool, which all aircraft manufacturers joined, enabling them to cross-license relevant patents.[5]

Freed from the patent straitjacket, a variety of engineers and entrepreneurs pushed forward. Structures evolved from wood and fabric to aluminum. Biplanes—aircraft with two wings stacked one above the other on each side—gave way to monoplanes—with one wing on each side. Improved engines allowed increased speed and the commercial airlines feasible. Jet engines, developed during World War II, proved their practicability in Boeing's 707 and Lockheed's Electra turboprop. Advances in electronics grew Sperry's 1919 wing leveler into sophisticated autopilots.

Helicopters received as much attention in the early days as airplanes. In 1877, an unmanned helicopter powered by a steam engine flew at about 40 ft over Milan for less than minute. In his excellent book, *The God Machine*,[6] James R. Chiles profiles the engineering and political history of helicopters and other rotary-wing aircraft after that. Efforts to use rotary wing designs to mimic the Wright brothers' success were stymied by poorly understood stability problems. A number of helicopters flew a few feet off the ground but had to be steadied by human attendants holding on to them. Eventually, a different rotary wing design was successful: the gyrocopter. The Spaniard Juan de la Cierva proved the design, based on an unpowered rotor that was spun and generated lift through the principle of autorotation as a conventional gasoline engine and propeller drove the aircraft through the air. Gyrocopters captivated the public, leading to predictions that they would become more common than the automobile, and causing the US Post Office to launch gyrocopter-delivered mail.

By the late 1930s a variety of European engineers and promoters were beginning to demonstrate useful helicopter designs with a single main rotor and a tail rotor. In Russia, Igor Sikorsky, inspired by a Wright Brothers' tour of Europe, turned his attention to the problems still confronted by helicopter designers. After unsuccessful attempts to make a coaxial helicopter (one with two main rotors, turning in opposite directions) practicable, he redirected his attention to designing and building improved airplanes.

Sikorsky came to the United States to escape the turmoil of the 1918 Russian Revolution, which was occupied with matters other than aviation advancement. He had reasonable success in building and selling a series of flying boats, but realized that the future of that airplane design was limited and turned his attention back to helicopters as World War II was heating up in Europe. Advocating a tail rotor design instead of coaxial rotors or side rotors, which were then the most popular designs and had proven their capacity to fly for short distances, he struggled with dissymmetry of lift, ground resonance, and instability once ETL was reached. Effective translational lift (ETL) is the aerodynamic phenomenon that increases lift generated by a rotor as it moves out of its own downwash. Dissymmetry of lift is the tendency of the blade moving forward to generate more lift than the blade moving backward on the other side, tending to roll the aircraft over as forward speed increases.

He finally realized that his failure to design an effective cyclic control resulted from a failure to understand gyroscopic effects. The cyclic control on a helicopter changes blade pitch asymmetrically, as the rotor turns. When he changed the phase of applying asymmetric pitch changes to the main rotor by advancing them 90 degrees, the cyclic control worked. By 1942, the armed services, the public, and a variety of commercial entities were growing excited about the helicopter's potential.

Enthusiasm for wider civilian use was thwarted by political opposition to what the general public perceived as a noisy and dangerous toy of the rich. The armed services pressed forward, however. They used a few helicopters, and more gyrocopters, at the tail end of the war for search and rescue.

The pace of design innovations and manufacturing accelerated after World War II. In the Korean conflict, US forces regularly used helicopters, mostly for search and rescue. As the Cold War developed, US strategists were preoccupied with strategic nuclear weapons, and the Army scrambled for new military doctrines that would prove its continued worth. The solution was greater mobility. Led mostly by army chief of staff Matthew Ridgway, the army reinvented itself around helicopters to move troops quickly, not only to rescue them after they got shot.

During the Vietnam War, helicopter reconnaissance and fire-control scouts, transports, and logistics were central to military operations. To make this possible, new, more specialized helicopter designs proliferated. After the war, military helicopters became available in large numbers to civilian operators, and discharged helicopter pilots were available to fly them. Scheduled helicopter passenger operations sprang up in large cities, though most failed because of a combination of unworkable business models and public concerns about safety and noise. Tourists flocked to helicopter sightseeing businesses that bloomed in Hawaii, near the Grand Canyon, and elsewhere. Offshore drilling in the Gulf of Mexico and the North Sea needed helicopters to carry crews back and forth to and from drilling platforms. Law enforcement agencies and TV stations relished what aerial observation and video images could do to enhance their missions and improve their competitive positions. The medical community embraced them to evacuate critically injured patients to hospitals. Helicopter flight schools sprang up.

2.3.2 Model aircraft

Microdrones emerged primarily from the model aircraft hobbyist community. How that happened requires an appreciation of the evolution of model aircraft. After World War II, a surge of interest in model airplanes was fueled by three phenomena: general public fascination with aviation's role in the war, a rich inventory of wartime breakthroughs in aeronautical engineering, propulsion systems, electronics and radio engineering, and a glut of young men and women who had come in contact with these technologies in wartime service.[7] Wishing not to leave their involvement with aviation altogether behind them as they reentered the civilian workforce, they took up aviation as a hobby.

Fabrication techniques for building workable wings and aircraft structures existed well before the war; indeed the same techniques were used to build full-scale aircraft during World War I and during the interwar period. A person could design a flyable airplane by starting with basic knowledge of aerodynamics and Newtonian physics, build the structure reflecting the design, usually with wood, and overlay fabric on top of it. That knowledge and those techniques could enable one to build a pretty good glider at almost any size he wanted.

The first missing link was an appropriate propulsion system. For a variety of reasons—cost, limited flying space, and more complex structures required for larger vehicles—hobbyists sought smaller vehicles, too small to carry available reciprocating engines.

Reciprocating engines were the target. Jet engines had barely achieved practical utility by the end of World War II and were far too complex, expensive, and heavy for hobbyists. Electric motors were widely used in industrial and transportation applications. But their weight and their requirement for a physical wired electrical connection or batteries excluded them as candidates for aircraft propulsion. Portable lead-acid batteries were much too heavy to attract any interest.

The breakthrough was the development of small two-cycle, one-cylinder gasoline engines, a subject considered in § 2.3.3. With their development, it was feasible and affordable for model airplane enthusiasts to build powered airplanes.

The next problem was controlling them. At first, the operator set whatever control surfaces existed while the airplane was still on the ground, spun the prop to start it, held the little airplane in place until full power developed, released it, and watched it take off. Part of the excitement was wondering where it would go. The operator and his friends would watch it follow an undetermined path, hoping it would not crash, and that it would not "land" in an inaccessible place. The experience was not unlike hitting a golf ball and hoping that it landed near the pin; once the ball leaves the club face, nothing more can be done.

Enthusiasts with even modest electronics knowledge set about to figure out radio control systems.

Radio-controlled model aircraft systems developed from the confluence of radio telegraphy, railroad and telephone switching systems, transistors, and integrated semiconductor chips.

As hobbyists were developing small radio-controlled model aircraft after World War II, some combination of protocols, radio receiving and transmitting equipment, and electronic coding and decoding was necessary to permit the ground operator to provide control inputs for pitch, roll, yaw, and thrust. Additionally, methods had to be developed to couple these control inputs to control surfaces on the model aircraft, making use of technologies being developed in parallel for onboard control systems in manned aircraft.

Even if aircraft of a useful size could be flown remotely, however, they could not do anything useful—except maybe execute kamikaze suicide attacks—unless they could take pictures and send them back to the ground or record them in some fashion so they could be retrieved after the aircraft landed.

§§ 2.3.4–2.3.7 explain the development of digital coding and modulation schemes through radio telegraphy, the development of digital logic implemented through electromechanical relays in the railroad and telephone industries, and signal multiplexing through tuned metal reeds. The challenge for the model aircraft engineers was how to turn these discoveries into useful control systems for their aircraft. And how to do it on a channel involving radio waves rather than electrical signals traversing a wire.

By the mid-1950s, enthusiasts had it pretty well figured out. A small airplane could be equipped with a radio receiver—a crystal-controlled receiver operating on only one frequency. Pulses were detected by the receiver and fed to a collection of relays that controlled surfaces on the airplane, sometimes just a rudder. An airplane with the requisite lateral stability will eventually enter a coordinated turn initiated with rudder deflection alone; ailerons are not strictly necessary. An initial skid caused by rudder deflection makes the outside wing move faster, generating more lift on that wing, and resulting in a roll in the direction of the turn. Initiating a turn in this way was possible even if there were only three rudder positions, left, neutral, and right. These could be effectuated by two-position switch for the ground-based operator. By turning the switch on, the operator would send a pulse over the radio channel to the aircraft. A stepping relay or escapement on the vehicle could move the rudder left with the first pulse, center it with the second pulse, move it right with the third pulse, and so on.

The same technique could be used for other control surfaces and throttling of engine thrust. Each relay and control surface actuator added weight, of course, but not as much weight as the radio receiver. Transistors were at least 10 years away and every radio receiver required at least one vacuum tube with an incandescent element that consumed lots of current. The associated battery weighed as much as the engine.

As late as 1961, a top-of-the-line model airplane had 1 to 10 channels of radio communication with the operator, who had a series of 2-3-position switches with which to issue commands for the various control services. Tuned-reed multiplexing of signals on the same frequency was the norm, given the limited number of radio frequencies available. One reed, and thus one audio frequency, for each control channel provided one channel for each control surface.

The results and capabilities for these model airplanes were exciting, but the flight profiles were clumsy. Only a limited number of fixed positions for each control surface and throttle setting were possible, and an operator had to be quick with his brain, and nimble with his fingers, to get an acceptable results.

Then, in 1964, a proportional control system that Phil Kraft been working on for 10 years reached the market.[8] While Kraft was experimenting with and publishing papers on successive proportional control designs, he and his business partners also were successful in selling reed-based-controlled model airplanes. They were so successful, in fact, that Kraft occasionally was tempted to be critical of the proportional control approach. Eventually, however, he realized that proportional control would dominate the future, and he quit his job to pursue its development full-time.

Under his innovation, the operator had not only switches but also potentiometers—rotating knobs that continuously vary a resistance and can impart the same variations to an electrical current or voltage. Potentiometers were already in wide use as volume controls on radios and televisions. To make use of the varying control signal, the vehicle required servo motors that moved control services in direct proportion to the input signal, reflecting potentiometer position. Transistors had just entered the consumer market, and Kraft's design made liberal use of them, dramatically reducing power requirements for transmitter and receiver, making a margin available for the servomotors.

Proportional control took the community by storm, making Kraft and his manufacturing partner Ace Model Airplanes, quite wealthy. Incremental improvements in control box design, servos, and coding schemes proceeded apace, but several important limitations remained:

- Only a limited number of frequencies in the low VHF band—40 MHz or so—were available, resulting in frequent interference with the control signal
- The servo motors were heavy and consumed lots of electrical power necessitating bigger and heavier batteries and
- The vehicles worked acceptably, but reasonable endurance required carrying enough gasoline, which impinged further on weight limits.

No one seriously considered trying to fit helicopter designs into these tight constraints. Full-scale helicopter designers were still working out the kinks which regularly resulted in fatal crashes due to phenomena that no one had thought about.

But technology was improving in other arenas. The transistor attracted consumer enthusiasm. Transistor-based radios fueled industrial research and development to make them smaller. Consumer electronics enterprises and the emerging space industry wanted more transistorized circuits but struggled initially to fit transistors, resistors and capacitors into smaller, integrated packages. The obvious utility of digital computers was energizing an intensive engineering effort to make them smaller and more powerful. Before long, the consumer electronics and computer industries were designing and manufacturing integrated circuits by the thousands, and they got smaller and smaller and required less and less power.

2.3.3 Propulsion systems

Before the turn of the twentieth century there had been experimentation with steam engines fitted to larger model airplanes, but no great demand developed for them. Miniature gasoline engines for model aircraft first became available in 1911.[9] They had tiny carburetors, and spark plug ignition systems, one cylinder and usually operated in a two-stroke cycle. They were heavy and unreliable. Ray Arden, an enthusiast for improving all kinds of toys, first developed a simplified ignition system for model engines (boats as well as airplanes) and then was responsible for crucial breakthrough, the invention of a glow plug engine.[10] The glow plug design simplified the engine by eliminating the need for an ignition system to deliver a spark to the cylinder at the right time in the cycle. Glow plug engines are started by connecting an external battery to the glow plug. After start, the heat of combustion keeps the element in the glow plug at a sufficient temperature to ignite the fuel-air mixture on the next cycle.

First marketed in 1947, glow plug engines galvanized the model airplane community and soon dominated model airplane propulsion, continuing to do so until high-torque electric motors became available in the early 2000s. In 2014 2.5-cc-displacement engines, delivering 1 HP, and weighing 275 grams were available for $50.

The possibility of electric propulsion for air vehicles remained in the background of research until the late 2000s. Available motors were too heavy, too difficult to control, and required huge batteries. DC motors offered the advantage that their speed could be varied continuously simply by changing the voltage applied to them. They paid for this advantage by the complexity, unreliability, and RF interference resulting from the necessary brushes and commutators.

AC motors existed from the earliest days of the practical application of electricity. The simplest kind of AC motors were, however, *synchronous*, operating only at multiples of the electrical supply frequency—60 Hz for power generated and distributed in the United States. Displacing DC motors required advances in electronic motor control systems, which, in turn, depended on advances in electronics. The necessary electronics as well as small motor design benefited from the microcomputer revolution, which needed small motors able to operate at high RPM to power disc drives and printers.

Nikola Tesla emigrated from Croatia to the United States to find a more hospitable climate for his ideas on alternating current as an alternative to direct current electrical power systems being promoted by Thomas Edison. Ironically, he initially went to work for Edison, but their growing disagreement over the best system for distributing electricity caused him to switch allegiances and to team up with George Westinghouse. In order to make his preferred AC current practicable for households and businesses, Tesla had to invent an AC motor. Otherwise, the AC power supply would have to be rectified to deliver DC current to motors, and that would have lessened its advantage over Thomas Edison's preferred DC distribution. Engineers already understood multiphase synchronous AC motors. Three sources of alternating current in different phase relationships with each other (a *three-phase* supply) were fed to a stator—the stationary part of the motor—causing the field of the stator to rotate. This induced an opposing current in the windings on the armature—the rotating part of the motor—and the armature rotated in sync with the frequency of the supply current. Torque varied with the amount of current supplied to the stator. The speed of the motor, however, was not adjustable. Tesla's *induction motor* was similar to a synchronous motor, in that a field current induced a current in the windings of the armature. But in an induction motor, the armature moved more slowly than the rotating field. The current induced in the armature was thus out of phase with the current in the stator. The result was rotation of the armature. When a load was applied to an induction motor, slowing the rotation of the armature shaft, the change in the relationship between the phase of the armature field and the stator field increased the torque until the motor reached its design speed again. Induction motors, unlike synchronous motors, did not require a three-phase power supply.

AC motor controllers developed to impart variable speed capabilities to induction motors, by varying the frequency supplied to them. The earliest controllers were rotating machines, which took the supply current as an input, rectified it into DC, and drove a variable-speed alternator. The output of the alternator drove the AC motor. Later controllers used specialized vacuum tubes known as thyratrons to rectify the AC supply into DC current. Thyratrons can handle larger currents than conventional vacuum tubes. Operators controlled current and frequency. This double conversion with vacuum tubes and associated transformers, capacitors, and resistors was inefficient. As § 2.3.7 explains, vacuum tubes are large, consume significant power, and burn out frequently. It was not until the 1980s, when speed controllers improved with the installation of miniaturized semiconductor components such as thyristors to replace thyratrons that AC motors could compete effectively with DC motors and became attractive for a wide variety of applications, including model aircraft.

The first battery-powered model aircraft was the Radio Queen, designed by LTC H. J. Taplin and sold in kit form by Electronic Developments Ltd. Beginning in 1957, Taplin was a prominent pioneer in RC modeling and had proposed in 1958 that beach lifeguards could be replaced by one his RC-controlled boats, which he demonstrated in newsreel footage posted on YouTube many years later.

Electronic Developments had been started in 1946 by recently discharged RAF and war production employees who were casting about for a way to make a living in the post-war world. It dominated the British model airplane marketplace for a while. The Radio Queen had a 7 ft wingspan and used two permanent magnet DC motors and a silver-zinc battery. Silver-zinc batteries commonly were used in torpedoes. The superiority of the small nitro-methane and gasoline powered engines then becoming available, however, relegated the Radio Queen to a short market life.

AC induction motors had the advantage that they did not need commutators to deliver reverse the electric field in the armature, but DC motors offered advantages of higher torque over a wider range of RPM. Brushless motors eliminated commutators and combined the advantages of both. Brushless motors were developed for the space program in the 1960s. They rely on semiconductor circuits to deliver pulsed current to the field windings, which induce the necessary currents in the armature. Because the pulsed current is neither pure DC or pure AC, they sometimes are called *brushless DC motors* and sometimes *brushless AC motors*.

Even before they became attractive for propulsion in competition with glow plug engines, small electric motors were used to power control-surface actuators and throttles.

2.3.4 Control systems

Safe and useful drone flight could occur only when it was possible to control aircraft electronically rather than requiring an onboard pilot to integrate kinesthetic, visual, and aural perception into hand and foot movements linked to throttles, pedals and sticks connected to engines, elevators, ailerons, and rudders by mechanical cables or rods.

Key developments with respect to aircraft control systems involved autopilots and fly-by-wire systems—control systems in which the pilot controls are not mechanically connected to the control surfaces, but rather through electrical circuits embodying digital or analog logic. Those developments, in turn, required improvements in lightweight electronics to implement digital logic, and in wireless data communications. § 2.3.2 explains how model aircraft hobbyists experimented with radio-control systems that delivered signals to onboard components such as relays, stepper motors, and servomechanisms to operate aircraft control surfaces. These developments drew on broader innovations in electro-mechanical devices and control theory.

Mechanical devices that translated Morse code signals into movement already existed. These devices were part of Samuel F.B. Morse's original patent. He had vision that a message in Morse code would be received, not by an operator listening to a pattern of clicks or tones, but by an operator reading a sheet of paper on which an electrically actuated pen had traced the pattern of dots and dashes.

Electrical relays were in wide use. They had been developed in the railroad industry to operate wayside signal systems as early as the turn of the twentieth century. Relays were all over the place by 1920. More sophisticated circuits involving relays to execute digital logic had been developed and were used for signaling in automated dispatch systems by the 1930s, making it possible, by 1950, to automate railroad traffic control on hundred-mile sections of railroads.

Relay-based digital logic also had been pushed forward for the voice telephone infrastructure, enabling automatic dialing to be offered in some parts of the country before World War II.

Servomechanism design leaped forward during World War II as the armed services funded centers like the Servomechanism Laboratory at MIT.[11] They were seeking improved gun sights for ground-based artillery and bombs delivered from the air, target acquisition systems for the just-emerging radar, and aircraft autopilots. The result was an improved theoretical understanding of control loops, and improvements in the design of vacuum-tube based amplifiers for those loops.

The airborne gunsight resulted from collaboration between a young USAF officer, Leighton L. Davis, and Charles Stark Draper, a young professor of aeronautical engineering at MIT. Draper had been working on aircraft instrumentation, including gyroscopic turn-and-bank indicators, altimeters, airspeed indicators, and compasses, when Davis sought his help in adapting ship-borne gunsights for combat aviation. The problem to be solved was computing the appropriate lead for a bomb or burst of gunfire from aircraft speed and angle of dive, despite rapid oscillations in aircraft orientation. By the end of World War II, the team had solved the problem with a combination of radar, gyroscopes, equations of motion, and rapid analog computation—techniques that represented the foundation for subsequent automatic control systems.

2.3.5 Communications links and hardware

Even the most sophisticated electronic control systems would not make drones possible, unless they could be controlled remotely, by radio. § 2.3.2 details the early efforts by model aircraft hobbyists to couple their aircraft to their human controllers by RF signals. Wireless control technology, in turn, required the development of signals and associated receivers and transmitters that could format information and interpret it without human intervention.

Radio (wireless) communication grew from the discovery that electrical signals could move through space without wires. The starting point was mathematical proof by James Maxwell in 1873 that electromagnetic forces (electrical currents that varied with time) could travel through free space. Fifteen years later, Heinrich Hertz conducted experiments validating Maxwell's theory.[12] The early work focused on induction—the capacity of a current in one wired circuit to induce a matching current in a nearby but separate wired circuit. Follow-on experimentation, mostly in Europe, demonstrated that higher frequency variations in the current could induce much weaker currents in circuits located at greater distances. By the turn of the century, Guglielmo Marconi demonstrated that an electromagnetic signal could be transmitted across the English Channel and, later, across the Atlantic Ocean. More important, Marconi's signal included information, in the form of Morse Code.[13]

Marconi's breakthrough combined *propagation* with *signaling*. Propagation is simply the movement of electromagnetic waves from one point to another not connected by wires to the first. Signaling is the superimposition of information on the waves. Propagation of radio signals depends on the frequency of an underlying carrier signal, and is possible only above certain frequencies. Signaling depends on the combination of a coding method and the process of combining the coded information and the carrier signal, called *modulation*.

Marconi's experiments used the Morse Code as the coding protocol, and his simple modulation technique involved nothing more complicated that turning on and then turning off the carrier signal. This method imitated wired telegraphy in which code was sent by turning on and turning off an electrical current in a wire.

Radio technology initially advanced through the twentieth century by embracing vacuum tubes. Vacuum tubes were combined with tuned electrical circuits to transmit more precise higher frequency carrier signals modulated by continuously varying currents representing audio information such as voice. The same phenomena used to transmit voice was extended, first, to still pictures, and then to full-motion video. Meanwhile, refinements in vacuum-tube and then semiconductor technologies allowed frequencies to increase, which, in turn, permitted information to be sent at the higher rates necessary for video.

Radio telegraphy was well-developed before the war, and indeed was the dominant means of military radio communication during the war, rather than radio telephony.

Radio telegraphy actually was a better conceptual foundation on which to build radio control systems than radio telephony. Telegraphy is inherently digital and it brings with it accepted coding and modulation schemes. Digital, in this sense, means that the basic signal is either on or off: a light or its absence; a buzz or silence; an electrical current or none. Analog signals, in contrast, vary continuously, to represent things like sound pitch.

The Morse Code was the telegraphic coding system, providing a system for translating the alphanumeric characters comprising textual messages into a series of dots and dashes in a unique pattern for each letter. The letter E, for example, is represented as a single dot; the letter O as a series of three dashes, and the letter S as a series of three dots. The letter H is represented as a series of four dots, and the letter P as a pattern of dot-dash-dash-dot.

The modulation scheme was simple; all it needed was a spring-loaded switch, which already existed—a telegraph key, already in use for about 100 years. The telegraph key was used rapidly to switch the RF carrier on or off intermittently. The brief presence of the carrier represented a dot, while a longer carrier pulse represented a dash. Any circuitry that could generate an RF carrier at a particular frequency sufficed for the transmitter. All a receiver had to do was to tune to the desired frequency, detect the interruptions in the carrier accurately, and provide an audible tone of the pattern of dots and dashes in the operator's headphones.

Radiotelegraphy required a human radio operator at the sending end to key Morse Code, and a human operator at the receiving end to decode it into the characters the dots and dashes represented.

Radio teletype, in use by 1930, was a seminal development because, for the first time, a machine could accept information in the form of alphanumeric key presses, automatically translate it into digital representations, and transmit it. On the receiving end, a similar looking machine could accept the digital pulses and translate them into letter or number bar strikes against a page.

The problem was finding enough distinct frequencies on which to transmit the information. If multiple streams of information could be transmitted on a single frequency fewer separate frequencies would be necessary. Alexander Graham Bell had pursued the solution—multiplexing—before he invented the telephone: the use of tuned metal reeds to generate audio tones to carry signals. By using a multiplicity of such reeds, multiple telegraph signals, one on each audio frequency, could be transmitted simultaneously over one telegraph wire. This was the *harmonic telegraph* that Bell was exploring when he stumbled across the capacity of the same circuit to carry voice signals.

Bell had been fascinated with the possibility of using electrical signals to transmit speech from an early age to help his deaf mother. At one point he invested considerable effort in teaching his dog to talk, by manipulating its tongue and lips while it growled. The canine experiments were unsuccessful, but the idea of sending multiple signals over a telephone line by encoding each at a different audio frequency resulted in US Patent No. 174465 A, "Improvement in Telegraphy," granted to Bell on March 7, 1876.

Telephony has a different purpose from multiplexed telegraphy, but relies on the same underlying capability of sending different audio frequencies over a wire.

Increasing the carrier frequency of remote control transmitters and receivers was desirable for several reasons. First, antenna lengths must be proportional to the wavelength. Wavelength diminishes as frequency increases. A dipole antenna for a 56 MHz channel is 8 ft long. By comparison, a dipole antenna for a 2.4 GHz signal is 2.5 in long. Second, higher frequencies allow for greater bandwidth. Shannon's theory declares that bandwidth requirements increase as the rate of information transmitted increases. As the bitrate of the information signal increased, the frequency of the carrier signal had to increase as well. Chapter 3 explains that a carrier must be at least twice the frequency of a modulating signal.

Increasing the frequency of transmitters and receivers depended on developing better hardware for oscillators and amplifiers. Oscillators generate a carrier signal at the requisite frequency; amplifiers make the signal stronger. Medium wave and HF communications were common during World War II, while low band VHF became commercially attractive by the late 1950s, with higher band VHF and UHF joining the offerings by the 1960s and 1970s.

Communication above 2 GHz awaited development of replacements for the conventional vacuum tube.

Microwaves are ideal for transmission of large quantities of information, such as that involved in video imagery, from point to point. The microwave portion of the radio spectrum usually is considered to lie between 1 GHz and 30 GHz, with corresponding wavelengths ranging from 30 centimeters at the lower end of the frequency spectrum to 1 centimeter at the higher end. The small wavelengths make highly directional antennas possible at sizes feasible for airborne use, and greater bandwidth is available to accommodate video signals than at lower frequencies.

Advances in radar technology during World War II made microwave communication, which had been demonstrated as early as 1931 by a US–French venture, commercially practicable. Particularly significant were invention of the Klystron oscillator and parabolic antenna designs. The Klystron (US Patent No. 2242275), invented mainly at Stanford University and perfected at MIT's Radiation Lab, overcame the limitation of conventional vacuum tubes and associated circuitry necessary to create oscillators at the small wavelengths. An oscillator is the starting point for any radio transmitter; it is what generates the RF frequency. Conventional vacuum-tube oscillator designs failed to work properly at higher frequencies because the higher frequency electrons traveling inside the tube moved between cathode and plate too rapidly. Internal capacitance and inductance of the tube essentially short-circuited the resonant part of the external circuit, dampening isolation as well. Klystron tubes reflected the crucial insight that the resonance necessary for oscillation should occur inside the vacuum tube rather than outside it in resistor-capacitor circuits. Magnetron tubes, developed at almost the same time as the Klystron, also generated acceptable levels of microwave RF, but their less stable frequency and phase limited their utility for communications and relegated them to radar and microwave-oven applications.

Beginning in the mid-1940s, microwave relay links using transmitters with Klystrons formed the backbone of the long-distance US voice telephone infrastructure, gradually being replaced by optical fiber and satellite links in the 1990s and afterward.

Klystrons are still used in high-power microwave applications, such as TV and ground-station satellite transmitters. For smaller, more weight- and power-sensitive applications such as in airplane, helicopter, or drone video downlink systems, they have largely been replaced by semiconductor devices such as *Gunn diodes*, transferred electron devices.

2.3.6 Wireless transmission of photographs and full-motion video

Most drone missions involve aerial photography, or closely associated infra-red, radar, or sonic imagery and profiling. To realize their full potential, drones must be able to send the resulting imagery to the ground.

The idea of aerial reconnaissance is older than flyable aircraft. Manned heavier-than-air balloons were used during the Civil War to observe lines of battle and artillery positions. Early work on drones by the armed services focused on reconnaissance roles, as did early helicopter use. Accurate reconnaissance depends on good aerial photography.

Sending pictures to a remote location by wire or radio relies on basic concepts understood as early as the mid-nineteenth century. As early as 1850, experimentation was underway on transferring simple images, such as line drawings, by means of electrical signals.[14] The basic ingredients were clear: some kind of sensor was required on the sending end to detect patterns of lightness and darkness. A light area could be represented by a one, and the dark area by a zero. The ones and zeros representing parts of the image could easily be coded as the presence or absence of a current or voltage, just as in telegraphy, which then could be sent over an electrical circuit.

Once radio was feasible, sending an image required no more than sending something resembling a Morse Code signal; the components of the image, like Morse Code, were represented by binary values. Each value could be represented by the presence or absence of a carrier. On the receiving end, the image could be replicated by a device, such as Morse's pen register, which applied ink to a moving surface when the carrier was present, and lifted off the surface when the carrier was absent. The challenge was in deciding how fine the granularity should be in the detection of the original and its replication on the receiving end, and, more importantly, how to synchronize the movements of the original and the replicated copy, including when to move to the next line.

The finer the granularity of the original image, the finer the granularity of the reproduction process needed to be. Speed of transmission was not particularly important; a lower rate of data transfer simply resulted in transmission of a particular image taking longer.

The simplest approach would use a pattern of dots and lashes representing Morse code. As long as the granularity was sufficiently fine to distinguish between different lines of code, the detection device would replicate the signal in the same way Morse's pen register did. The only difference is that the signal would originate with a scanned image instead of with a human operator manipulating a telegraph key. On the other hand, if the original image contained greater detail, such as lines of alphanumeric characters or the light and dark areas of a photograph or drawing, the granularity needed to be finer.

This basic system for transmitting an image has remained intact for more than 150 years. Strips of the original are scanned left to right by a moving sensor or by moving the image under the sensor. Then, either the image or the sensor is incremented down and scans the next strip in the same way. The resulting pattern is known as a raster. Obviously, the transmitting system and the receiving system must use the same raster definition, in terms of the width of the strips and their length and the size of the increment that reflects the spacing of the strips. The size of the sensor on the sending size and whatever replicates the light and dark areas on the copy must be proportional to the smallest significant elements of the image. Standardizing rasters is mainly a matter of conceptualization and agreement.

The real challenge is finding a way to synchronize the two. The earliest effort used pendulums on both the sending and receiving side. Then clocks were used at both ends. Eventually a system of synchronization electrical pulses was adopted. The data structures had fields that allowed the receiver to distinguish synchronization pulses from pulses representing content. One type of synchronization pulse signified the beginning of a line, another signified the end of the line, and a third signaled the receiver to increment by one line.

By 1935, the Associated Press, Western Union, and AT&T were offering photofacsimile services to their members. The original was placed on a rotating drum and scanned with a light-sensitive sensor. At the receiving end, a drum of the same size rotated under some kind of pen. The speed of rotation of the drums and the size and timing of the line increments were precisely synchronized by synch pulses embedded in the standard. At first, dedicated telephone lines were required. By 1940, smaller machines could transmit an image over ordinary voice lines.

Xerox Corporation adapted its basic technologies for photocopying and introduced LDX (long distance Xerography) to the commercial marketplace in 1966, with its Magnafax Telecopier which enabled anyone to send images over ordinary telephone lines at a rate of a page every six minutes. The Magnafax had strong appeal, because at 46 pounds, it was cheaper and less cumbersome than Western Union's Deskfax machine, introduced in 1948. Faster speeds soon resulted as competitors dispensed with the rotating drum design and brought to bear digitization and data compression technologies developed for other purposes.

The same signal logic used by fax machines applies to all image transmission, including modern-day television and other video transmission. Only the granularity has changed. Organizations provide values for number of lines for screen. For example, Standard-Definition (SD) video is standardized in the US as 480 lines by 1080 pixels per line. HD video is either 720 lines by 720 pixels per line or 1080 lines by 1080 pixels per line.

By comparison, early fax standards specified 96 scan lines per inch. Current standards, associated with digitization of the scan, allow from 100-400 scan lines per inch.

Refinements involved mechanical engineering advances so that the movements of the sensor and pen could be smaller and more precise. Advances in coding, compression, and modulation schemes were needed to reduce the time required for transmission.

The developments in digital photography and videography to make drone-based imagery—particularly microdrone-based imagery—attractive came much later, well after the PC and Internet revolutions were underway.

2.3.7 Computing power and miniaturization

Many of the seminal advances in specific technologies for unmanned aircraft improvement occurred only when digital logic circuits in small packages became available. Miniaturization required the replacement of vacuum tubes and mechanical or magnetic memory devices with transistors and the development of manufacturing techniques to fabricate transistors and associated circuit components in smaller sizes.

The earliest digital computers relied on vacuum tubes acting as amplifiers, switches, and relays as memory components. Performance of practical logical operations required a large number of vacuum tubes and relays, which sparked, burned out, and consumed substantial amounts of electrical power.

Two technological developments, occurring more or less in parallel, overcame these limitations, resulting in significantly improved and more complex logic circuits. IBM, in the late 1950s, introduced *core memory*.[15] Each memory element was built around a small core of a magnetizable material. The core was wrapped with a coil of wire through which a current could flow. Depending on the direction of the current, the core would be magnetized in one direction or the other. One direction represented a zero; the other represented a one.

The second development, eventually eliminating core memory, was the invention of the transistor.

A transistor comprises two or more layers of silicon, each of which has a different chemical composition resulting from doping it with another chemical. The sandwich of layers functions much like a vacuum tube. A current can be applied to the layer on one side. When a signal is applied to the intermediate layer, the output current extracted from the final layer is an amplified, accurate, representation of the signal. In this mode of operation, the transistor serves as an amplifier. If the current applied to the intermediate layer reaches a cut off value, the transistor also, like a vacuum tube, serves as a switch.

Two transistors can be wired together to form a flip-flop, a simple device capable of holding state. As one or the other transistor applies output that holds the other transistor in cut off, the flip-flop holds a one. The state can be flipped to represent a zero, by pulsing the flip-flop.

Transistors consume far less power than tubes, because nothing inside them needs to be heated up. They never burn out, under normal operation. They also can be made arbitrarily small, limited only by fabrication techniques.

Tiny miniaturization now was possible. As researchers worked on miniaturizing transistors and their associated components, it became obvious that it would be convenient if the combination of components comprising commonly used circuits such as amplifiers, oscillators, and flip-flops could be delivered in ready made packages—integrated circuits. A designer could draw his logic flow diagram at a higher level of abstraction and simply select from available functional components, without having to worry about matching specific transistors, resistors, and capacitors to design the desired circuit at a more specific level of abstraction.

And, the same concept of integration could be applied at even higher-level abstractions. Not only would a system designer be able to buy a complete oscillator package, he could buy a larger, more fully integrated package that comprised a complete radio transmitter or receiver. Not only could he buy a flip-flop; he could buy a complete central processing unit (CPU) for a computer, or a package of accelerometers and associated circuitry to deliver to deliver velocity and acceleration information, or a complete GPS package that would deliver latitude and longitude data.

For fifty years after 1960s, progress took place in the form of greater miniaturization and larger scale integration, with associated improvements in fabrication technology, material science, and dissipation of heat.

The first integrated circuits, developed in the late 1950s, contained one transistor and a few resistors and capacitors. Since then, their density has doubled approximately every two years, with the 2005 Intel Itanium series containing 500 million transistors and associated components.[16]

2.4 Military drones

As military and naval aviation flourished, operators began to consider how they might reduce flight crew costs and the hazards to pilots, gunners, and passengers. One obvious possibility was to get rid of the pilot for certain reconnaissance and combat missions (transport was exempt from this quest because of the presence of passengers). The potential of unmanned aircraft, however, had to await better technologies for aircraft control systems and for radio links to pass information back and forth between the aircraft and the ground operators, subjects explored in § 2.3.

Drones have been flown by armed services for at least 50 years. Only recently, however, have digital wireless technologies and video technologies progressed to the point that they are useful in the civilian world. Until 2010 or so, all military drones had fixed wings. Now, development is well underway on rotary wing drones for the military services and intelligence agencies.

Work on remotely controlled aerial weapons began toward the end of World War I[17] and continued through World War II.[18]

After airplanes supplanted balloons as observation platforms for tactical battlefield intelligence collection and artillery spotting—"Why all this fuss about airplanes for the Army—I thought we already had one"[19]—interest began to grow in flying airplanes without pilots. The idea apparently originated with Nikola Tesla, famous for promoting alternating current as the means for distributing electric power, and opposed by Thomas Edison's preference for direct current. Tesla reportedly speculated about the idea with Elmer Sperry, who received a Navy contract in 1917 to use unmanned aircraft to launch torpedoes. Sperry and Glenn Curtis had already demonstrated the use of a gyroscope to keep the wings of an airplane level in flight.

Sperry elaborated his early automatic pilot experiments by an unsuccessful effort to develop an unmanned flying bomb, programmed to be launched by a kind of catapult, fly to a pre-designated position and then dive into the ground.

Contemporaneously, Charles Kettering, who developed a self-starting automobile engine, built the "bug" for the army, also a flying bomb. Anticipating Sperry's failure, the bug counted propeller rotations and, after a certain number, dived into the ground. Meanwhile the British were experimenting with unmanned aircraft in the form of flying bombs and as artillery targets. One, the Larynx, flew more than 100 miles.

As World War II approached, the armed services, particularly the navy, used drones for anti-aircraft target practice. The navy went further, equipping a drone with a torpedo and a primitive television camera. The operator watched imagery collected from the drone in order to guide it to a torpedo run on a ship. By the end of the war, the army and navy had purchased some 8,000 small drones, including some with wing spans of 12 ft, 22 HP engines, and capable of speeds up to 140 MPH. All of them were used for target practice, though the army began to experiment with their use for reconnaissance after the war.

Toward the end of the war the Army Air Corps experimented with remote control of bombers loaded with explosives, after their crews bailed out. The results were mixed.

Throughout the remainder of the 1940s, and into the 1950s and 1960s, the services continued to explore expanded uses of drones for reconnaissance and specialized tasks such as laying wire. One research focus involved radio transmission of drone-captured photographs.

Lightning Bug reconnaissance drones were used by the Air Force during the Vietnam War, but intensive drone usage first occurred during the NATO intervention into the Kosovo Conflict in 1999.[20] Eight Army Hunters and five Air Force Predators, joined German, French, and British drones early in the conflict.[21] They were used not only for reconnaissance, but also for coordination of fires from manned aircraft that were restricted to high altitudes by the rule of engagement.[22] Serbian forces sought to shoot down the drones, in one instance launching a Serbian helicopter to fly alongside a Hunter drone and have the helicopter door gunner shoot the drone with his 7.62 mm machine gun.[23]

Drones have played a very public role in finding and killing leaders of terrorist groups in Afghanistan, Pakistan, and the Middle East. The FY 2012 inventory ranges from the Navy's 44,600-pound aerial combat X-47B and the Air Force/Navy 14,900 pound Global Hawk to the Army's 4.2-pound, hand-launched Raven.[24] In 2010 the Air Force had 192 drones, the Army 364, the Navy 15, and the Marine Corps 28, in addition to some 6,200 small unmanned aircraft, which require less support.[25]

Fixed-wing configurations dominate the military and foreign-intelligence inventories for several reasons. First, unarmed reconnaissance missions require long endurance. With existing powerplant technology, this is more easily achieved with piston-engine or turbine-engine power plants than with electric propulsion. Fuel weight for long endurance times with a given thrust are considerably less than battery weight for the same capability.

For delivery of munitions, payloads, and therefore drone size, must be significant. It has been easier to adapt manned-fixed-wing aircraft technologies for these applications than to scale up rotary wing technologies. The substantial amount of pilot inputs required to fly rotary wing aircraft have been more difficult to replicate in autonomous or remotely controlled flight control systems. Large-scale rotary wing drone technology lags behind large-scale fixed-wing drone technology.

The aviation industry has recently made significant progress, however, on rotary wing technology for drones. The Navy is in the carrier landing proof stages for the MQ-8C Fire Scout, an unmanned adaptation of the Bell 407, and expected to deploy some at sea, beginning in 2014.[26] The Fire Scout is 23 ft, 11 in long, has a rotor diameter of 28 ft, a maximum gross weight of 6,000 pounds, and a maximum sling load of 2,650 pounds. It uses the Allison 250-C47B turbine engine, has a maximum cruise speed of 140 knots, a service ceiling of 17,000 ft, maximum endurance of 14 hours (11 hours with a 600-pound payload). It currently costs $18.2 million per aircraft.[27]

Rotary-wing drones recently extended their missions from pure reconnaissance and delivery of fires to logistics support. An unmanned, remotely-controlled version of the Kaman K-MAX has been used by the Marine Corps to deliver 3,500 pounds of food and supplies to troops in Afghanistan.[28]

2.5 Intelligent missiles

Some of the technology development in control systems and wireless communications occurred in connection with guided missiles. Guided missile development began in Germany with efforts to add wireless guidance to the V-1 cruise missile (popularly known as the *buzz bomb*).[29] Most of the V-1 and V-2 missiles used to attack England during the Blitz, however, were unguided. Efforts to improve guidance systems continued in the United States after the war. No stone was left unturned. In one unusual approach, behavioral psychologist B. F. Skinner received modest research funding for a system that relied on three specially-trained pigeons. They would guide a missile by pecking on different areas of a video monitor inside the missile to correct its course toward the target.

While external human guidance is common in missiles aimed at moving targets such as aircraft and ground vehicles, the preferred approach is to rely on internal guidance systems based on inertial measurement units, GPS navigation, heat sensing, onboard radar.

Internal guidance eliminates the possibility that an enemy will corrupt or jam the control link.

2.6 Space systems

The challenges of unmanned aerospace vehicles and the challenges of controlling them and flying them safely to accomplish their missions are not new. They have been a reality since the Soviet Union launched Sputnik in 1957, and the United States launched Explorer six months later. In the ensuing half-century, hundreds of unmanned space vehicles have been launched and performed their missions. GPS satellites, military reconnaissance, and telecommunications satellites blanket the globe.

Space probes fly to Mars and collect samples on the ground; deep-space probes penetrate the outer reaches of the galaxy, sending photography back to Earth. None of these activities can be successful without robust control links and other data links. The demands of such datalinks have caused NASA to project the need to multiply available bandwidth by a factor of ten every ten years.[30] NASA is embarked on an aggressive R&D project concerning optical and laser links, new modulation techniques, and new types of antenna arrays.[31]

In comparison to the demands of space-vehicle-control technology, the requirements for terrestrial drones are modest.

R&D for space systems consistently has blazed new trails for technologies eventually adapted to civilian uses. Composite materials, structural analysis software, inertial guidance systems, new techniques for efficient, long-distance radio communication, solar cells, analysis of photographic and infrared imagery, and, of course, autonomous flight vehicle operations, all were developed or enhanced for the space program before they migrated to civilian aviation.

2.7 Other remotely controlled vehicles

The technology improvements necessary for drones have not been confined to the aviation community. Radio control is interesting for all kinds of vehicles. Radio controlled motor vehicles and boats were available for hobbyists long before radio controlled airplanes. Indeed, the earliest demonstration of radio control technology involved such vehicles.

Railroads long have been interested in remotely-controlled locomotives to handle switching of freight cars in rail classification yards.[32] Allowing the supervising crew member on the ground (the "conductor" or "foreman") to control the switching locomotive eliminates the need for an engineer aboard the locomotive. This saves crew costs, and also reduces delay and error associated with voice-radio instructions from the conductor to the engineer. As might be expected, the unions representing affected employees have been opposed. Despite the opposition, their use is expanding because of the obvious advantages.

In February 2014, the National Highway Traffic Safety Administration announced its intention to require vehicle-to-vehicle data communications, functionally equivalent to ADS-B, for automobiles.[33] Chapters 3 and 12 explore more fully the implications of self-driving ground vehicles for drone operations.

2.8 Bridges to the civilian and commercial worlds

As the introduction explains, microdrones are derived from—and in many cases, are identical to—products popularized by the model aircraft hobbyist community. For them the question is not how to adapt them to civilian applications; they started out as civilian. The question is how to adapt them for *commercial* use.

Often little adaptation is necessary to earn money for the same mission that can be flown with the same equipment on a volunteer basis—capturing aerial video of a wedding, for example. The market has rushed to embrace microdrone products initially marketed as toys.

Macrodrones derived from military drones present more daunting challenges. The required adaptation starts with revisiting the basic design criteria:

- enough useful load to carry and launch weapons
- sensors designed for target detection and tracking to enable successful weapon launches and
- endurance times and ceilings designed for tactical military reconnaissance and avoidance of detection or vulnerability to attack.

These are not the most relevant criteria for civilian applications. The most interesting civilian applications—in law enforcement, electronic news gathering (ENG), surveying, and utility infrastructure patrol—do not require as much endurance as military and foreign intelligence missions. So, considerably smaller-sized drones are interesting for these applications.

Law enforcement patrol, ENG, and utility patrol require slow speeds, relatively low heights above ground level and frequent hovering. Rotary-wing aircraft can do this better than fixed-wing aircraft; indeed only rotary wing aircraft can hover. That is why helicopters predominate over fixed wing aircraft in law enforcement support and ENG. Rotary wing drones have the distinct advantage over fixed-wing drones, because they do not need a specially prepared place to operate from. In addition, depending on their basic design, they may have much more simplified control systems.

The question for existing or would-be macrodrone designers and manufacturers is whether they are better off adapting military designs or starting on a clean slate. Military designs usually involve prices and elaborate support infrastructures that make adoption infeasible in civilian commercial markets.

Notes

1 Leonardo da Vinci, *The Helicopter*, 2015, www.leonardodavincisinventions.com/inventions-for-flight/leonardo-da-vinci-helicopter/ (early helicopter design concepts).
2 David McCullough, *The Wright Brothers*,102–103 (2015) (describing the first powered flight) [hereinafter "McCullough"].
3 McCullough at 100 (describing Langley's second failure).

4 US Patent No. 821393 (May 23, 1906).
5 Dustin R. Szakalski, *Progress in the Aircraft Industry and the Role of Patent Pools and Cross-Licensing Agreements*, 15 *UCLA J. L. & Tech*. 1, 5–10 (2011) (explaining the problem and the patent-pool solution); *The Wright/Smithsonian Controversy: The Patent Pool* (2014), www.wright-brothers.org/History_Wing/History_of_the_Airplane/Doers_and_Dreamers/Wright_Smithsonian_Controversy/08_The_Patent_Pool.htm (describing the controversy and US government pressure to establish a patent pool).
6 James R. Chiles, *The God Machine* (2007), Bantam Reprint.
7 Pete Carpenter, *Model flying – an overview*, www.rc-airplane-world.com/model-flying.html (early model airplane design, concepts, and history).
8 Radio Control Hall of Fame, *First Kraft Proportional*, www.rchalloffame.org/Exhibits/Exhibit36/ (describing the first feasible proportional control system for model aircraft).
9 Bill Mohrbacher, *History of Model Engines*, www.modelaviation.com/enginehistory (1911 Baby engine design and history).
10 Bill Mohrbacher, *History of Model Engines*, www.modelaviation.com/enginehistory (Ray Arden commercialized the glow plug engine).
11 Lincoln Laboratory, Massachusetts Institute of Technology, *Lincoln Laboratory Origins*, www.ll.mit.edu/about/History/origins.html (summarizing the history of MIT Servomechanisms Lab and its work on gunsights).
12 *Scientists and Electromagnetic Waves: Maxwell and Hertz*, science.hq.nasa.gov/kids/imagers/ems/consider.html (Maxwell and Hertz theory on electromagnetic waves; Hertz applied this theory to radio waves).
13 Guglielmo Marconi, www.history.com/topics/inventions/guglielmo-marconi (long distance wireless telegraph and broadcast of transatlantic radio signal).
14 FaxAuthority, *The History of Fax – from 1843 to Present Day*, www.faxauthority.com/fax-history/ (explaining the origins of practical image transmission).
15 Computer History Museum, *Magnetic Core Memory*, www.computerhistory.org/revolution/memory-storage/8/253 (reviewing history and significance of core memory).
16 Harry K. Charles, Jr., *Miniaturized Electronics*, 26 *Johns Hopkins APL Technical Digest* 402, 402 (2005).
17 Lockheed Martin, *No Pilot Seat Necessary*, www.lockheedmartin.com/us/100years/stories/uavs.html (briefly describing the history of drones, beginning with the 1918 remotely controlled aerial torpedo).
18 Much of the history in this section is drawn from John David Blom, *Unmanned Aerial Systems: A Historical Perspective*, Occasional Paper 37 (Combat Studies Institute Press, US Army Combined Arms Center, Fort Leavenworth, Kansas 2010) [hereinafter "Blom"].
19 Blom at 6 (quoting alleged comment by member of Congress).
20 JD R. Dixon, *UAV Employment in Kosovo: Lessons for the Operational Commander* at 2 (2000) (Naval War College Paper), www.google.com/search?q=drones+in+kosovo&rlz=1C1LENN_enUS443US443&oq=drones+in+kosovo&aqs=chrome..69i57.2460j0j4&sourceid=chrome&espv=210&es_sm=93&ie=UTF-8 [hereinafter "Dixon"].
21 Dixon at 7–9.
22 Dixon at 6.
23 Dixon at 10.
24 DOD, *Report to Congress on Future Unmanned Aircraft Systems Training, Operations, and Sustainability* (Apr. 2012), (RefID:7-3C47E5F) at p. 2.
25 Government Accountability Office, Unmanned Aircraft Systems Table 1 at p. 6 (Mar. 2010), www.fas.org/irp/gao/uas2010.pdf.

26 *Unmanned Pair*, Aviation Wk & Space Tech., Feb. 17, 2014 at p. 13 (reporting additional flight testing of MQ-8C and expectations for deployment on destroyers by end of 2014).
27 www.northropgrumman.com/Capabilities/FireScout/Documents/pageDocuments/MQ-8C_Fire_Scout_Data_Sheet.pdf; www.navair.navy.mil/index.cfm?fuseaction=home.display&key=8250AFBA-DF2B-4999-9EF3-0B0E46144D03 (Naval Air Systems Command program description).
28 en.wikipedia.org/wiki/Kaman_K-MAX#Unmanned_remote_control_version.
29 *The Fieseler FI103 (V1) German "Buzz Bomb,"* www.museumofflight.org/exhibits/fieseler-fi-103-v1, (history of the German "buzz bomb" WWII).
30 NASA, Deep Space Communications, scienceandtechnology.jpl.nasa.gov/research/research-topics-list/communications-computing-software/deep-space-communications (projecting a ten-fold increase in requirements in each decade).
31 NASA, Deep Space Communications, scienceandtechnology.jpl.nasa.gov/research/research-topics-list/communications-computing-software/deep-space-communi cations (listing research areas).
32 Federal Railroad Administration, *Remote Control Locomotive Operations: Results of Focus Groups with Remote Control Operators in the United States and Canada* (2006) (reporting on focus group discussions of remotely controlled locomotives).
33 www.nhtsa.gov/About+NHTSA/Press+Releases/USDOT+to+Move+Forward+with+Vehicle-to-Vehicle+Communication+Technology+for+Light+Vehicles.

3 Technologies

3.1 Introduction

Drones are possible because of advances in specific technologies. Chapter 2 explains the critical advances in aircraft control systems and wireless data communications that enabled remote control of aircraft. It also explains advances in propulsion system technology, especially those for electric propulsion systems that made microdrones a reality.

The drone revolution in aviation technology opens up the opportunity for aircraft designers to revisit many things they usually take for granted. As Space-X and Tesla Motors entrepreneur Elon Musk said in an October 2014 symposium celebrating the centennial of MIT's Aero and Astro Department, "Airplanes have many unnecessary things, like tails, rudders, elevators, and ailerons. Rockets have gimbaled motors to control them. Why don't airplanes? You just gimbal the fan." A gimbaled engine controls vehicle orientation and therefore flight path by varying the direction of thrust pursuant to commands from the control system. With this means of adjusting vehicle orientation, control surfaces such as ailerons, elevators, and rudders—explained in § 3.4—are unnecessary.

Designing new microdrones and macrodrones allows engineers to follow Musk's advice: to reexamine conventional wisdom about the basic configuration of aircraft—such as whether control surfaces for airplanes must be mounted on the tail and whether rotorcraft must have tail rotors. Both have substantial parasitic effects: a rear mounted horizontal stabilizer subtracts from the lift generated by the wing; a tail rotor on a helicopter consumes about 30 percent of the power developed by the engine. Neither adds anything to performance beyond countering other forces.

Aircraft design is the art of making trade-offs. One can design a fixed wing drone to be stable about its yaw, pitch, and roll axes, but it cannot be stable about all three and still be stable in the combination of roll, pitch, and yaw that results in a death spiral like the one that killed John F. Kennedy, Jr. An unmanned helicopter with a 100 ft diameter rotor would have much better rotor efficiency and desirable autorotation characteristics than one with a smaller rotor, but it would be hard to find a hangar big enough to hold it.

Trade-offs are a function of the technologies available and the economics when the design decisions are made. Technologies change. The question in the background for any aircraft design is how many past design decisions should be reconsidered.

Regulatory impediments usually produce the answer: "not many." Empirical results and customer acceptance are well-established for existing designs, and airworthiness certification is far less demanding when a proponent approaches the FAA with a modification of an existing design rather than something completely new. Nevertheless, the evolution of technology enables new designs, some driven by long-wished-for mission capability; some driven by what the new technologies can do that never has been feasible before.

This chapter explains the state of drone technology and pinpoints the most interesting further paths of technological development that will increase the attractiveness of both classes of drones in the future, in terms of performance and cost.

The chapter focuses on technology that is most relevant to drone flight. It concentrates on those most recently commercialized, including electric propulsion systems, the choice for microdrones; on sensors, image capture, and image processing, the most likely mission oriented technologies for microdrones and macrodrones alike; and on autonomous navigation systems, control links and image downlink systems. To the extent necessary, it provides a theoretical foundation to understand the more specific analysis of hardware, software, subsystem performance, and likely improvements as technology develops further.

Throughout, it highlights critical technology developments that will determine the future of drones.

Most drone designs delegate to complex electronic systems many of the functions that used to be performed by human pilots and mechanical systems meeting detailed certification requirements under the watchful eye of regulators. Now, as chapter 9 explains, drones can be designed and manufactured to be law-abiding; reducing the need for detailed regulatory oversight. As responsibility for compliance with safety standards shifts to autonomous systems, however, regulation must develop alternative ways to ensure that these electronic systems perform as advertised. Chapter 9 and § 3.15 of this chapter explore how reliability engineering can be adapted to the task.

3.2 What do drones look like, and how do they work?

The basic configuration choices for drones are essentially the same as for manned aircraft: fixed wing or rotary wing. Rotary wing designs can feature a single main rotor, as for a helicopter, or multiple rotors. Drones of any size can be built in any of these configurations. But design choices for microdrones have clustered around the multirotor concept, and design choices for macrodrones concentrate on fixed wing concepts, with an occasional unmanned helicopter design thrown in.

The dividing line between microdrones and macrodrones is statutorily defined as 55 pounds gross weight. As regulatory regimes evolve, a finer set of weight- and performance-based distinctions are likely, but for now, 55 pounds marks the distinction.

3.2.1 Microdrones

Most microdrones are electrically powered multicopters, typically with an even number of motor and rotor combinations to facilitate pairing the thrust of directly opposed rotors on opposite sides of the aircraft for control. *Quadcopters* have four rotors, *hexacopters* have six, and *octocopters* have eight. A few *decacopters,* with ten rotors, also have been marketed. All of them use differential rotor thrust to provide control about the three axes of flight.

The most popular multicopter model as of the publication of this book is the DJI Phantom, a quadcopter, which comes in several models. They offer:

- Take-off weight: 2.8 pounds
- Maximum flight speed: 31.1 knots
- Diagonal length (including rotors): 23.2 inches
- Flight time: 25 minutes
- Battery type: 2S LiPo
- Battery capacity: 6,000 milli-ampere hours, at 7.4 volts
- Control link frequency: 2.4 GHz
- Communication distance: 1.1 nautical miles
- Autonomy: automatic takeoff, hover, landing, and return-to-home; automatic waypoint navigation; geo-fencing

Priced at around $1,000 depending on the specific model, the Phantom straddles the line between hobbyist and professional user. It is inexpensive, requires no significant assembly, and yet can capture high-quality video imagery.

Phantoms carry either a user-supplied GoPro camera or a similar but more advanced camera installed by DJI. Depending on the particular model of Phantom, the video captured by the drone is stored on board and downloaded from a micro-SD card memory chip when the drone lands, or it can be downlinked in real time.

More sophisticated microdrones have higher gross weight and more power, usually delivered through more rotors, to allow them to carry larger payloads. The larger payloads can include bigger cameras, more sophisticated gimbals, and more advanced wireless links and control and navigation systems. They cost between two and ten times as much as the Phantom, often require some assembly by the purchaser, and need a good deal more testing of components before the vehicle is flight ready.

The drone operator (DROP) controls any of these microdrones by moving sticks and altering the state of switches on a drop console (DROPCON) (or by tapping and swiping a touchscreen) held in his hands or suspended by strap worn around his neck. The DROPCON allows the DROP to mount a smart phone or tablet computer on top of the DROPCON, which displays imagery captured by the Phantom's camera. The display also allows the DROP to read telemetry values for speed, height, and direction of flight.

Microdrone navigation systems provide substantial autonomy. At the option of the DROP, they can take off and land autonomously, and they can be programmed to fly from one waypoint to another and to hover over or orbit around selected waypoints defined by tapping on GPS-derived map, correcting for wind.

Almost all of them have automatic return-to-home capability, which can be triggered by the onboard navigation system or by the DROP. When triggered, the systems arrest the drone's mission and cause it to navigate back to its launch point and to land autonomously.

Lithium polymer batteries are the norm for the main power supply, capable of producing flight times from 15 minutes to nearly an hour. Microdrones typically fly at up to 35 knots or so and have service ceilings of several thousand feet above the ground. Their range is determined more by the range of the control link and the video downlink wireless signals than by battery life.

Control links use Wi-Fi coding and frequency-hopping spread-spectrum modulation schemes on the 2.4 GHz band, and allow sensor imagery download on either the 2.4 GHz or the 5.7 GHz band, using similar coding and modulation technologies.

Not all microdrones are multicopters. The Delair-Tech DT18 is a 2 kilogram (4.4 pound) electrically powered fixed wing microdrone with a range of about 15 miles and endurance of several hours. It is capable of fully autonomous operation or flying under DROP control. It has been approved by the French aviation agency and is in operational use for aerial surveying.

3.2.2 Macrodrones

The market for macrodrones is developing more slowly than that for microdrones, and so portrayal of a typical design is murkier. Designers, manufacturers, and vendors, apparently are reluctant to commit themselves to a type certification process which is reported to cost up to $20 million for a new general-general aviation aircraft type, though a variety of military drones have been adapted for civilian use.

The high end is represented by Northrop-Grumman's fixed-wing Global Hawk. Not yet marketed in commercial adaptations, its military version offers:

- Wingspan: 130.9 ft (39.9 m)
- Length: 47.6 ft (14.5 m)
- Height: 15.4 ft (4.7 m)
- Gross take-off weight: 32,250 lbs (14,628 kg)
- Maximum altitude: 60,000 ft (18.3 km)

- Payload: 3,000 lbs (1,360 kg)
- Ferry range: 12,300 nm (22,780 km)
- Loiter velocity: 310 knots True Air Speed (TAS)
- On-station endurance at 1,200 nm: 24 hrs
- Maximum endurance: 32+ hrs

At the smaller end, are the Institu Scan Eagle and Aeroenviron's Puma AE (MQ-20)., Both arguably are big microdrones rather than macrodrones. Both aircraft were granted limited certification by the FAA for operational flights in Alaska. The Scan Eagle is a turboprop (optionally powered by a gasoline engine) with the following capabilities:

- Length (EO dome): 5.1 ft (1.55 m)
- Length (EO900/MWIR dome): 5.6 ft (1.71 m)
- Wingspan: 10.2 ft (3.11 m)
- Empty structure weight: 30.9–39.68 lbs (14–18 kg)
- Max takeoff weight: 48.5 lbs (22 kg)
- Max payload weight: 7.5 lbs (3.4 kg)
- Endurance: 24+ hours
- Ceiling: 19,500 ft (5,950 m)
- Max horizontal speed: 80 knots(41.2 m/s)

The Puma AE is a fixed-wing, electrically powered configuration, with the following capabilities:

- Endurance: 2 hours
- Sensors: Photographic and infrared
- Control link range: 8.1 nm
- Range: 8.1 nm
- Speed: 20 to 45 knots
- Operating altitude: 500 ft (152 m) AGL
- Wing span 9.2 ft (2.8 m)
- Length 4.6 ft (1.4 m)
- Weight 13.5 lbs (6.1 kg)
- Launch method: Hand-launched, rail launch (optional)
- Recovery method: Autonomous or manual deep-stall landing

The Scion UAS SA-400 Jackal is a turbine-powered, optionally piloted helicopter, built for the US Navy's Naval Research Laboratory. It has completed its flight test program. Its specifications are:

- Height: 82.1 inches
- Width: 51.5 inches
- Length: 232 inches
- Rotor diameter: 250 inches

- Gross weight: 1200 lbs
- Empty weight: 700 lbs
- Useful load: 500 lbs
- Fuel capacity: 60 gallons
- Max speed: 95 knots
- Cruise speed: 82 knots
- Endurance: 4+ hours

3.3 Finding the right design philosophy: mission requirements or technology's possibilities?

Sometimes new mission concepts lead to completely new designs. This was the case when the need for high-altitude reconnaissance led to the U-2; demand for intercontinental ballistic missiles to deliver multiple independently retargetable warheads led to new kinds of inertial guidance systems; and the goal of putting a man on the moon led to the multistage Saturn rocket with a small orbiter payload.

Other times, the obvious potential of new technologies has caused people to dream up new missions. The helicopter, which first became practicable after World War II, is an example. It caused people to think, "what could we do if we didn't need runways and could remain stationary in the air?" Sometimes the mission-driven process prevails; at other times the technology-driven process prevails.

In a market economy, neither the new-mission-invites-new-design nor new-technologies-inspire-new-missions process is unidirectional. Engineers and entrepreneurs come up with new design ideas all the time. Operators go shopping for systems that can perform new missions. When their ideas resonate with those of the engineers and of other operators, something new happens, and the market either accepts or rejects it.

In the drone world, mission will drive design for macrodrones, and technology will inspire new missions for microdrones.

The following sections survey the technology categories most relevant to drone development. They begin with basic aircraft technologies. The same basic principles that allow airplanes and helicopters to fly also enable drones to fly. But drones also have significant differences from manned aircraft. The technology sections highlight those differences and explore the technologies that accommodate the differences and thus are more important in the drone world than in the airplane and helicopter world.

3.4 Lift and drag

Basic aerodynamic principles of lift and draft determine how any aircraft—manned or unmanned—can fly and how it can be controlled by pilots on board or DROPs on the ground linked to the vehicle by radio communication.

3.4.1 Getting aloft and staying in the air

Aircraft must have a means to oppose gravitational forces so they can fly and to generate velocity so they can go somewhere. Vehicles intended to operate in space have a rocket engine to perform both functions. Vehicles intended operate in the atmosphere have wings or rotors to oppose gravity and internal combustion engines, or electric propulsion systems to generate velocity.

The rotors on a 3 pound Phantom quadcopter generate lift under the same principles as the rotors on a 5,200 pound Airbus AS350B3 helicopter do. The wings on a Global Hawk macrodrone generate lift under the same principles as the wings on a Boeing 787 do.

Wings, control surfaces, and rotor blades are *airfoils*, signifying a shape that generates lift as the structure moves through the air.

Figure 3.1 depicts a simple airfoil. As the airfoil moves through the air, it generates lift because of a phenomenon known as the *Bernoulli's principle*, which says that when a fluid, such as air, moves faster, its pressure drops proportionately. When an airfoil is generating lift, the air moving over the top of the airfoil has farther to go than the air moving over the bottom. It must travel faster to meet up with the corresponding air crossing the bottom. As it moves, its pressure drops according to Bernoulli's principle. Thus the pressure on the upper surface is lower than the pressure over the bottom surface. The pressure differential creates a force known as lift.

The airflow over the upper surface can be forced to travel longer distances and thus move faster in one of two basic ways. First, the wing can be angled upward with respect to the airflow so that the leading edge is higher than the trailing edge. The size of the angle is known as the *angle of attack*. Second, regardless of the angle of attack, the air may be forced to travel farther over the top by the basic shape.

The equation of lift universally is represented as

$$L = \rho/2 \text{ x } C_L \text{ x } S \text{ x } V^2$$

The Greek letter ρ (rho) signifies the air density. Its placement in the equation indicates that more lift will be generated in higher density air, such as air close to sea level; while less lift will be generated in thinner air, such as air at higher altitudes.

The *C* with the subscript *L* is the *coefficient of lift*, a value usually determined experimentally in a wind tunnel. It depends on the basic shape of the airfoil and the angle of attack. As the angle of attack increases, the lift coefficient increases proportionally.

S is the surface area of the wing. A longer chord (the distance from leading to trailing edge) increases the area, as does a longer wingspan. A tapered wing has less surface area than a rectangular one with the same span and chord.

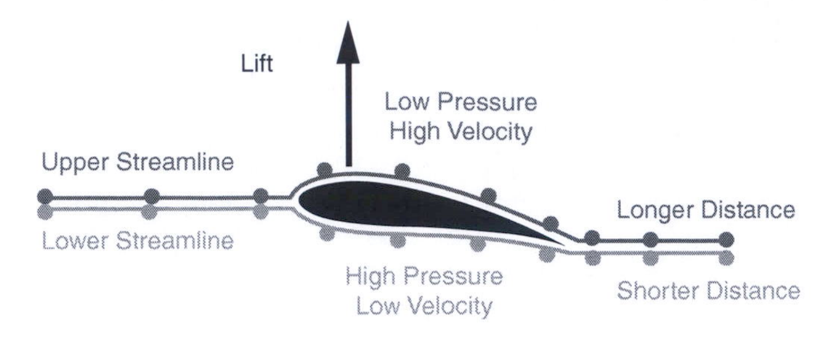

Figure 3.1 Airfoil

At the *critical angle of attack*, usually between 15 and 20 degrees, the flow of air traveling over the upper surface of an airfoil breaks away from the wing and becomes turbulent. When this happens, the lift generated by the airfoil is sharply reduced, usually to 0 or close to it. This phenomenon is known as a *stall*. It has nothing to do with the amount of power being generated by the engine or its rotational speed.

An uncorrected stall leads to the fixed-wing aircraft crashing to the ground because it no longer has lift to oppose gravitational forces.

In a rotary wing aircraft, the rotor blades are airfoils. The amount of lift they generate is, as with fixed-wing aircraft, varied by changing their angle of attack or their speed through the air. The angle of attack is increased or decreased by changing the pitch of the blades with respect to the hub where the blade roots are attached. Airfoil speed through the air varies with the RPM of the rotor.

Airfoils and the aircraft bodies to which they are attached generate drag as well as lift. The drag equation is similar to the lift equation

$$D = \rho/2 \text{ x } C_D \text{ x } S \text{ x } V^2$$

The only difference is the use of a drag coefficient instead of a lift coefficient.

Lift can be increased by:

- Flying at a lower altitude (where the air is denser)
- Increasing the angle of attack
- Flying faster
- Flying with a different airfoil

Drag can be decreased by:

- Flying at a higher altitude (where the air is less dense)
- Decreasing the angle of attack (which reduces induced drag)
- Flying slower (which reduces profile drag)
- Flying with a different airfoil (which improves the lift-to-drag ratio)

All of this was understood, though not yet reduced to mathematical equations, by the late eighteenth century. The Greek philosophers Aristotle and Archimedes had explored the relationship between moving fluids and the forces on objects moving through the fluids, but it was the English engineer Sr. George Cayley who first crystallized the relationships among weight, lift, drag, and thrust, in 1799, when he was 26 years old. Ten years later, he flew the first manned glider.

As chapter 2 explains, it was the Wright Brothers who first demonstrated how onboard pilots could control flying machines.

3.4.2 Controlling flight

The pilot of a manned aircraft or a DROP may change lift while in flight by changing the angle of attack, by flying faster or slower, and by changing the airfoil shape.

In fixed-wing drones, aircraft control possibilities are the same as for airplanes. The DROP can change the shape of the airfoil by extending flaps, or by moving control surfaces such as ailerons, elevators, and the rudder. *Flaps* are surfaces built in to the wing near the trailing edge that can be lowered or raised. As they are lowered, they generate additional lift at any given angle of attack. *Ailerons* are used to roll the aircraft—to cause it to rotate about an axis represented by a line drawn from the nose to the tail. They function by causing one wing to generate more lift and the other to generate less lift.

Elevators are similar devices mounted to the trailing edge of a fin attached to the rear of the aircraft known as the *horizontal stabilizer*. They make the nose go up and down with respect to the tail (*pitching* up or down) by exerting positive or negative lift as the elevators move to change the curvature (*camber*) of the horizontal stabilizer.

The *rudder* is a similar device intended to make the nose go left and right with respect to the tail (*yawing*) by changing the camber of a vertical fin known as the *vertical stabilizer*.

Control of rotary-wing drones depends on whether they are unmanned helicopters or multicopters. Control of unmanned helicopters works the same way that it does for manned helicopters: symmetrical and asymmetrical pitch changes on the main rotor blades change the lift generated by the blades and cause the aircraft to go up, down, forward, backward, or sideways, and to speed up or slow down.

Changing the pitch of blades that are rotating at 400 RPM is an engineering challenge. Typically, control inputs from the pilot are applied through three control rods to a lower *swashplate* which does not spin with the rotor. The swashplate idea is attributed to French engineer Étienne Edmond Oehmichen, who studied dragonflies and in 1930, used his experimental results to obtain US Patent No. 1828783 A for the concept; which he incorporated into one of the first successful manned helicopters.

The lower swashplate tilts forward, aft, left and right, at arbitrary angles, depending on the forces applied to the control rods. An upper swashplate rests directly on top of the lower swashplate, with some kind of bearing between the two plates. The upper swashplate rotates with the rotor blades. It has pitch links protruding upward to levers at the roots of the blades. As a pitch link moves up, it raises the lever, thereby reducing the pitch of the blade; as it moves down it does the opposite.

The swashplates working together enable pilot control forces to vary the pitch of the blades in the correct position as the rotor turns.

A peculiarity of single-rotor helicopters is that the engine tries to spin the fuselage to the right as the reaction to its intended function of spinning the main rotor counterclockwise (French, Russian, Polish, Chinese, and Indian helicopter rotors spin clockwise, so the tendency is for the fuselage to yaw to the left). The tail rotor opposes this tendency by applying thrust in a counter-clockwise direction, variable so as to control the aircraft's yaw.

Foot-pedals change the pitch of the tail rotor and thus its thrust. When the pilot pushes the left pedal on an American helicopter, the tail rotor pitch increases, yawing the nose to the left. When he pushes the right pedal, the tail rotor pitch decreases, allowing the torque of the engine to yaw the nose to the right.

Multicopter control works differently. They do not need tail rotors, because yaw can be adjusted by varying the torque generated by the different motors. Varying the thrust generated by their rotors differentially causes them to climb, descend, go forward, backward, or sideways. If the aft rotors generate more lift than the forward rotors, the vehicle dips its nose and goes forward. If the rotors on the left generate more lift than the rotors on the right, the vehicle flies to the right.

3.5 Propulsion systems

Drones use the same basic propulsion systems as manned aircraft. Aircraft, except for gliders and balloons, require propulsion systems to move them through the air and, in the case of rotary-wing aircraft, to cause their rotors to generate lift. Balloons rely on the lower density of hot air, or the lower density of helium or hydrogen to generate lift, not needing propulsive power to spin rotors or pull the wings through the air. Gliders use the propulsion system of a towing airplane to get in the air and thereafter take advantage of updrafts to remain aloft.

Two basic classes of aircraft propulsion systems are in wide use: those built around internal combustion engines, such as reciprocating engines and gas turbines, and electric propulsion systems comprising batteries and electric motors. Electric propulsion is the system of choice for microdrones. Macrodrones are more likely to use internal combustion engines than electric propulsion because of better specific energy of fuels, compared with batteries. The size boundary, however, for electric propulsion, will increase as battery technology improves.

3.5.1 Internal combustion engines

Appreciating why internal combustion engines are the propulsion system of choice for bigger drones and a poor choice for smaller drones requires an understanding of how they work. The efficiency of internal combustion engines improves as size increases; the efficiency of electric motors gets worse as size increases. The Second Law of Thermodynamics says that heat engines such as reciprocating gasoline engines or gas turbine engines cannot extract all the heat energy supplied to them in the form of fuel. This is because all of them generate some waste heat, exhausted to the atmosphere; they cannot turn all their fuel's energy into work in the form of thrust or torque. The proposition was first articulated by a French military engineer for Napolean, Sadi Carnot, who set forth the theory in 1824. Steam engines were proliferating at the time, and their designers struggled to understand what was necessary for them to work efficiently.

The *Carnot Cycle* expresses this limitation as the difference between the heat energy harnessed to do work and the heat exhausted to the atmosphere, divided by the maximum heat energy. Thus an engine with a combustion temperature of 700 degrees and an exhaust temperature of 200 degrees has a theoretical maximum efficiency of 71 percent. Other energy losses in shaft bearing friction, less-than-optimal flow in inlets, cylinders, burner case, compressors, and turbines, lower the practical efficiency to levels of 25-30 percent.

Many macrodrones use reciprocating (piston) engines, characterized as *Otto Cycle* engines (because simultaneous changes to pressure, temperature, and volume occur during combustion). The Otto Cycle is named for Nikolaus August Otto, a self-taught German engineer. Otto, son of a farmer and postmaster, quit his job in a business office, and designed the first successful commercial internal combustion engine in 1875, which ran on illuminating gas made from coal. His four-stroke cycle continues to dominate internal-combustion engine design, and was first patented in the United States in 1877 under US Patent No. 194047 A.

Reciprocating engines develop power from the expansive force of a burning fuel-air mixture acting against one or more pistons that move back-and-forth in sealed cylinders. When they are not developing power, the pistons and cylinders act as pumps, sucking the requisite fuel air-mixture in on the down stroke of the piston, compressing it on the following upstroke; then being driven down after ignition, on the power stroke; and finally pumping the combustion products out during the exhaust stroke. They are equipped with valves that allow the fuel-air mixture to be introduced and the combustion products to be pushed out at the appropriate times. The force they produce during the power stroke is collected by means of a crankshaft to which each piston is attached by an eccentric bearing.

Diesel engines, unlike other reciprocating internal combustion engines, have no spark plugs. In diesels, ignition occurs when much higher cylinder compression raises heat to the ignition point of the fuel-air mixture.

Reciprocating engine technology is mature. Principal advances will relate to greater use of computerized fuel management systems to increase efficiency and reduce emissions

Turbine engines are *Brayton Cycle* engines, named after Boston mechanical engineer George Brayton, who developed a constant-pressure internal combustion engine in 1872. Though his design lost the competition with the Otto-cycle engine, his principles became the basis for gas turbine engines.

In Brayton-cycle engines, combustion takes place at constant pressure. Modern designs have three basic components: compressors, burners, and turbines. The compressor increases the pressure and temperature of free stream air. The burner, often called the "combustor," adds temperature by burning fuel, and the turbine extracts energy from the high pressure and high temperature air to generate torque or thrust. Some aircraft turbines are intended mainly to supply mechanical energy to a shaft which drives a propeller or helicopter rotors. These are called *turboshaft* engines. Others are intended to supply thrust directly by the momentum they impart to the exhaust of the engine. These are *turbojets* or *turbofans*.

Turbine engines cost more than reciprocating engines with equivalent power, but they are considerably lighter and their machinery is simpler.

The frontier for turbine engines relates mainly to material science, replacing aluminum and titanium components with composites and ceramics that are lighter, can be fabricated with greater flexibility, and have higher heat tolerance.

A rotary engine, of which the Wankel engine is the only commercial example, uses a three sided rotor inside an eccentric elliptical cylinder. The points of triangular rotor have seals that ensure a tight fit against the cylinder walls. Intake and exhaust ports and spark plugs are arranged around the sides of the cylinder.

As the rotor rotates, the space between each of its sides and the cylinder walls expands, sucking the fuel air mixture into the cylinder through the intake port during the intake cycle, and then contracts, compressing the fuel-air mixture. The ignition of the fuel-air mixture by the spark plugs, forces the space to expand, imparting a force to the rotor.

Because the motion is rotary instead of back-and-forth, and because it occurs in only one direction, mechanical stresses are much less than on a reciprocating engine, whose pistons and crankshafts that must accelerate, stop, and re-accelerate two or four times for each power cycle. Wankel engines, though they use the Otto cycle, have higher power-to-weight ratios than reciprocating engines. Macrodrone designers are interested in them for that reason.

Their disadvantages involve relatively high emissions and the need to keep the seals tight over economically attractive maintenance cycles.

3.5.2 Electrical propulsion

Practical electrical propulsion systems can be much smaller than either piston engines or turbine engines, and their control can be fine tuned with inexpensive electronic control technology.

For these reasons, electrical propulsion systems predominate in microdrones. The main reason electrical component propulsion systems are not in wider use in manned aircraft and in larger drones is battery weight.

All electric motors have the advantage that they deliver maximum torque at start up and more constant torque over their RPM range than can an internal combustion engine. The torque of internal combustion engines is 0 at start up and tends to increase with RPM. Electric motors thus offer advantages for aircraft applications, which require relatively constant torque over a wide range of RPM and which cannot tolerate the weight and complexity of clutches and gear shifts, common in ground vehicles.

An electric motor generates torque by varying magnetic forces between a stator and a rotor. The rotor, as its name suggests, rotates. The stator is stationary. The coils of wire wound around the stator are known as *field* coils. The coils of wire wound around the rotor are known as *armature* coils.

Chapter 2 explains the evolution of electric motor technology from DC motors with brushes to modern brushless AC motors. Microdrone propulsion systems are dominated by brushless AC motors. They pack considerable power into small lightweight packages. For example, the motors for a mid-level octocopter can utilize 30 amperes or more, generate 900 watts of power (1.2 horsepower), and weigh only 150 grams (1/3 of a pound).

A typical lightweight brushless AC motor has a stator composed of field windings and an armature comprising fixed magnets. Because the windings are static, and electricity need not be delivered to the armature, brushes and commutators are not necessary. Instead, the electronic control subsystem delivers pulses of electricity in a sequence and with the necessary polarity to make the fixed magnets in the rotor spin at varying RPM. The pattern of current applied to the windings is not a smooth sinusoid generally associated with AC current. Instead, it is a train of pulses, reversing polarity every few milliseconds.

State-of-the-art electric motors can transform 80 to 90 percent of the electrical energy they receive into mechanical energy. The comparable figure for internal combustion engines is 50–60 percent, at most.

The practicability of electric propulsion systems for aircraft depends as much on battery technology as it does on electric motor technology. Batteries are heavy, and discharging and recharging them take significant amounts of time.

Batteries use an electro-chemical process to store and release electrical energy. They comprise multiple cells, each of which generates a fixed voltage associated with a particular chemical process. Voltage is a measure of the electromotive force that can be exerted by the battery. For example, the voltage generated by a lead-acid battery cell is 1.5 V, that generated by a lithium polymer cell is 3.65 V, a lithium polymer cell generates 3.65 V, and an alkaline battery common in consumer electronics, generates 1.5 V. An arbitrary number of battery cells can be wired together in series to produce the voltage needed by a particular circuit. The series wiring can be done mechanically, as when one puts multiple batteries in a flashlight with the positive terminal from each cell contacting the negative terminal of the next one. Or, the series wiring can be accomplished internal to the battery so that multiple cells comprise a single package.

Current, measured in amperes or fractions such as milliamperes (1/1000 amperes), is the measure of the amount of electricity that flows through a circuit. Ohm's Law teaches that current equals voltage divided by resistance. Power is voltage multiplied by current, and energy is voltage multiplied by current multiplied by time. Battery energy typically is expressed in ampere hours or milliampere hours.

Lead acid batteries found in automobiles typically store on the order of 10 ampere hours, and are optimized to provide high currents for relative short starter operation—approaching 500 amperes. A Concorde RG-350 lead-acid battery in an AS350 helicopter is rated at 17 ampere hours and can deliver the 150 amperes required by the starter motor for about 7 minutes.

Batteries for microdrones on the market in 2015 range in capacity up to 20,000 mAH, at a weight of 2,500 grams (5.5 pounds). For microdrone operation, peak currents are less important than steady-state current.

The most important development frontiers for electric propulsion are improvements in the specific energy of batteries—the amount of stored energy per pound. Significant improvements in battery technology dwarf the effect of any incremental improvements in motor technology.

3.5.3 Scaling propulsion systems up and down

Improvement in specific energy of batteries will shift the threshold at which electric propulsion systems become more attractive than internal combustion propulsion systems.

But this assumes that the competing propulsion systems scale linearly. They don't. That's why few large aircraft are powered electrically and why few microdrones are powered by internal combustion engines.

The power available from a piston engine depends on the volume (displacement) of the cylinders and the compression ratio (cylinder pressure).

If one scales a reciprocating engine downward, the weight of the cylinder necessary to withstand the pressure increases relative to the horsepower, for any given cylinder material.

If the pressure is kept constant, and the diameter of the cylinder reduced, the power generated diminishes with the square of the radius of the cylinder. The strength of the cylinder wall necessary to withstand the pressure is proportional to its thickness. As the diameter diminishes, the thickness of the wall needs to remain constant to withstand the same pressure. The weight of the cylinder wall is equal to the density of the material from which is made multiplied by its thickness, multiplied by the volume that it encloses. Dividing the power equation by the weight equation shows that the power to weight ratio of increases with size rather than remaining constant.

Turbine engine power depends on the operating temperature and the volume of air moved through the engine, which depends on engine volume. For the same reasons that reciprocating engines lose specific power as their size is reduced, the same thing happens with turbine engines. Nevertheless, a number of miniaturized turbine engines are on the market for model aircraft.

The power generated by an electric motor is proportional to the current flowing through its windings and to the number of windings. If wire diameter is reduced to fit more windings into the same volume, the resistance increases, which reduces the current. Assuming that a designer maintains the wire diameter to avoid this and simply increases the windings, the length of the wire used for the windings must increase. That increases its total resistance, and it also increases its total weight, and proportion to the length and format. The result is that the specific power of an electric motor increases somewhat less than linearly as its size increases.

In other words, internal combustion engines are better in large sizes, and electric motors are better in small sizes.

3.6 Specific energies

The flexibility of electric propulsion driven by batteries compared to internal combustion engines, has pressed technology development for space-systems and submarine batteries. The benefits are increasingly useful for terrestrial aircraft, especially microdrones. In 2015, the battery technology of choice was lithium polymer chemistry, which has a much higher specific energy than the lead-acid chemistry used in batteries for automobiles. It has much less, however, than the specific energy of common fuels.

Not only that, batteries are bulky, take far more time to charge than is required to fill a tank with gasoline or jet fuel, and have a limited number of recharge cycles—on the order of 300–500 for lithium polymer batteries. Their voltage declines below useful levels before the battery is fully discharged; fuel burns until it is depleted.

Table 3.1 Specific energy of different sources of energy

Type of energy source	*Specific energy: watt-hours/pound*
LiPo battery	817
Lead-acid battery	278
Ethanol	18,947
Automotive gasoline	40,835
Aviation gasoline	70,399
Jet A	74,809
Diesel	73,933
Hydrogen	231,942

3.7 Transforming mechanical energy into thrust: jets, propellers, rotors, and fans

All propulsion systems rely on increasing the momentum of the airflow. The thrust of any air propulsion system, ignoring efficiency losses, is the product of the mass of air that it moves and the amount of a momentum it imparts. A turbojet creates thrust by imparting momentum directly to the airstream that flows through the engine and exits the rear at higher speeds than ambient air. Pure jet engines are most efficient if they extract only enough mechanical energy from the air to drive the compressor.

Thermodynamics principles say that imparting a lower increment of acceleration to a larger mass of air results in greater efficiency than imparting larger acceleration to smaller arm mass of air. That makes larger diameter propellers or rotors more efficient than smaller ones. It also makes turbofans more efficient than pure jet engines. Turbofans differ from turbojets in that a fan larger in diameter than the first compressor stage is mounted adjacent to the compressor on the same shaft, and rotates within a shroud that is larger in diameter than the case for the rest of the engine. Turbofan engines are more efficient than turbojets because they accelerate a larger mass of air to smaller velocities. The amount of air moved is proportional to the diameter of the propeller or rotor disc, turbofan, or inlet diameter of a pure jet engine.

Most propulsion systems for drones rely on props or rotors driven by the rotary motion of the engine. The pitch (angle of attack) of propellers or rotors must turn as much of the rotating energy as possible into thrust by accelerating the air.

Any airfoil experiences loss of efficiency in the form of greater drag because of the turbulence generated around its tips (wing-tip vortices). Longer and narrower airfoils have better efficiency in this regard. Vortex effect can also be reduced by housing a propeller or rotor in a shroud. Shrouded props and shrouded rotors represent an alternative to a longer blades to produce the same efficiency.

3.8 Transmissions

Pure jet engines do not require transmissions; other types do. In fanjet engines, the transmission is simple: the same shaft that connects the compressors and turbines also connects the fan, in the same engine housing. Propulsion system designs relying on propellers or rotors require more elaborate transmissions.

Any type of drone engine can turn a propeller or rotor mechanically, or drive electric motors which turn the rotor or propeller. Internal combustion engines turning generators and driving electrically powered rotors are propellers are known as *hybrid propulsion* systems.

The shaft of a drone engine may be connected directly to the propeller that it drives (*direct drive*). Electric motors usually employ direct drive, because motor RPM can be matched to the most efficient propeller or rotor RPM for any particular condition of flight. *Reduction drives* are more common for internal combustion engines to match engine RPM to efficient propeller or rotor RPM. Internal combustion engines develop more power at higher RPM, while the RPM of propellers must be limited to keep their tips below the speed of sound. Otherwise, drag and vibration increase above tolerable levels. Reduction drives typically involve gearboxes, which convert high-RPM, relatively low torque, output from the engine to lower RPM, higher torque input for the propeller.

Fixed-wing performance can be increased by increasing the pitch of the prop as airplane speed increases. Fixed-pitch props suffer from the disadvantage that the relative angle of attack decreases as aircraft speed increases. The pitch can be set for relatively low-speed takeoff and climb, or for higher-speed cruise, resulting in loss of efficiency at speeds other than those for which the prop is optimized. The solution is a *constant-speed* prop, in which a governor—usually powered by engine oil pressure—keeps RPM constant, by varying blade pitch. At a given power setting, increasing blade pitch can extract more power from the engine.

Helicopter drives are more complex. Not only must they reduce engine RPM to drive the lower-RPM rotor, but they also must drive the higher-RPM tail rotor. In addition, they usually must change the orientation of the rotational motion from a typically horizontal engine shaft to the vertical main rotor shaft. They do so with a combination of helical, bevel, and spur gears, with ratios necessary to reduce engine RPM and orientation to change the axis of the torque. Some light helicopters, such as the popular Robinson R22 and R44, also use belts to transfer power from the engine shaft to the main drive shaft.

Tilt-wing configurations, of course, must have more elaborate transmissions, because the angle of the rotor shaft is intended to be changeable.

Electric propulsion systems offer simpler drivetrains. Because electrical motors have more favorable torque output over a wide range of RPM, rotor thrust can be varied efficiently by changing RPM, obviating the need for blade-pitch variation.

Electrically powered multi-rotor aircraft designs, which dominate the microdrone field, also avoid the loss of efficiency resulting from tail rotors, explained in § 3.4.2. A tail rotor is not necessary if multiple main rotors spin in opposite directions. This can be done by mounting them coaxially, by having two, laterally or longitudinally displaced from each other, or by having more than two symmetrically arranged in a rectangle or octagon. The disadvantage of such a configuration is that main rotors in conventional helicopters have complex mechanical linkages so that the pitch of the blades can be adjusted as they rotate. Equipping multiple rotors with these intricate control systems adds weight and complexity. Electric drives eliminate this problem.

A rotary-wing drone with an internal combustion engine and conventional controls must have a means for translating the control inputs received from the DROP over the control link into mechanical forces applied to the pitch rods for the main and tail rotors, either directly or through hydraulic servomechanisms. Electrically powered drones, however, can have much simplified control systems. A drone with electric propulsion does not need servomechanisms to move control rods; it only needs logic programmed into control systems that vary the current to the motors driving the rotors.

Most helicopters keep main- and tail-rotor RPM constant and control thrust of the rotor by varying blade pitch. Constant RPM is desirable mainly because it permits the rotor to operate in the relatively narrow peak torque available from a piston or turbine engine.

Constant RPM is not needed in rotorcraft with electric propulsion systems. Electric motors produce relatively constant torque over a wide range of RPM. Moreover, in rotorcraft with multiple main rotors, control over all three axes can be produced by differential rotor RPM, obviating the need for pitch control rods and the associated complexity and weight necessitated by them.

Brushless AC motors require a special kind of power input, one that simulates alternating current, although the battery delivers direct current. An electric drivetrain requires a voltage controller to deliver a voltage appropriate to desired rotor RPM, an inverter to convert the direct current coming from the battery into alternating current at the right voltage and frequency for the brushless electric motors driving the rotors, and wires to connect the subcircuits and the motors. Because the vehicle is controlled by differential thrust resulting from differential rotor RPM, each motor must have its own controller and inverter.

How much power is delivered to each motor depends upon signals generated by the flight control system (FCS). The FCS has logic in its hardware and software that translates signals received from the DROPCON into differential current instructions for the power distribution board (PCB) resulting in differential currents for each motor. For example, if a DROP moves the collective forward, the FCS tells the power distribution board to increase the current on all of the motors by the same amount, resulting in greater total thrust. When the DROP moves the cyclic forward, the flight control system tells the power distribution board to decrease the current on the forward motors and to increase it on the aft motors, thereby tilting the total lift vector forward and causing the vehicle to move in a forward direction. When the DROP moves the cyclic control right, the FCS tells the power distribution board to increase the increase to thrust of the left motors and to decrease the thrust of the right motors, resulting in a roll to the right, followed by movement of the vehicle to the right. The same process can command backwards and leftwards flight.

The PCB contains an Electronic Speed Control (ESC) chip for each motor. PCBs have two inputs: battery voltage direct current from the main battery, and a digital signal from the Flight Control System (FCS) that tells it how much current to deliver to that particular motor to generate the lift necessary to balance the vehicle in a hover or to tilt its lift vector for forward, backwards, or sideways flight. The ESCs execute the FCS instructions by varying the duration of a series of pulses—a process known as *pulse-width modulation*.

As all this suggests, the FCS contains all the smarts—it's the coach. The PCB routes the electricity according to what the coach says to the muscles represented by the motors and rotors.

3.9 Wireless communications

Because drones have no pilot aboard, they must have wireless communications links to connect them with their DROPs. Even drones with substantial autonomous flight capabilities usually need some human input to navigate around objects and to specify the course to be flown.

The drone also must send data back to the DROP so that he knows where the drone is. For microdrones intended to be flown only within the sight of the operator, the data sent from the drone to the DROP—video imagery or numeric values for position, speed, and orientation—is intended only as a secondary aid for his control inputs. Primarily he relies on what he sees. For longer-range drones, however, more complete data from drone to DROP is necessary. These vehicles typically send high definition imagery from a wide angle of view and more complete telemetry.

Imagery to enable the DROP to fly the drone is known in the model airplane world as *first person view (FPV)*, a term that has spread throughout the microdrone community.

For short-range microdrone flight, control link data usually is transmitted in unlicensed slices of spectrum set aside for Wi-Fi. Although the spectrum available for Wi-Fi is divided into frequency channels, Wi-Fi does not assign frequencies to particular transmitters and receivers; instead it uses a modulation technique known as *spread spectrum,* in which a signal, regardless of whether it comprises data or imagery, is sliced into short pieces that are transmitted and received on multiple frequencies, usually more than 100. In this form of frequency division multiplexing, receiver and transmitter are synchronized so that the receiver tracks the frequencies used by the transmitter at each instant. The receiver collects the slices of bitstream from the different frequencies and reassembles them.

Any data communications channel—wired, optical, or wireless—can experience errors such as loss of bits, loss of entire packets, or variable delay, called *jitter*. Quality of service algorithms mitigate these errors, frequently by buffering the received bitstream to allow missing packets to be resent or to correct for timing variation. Such buffering adds to latency, however.

In the typical microdrone control link implementation, the drone has a small transceiver powered by its own battery or by the main drone battery tuned to the requisite block of frequencies. The transceiver installed on the drone is about the size of a half a pack of cigarettes. It has either a simple wire as an antenna or more sophisticated Yagi (multi-element) or helical antennas, which are directional. The DROPCON is similarly equipped, though it more likely to have directional antennas because weight does not matter as much.

Video imagery from the drone often is transmitted on a different spectrum slot to avoid interference between it and the control link. For example, many microdrone applications use 2.4 GHz for the control link and 5.7 GHz for video.

A critical technology for microdrones involves extending the range of wireless links—both of the control link and of the video downlink. One reason that microdrones are flown within the line of sight of the DROP is that the DROP controls the vehicle and avoids obstacles and other aircraft by looking at it. But the photog cannot get good imagery unless he knows what imagery the drone is capturing as the flight proceeds. That imagery also is available to the DROP to assist in controlling the aircraft. Extending the range of the wireless links will open up the possibility for longer flights, thus reducing the nonproductive cost of the DROP's needing ground transportation to the mission site.

Of course it also would activate new concerns about safe operation out of the line of sight of the DROP, especially if the rules have been designed around a line of sight assumption.

Development frontiers for drone radio communication relate to improvements in software for adaptive antennas, development of feasible concepts for satellite and ground relay of LOS signals, mitigation for lost-link circumstances, and navigations fallbacks for lost GPS links. Expansion of microdrone package delivery advocated by Amazon, Google, and DHL, will depend upon the development of a sophisticated neighborhood access networking technology, such as discussed in § 3.13.

3.9.1 Signals

3.9.1.1 Frequencies and propagation

Only those electrical signals above a certain frequency are capable of traveling through space, because extremely low frequencies cannot affect or *excite* the atoms comprising physical circuits. Signals at frequencies high enough to be propagated through space as electromagnetic radiation are called *RF* signals. Accordingly, input signals from DROPCONs, and onboard sensors such as cameras (*baseband signals*) are superimposed on carrier signals at the right RF frequency for propagation through a process called *modulation*. Modulation shifts the frequencies involved in a baseband signal to higher frequencies appropriate for transmission.

The range of control links and mission downlinks depends on transmitter power and the propagation characteristics of its frequency.

The useful radio spectrum extends from about 50 Hz to about 50 GHz. Signals at different frequencies in the spectrum move through space (*propagate*) quite differently. Very low-frequency (VLF) signals in the spectrum from 3 KHz to 30 KHz at sufficiently high power are capable of propagating a certain distance into the earth and suffer very little attenuation in the atmosphere. Because of this characteristic, the United States Navy uses them to communicate with submarines while they are submerged.

Medium frequency (MF) signals such as those of the US AM broadcast band—540–1610 MHz—suffer more attenuation as they move through the earth but follow its curvature for distances of 100–200 miles.

High-frequency (HF), also known as *short wave*, signals—1.6 MHz to 30 MHz—are refracted and eventually reflected by higher levels of the ionosphere so that, depending on the time of day or night, they appear to skip. They can be received at distances hundreds or thousands of miles, but not over intermediate distances.

Very high-frequency (VHF) signals—30–300 MHz—Ultra High Frequency (UHF) signals—300 MHz–3 GHz—and microwave frequencies—above 3 GHz—do not skip; they penetrate the ionosphere and go out into space. They are, thus, suitable only for line of sight communication. Drones—like helicopters and airplanes—increase line of sight in proportion to their altitude above the ground. One of the many reasons that helicopters are good for ENG is that their antennas can find a line of sight free of obstacles.

Beginning at this point in the spectrum, atmospheric attenuation becomes important, increasing with frequency. Atmospheric attenuation means that the air molecules diminish signal strength. These higher frequency signals also experience attenuation by ordinary terrestrial objects—buildings and other structures, foliage, dust particles, raindrops, and snowflakes. Fog, rain, snow, and other atmospheric obscuration become a problem.

Other than different propagation characteristics, radio frequencies also have dissimilar antenna requirements and different capabilities for accommodating signals of different bandwidth.

3.9.1.2 Bandwidth

Wireless communications often are talked about in terms of the "speed" of a channel. This is a useful metaphor for how things appear to work, but it is not accurate; speed and *bandwidth* are distinct concepts. Technically the propagation speed of a wireless signal does not change much. Electromagnetic signals move from one point to another in a vacuum at 186,000 miles per second. They move slightly slower though air, water, wires, optical fibers, and other materials.

A core principle of communications theory, derived from Shannon's law, says that the bandwidth of the signal is proportional to the rate of information transfer. Under Shannon's law, when more information is transmitted in a limited amount of time, more bandwidth is used. In other words, the signal is wider, not faster. When someone wants to get more water through a hose, she can use a bigger hose and keep the speed the same. Fans exiting a football stadium can get out more quickly if they move faster, or if they move slowly through a wider gate. The width of the gate is analogous to bandwidth; how fast they walk is analogous to speed of propagation.

A Morse Code CW signal has a bandwidth of only 50–100 Hz, because the rate of information transfer is fairly low—less than 35 words per minute. AM voice communications signals require bandwidth on the order 3.0 KHz. High-fidelity sound requires about twice that. Broadcast television signals require about 6 megahertz, and digitally encoded signals, whatever their content, require bandwidth proportional to the bitrate, which depends on the modulation scheme—8 bits per symbol is about the best currently achievable. So 6 MHz bandwidth can handle, at most, 25 MB/s. The difference between bandwidth and data rates arises because bandwidth depends on *symbol rate*; one symbol can represent more than one bit in advanced modulation schemes.

Lower frequencies do not provide enough room for these high-bandwidth signals even if everyone presently using that part of the spectrum were pushed aside. One cannot fit 6 MHz bandwidth between 550 KHz and 1500 KHz. The bandwidth required is 5 times bigger than that slice of spectrum.

Higher frequencies, in the microwave range, provide enough radio spectrum for much higher bandwidth signals. Full-motion video signals contain a huge amount of information, and therefore require much greater bandwidth than, say, aviation voice-radio communications. Accordingly, signals requiring large bandwidth must be relegated to higher frequencies.

Because of the radio line of sight limitation, long range control links and imagery downlinks require radio repeaters. Repeaters maybe ground-based or satellite-based.

Most microdrone communication occurs on low microwave shared frequencies in the 2.4 GHz and 5.8 GHz unlicensed bands. Most macrodrone communication occurs on assigned frequencies in various microwave bands, often requiring repeaters, except when high macrodrone altitudes permit communication with ground stations at greater distances.

3.9.1.3 Coding, decoding, and modulation

To be transmitted wirelessly, information has to be put in a form that can be carried by radio waves. Information from the position of the cyclic stick on a DROPCON is analog: the physical position of the stick usually is translated into an electrical current, voltage or into an electrical property such as capacitance or resistance. This analog value is then digitized and encoded. For example, if the position of the cyclic stick is represented by 0 when it is fully back and 10 when it is fully forward, the digitization process (sometimes called *quantization*) might represent the intermediate values in increments of 0.25. Then, the encoding process translates these numeric values into digital binary values, in which an 8 is represented as 1000, a 1 as 0001, and a 6 as 0110. The three digital positions would comprise a data structure with three fields. A piece or hardware or software known as a *codec* performs the encoding and decoding functions.

Often, the coding process also involves *compression* as under the popular mp3 standard for music and mp4 and PNG for video. Compression algorithms typically represent changes rather than steady state values. For example, a compressed image with lots of white space might be encoded so that only the transition from the white background to the edge of an object in the foreground is represented.

Compressing digital video signals is especially desirable because of the large number of bits associated with raw video. Compression takes advantage of the fact that, in most images, the majority of the background details do not change much from left to right or top to bottom. Even some aspects of foreground figures exhibit a constancy over significant areas of the frame. Moreover, in full-motion video with a frame rate of 30 frames per second, for instance, little changes from one frame to the next, unless the subject is moving rapidly. All video compression algorithms take advantage of these characteristics of still images and full-motion video.

Compression of a still image—the same thing as a single frame in full-motion video—divides the image up into macroblocks varying in size from 4 x 4 to 16 x 16 pixels. The algorithm compares adjacent macroblocks and encodes only the changes in brightness and color from one block to another. The more compression that occurs, the greater the likelihood of pixelation: obvious boundaries between blocks. That undesirable artifact of compression can be mitigated by making a smoothing procedure part of the algorithm.

For full-motion video, compression algorithms such as the popular H.264 periodically transmit *index* or *reference* frames, which include all of the detail resulting from frame compression. In between the index frames, interframes of two types are interspersed. P frames predict the next from the preceding index frame; B frames interpolate between the preceding and following index frames. The associated group of index and inter-frames is known as the *group of pictures* (GOP). The frequency with which index frames are sent is an adjustable parameter in different implementations, as is the sampling rate for P and B interframes. Obviously, sampling more frequently than the frame rate is not useful, because nothing changes between frames.

Any compression process requires related macroblocks and frames to be stored in a buffer while they are being compared or constructed. Buffer size limits efficiency, and the processing steps required introduce delays, known as *latency*.

Compression is accomplished by a coder and decoder (codec) embedded in software or, more often, in hardware. The coder performs the compression; the decoder restores a facsimile of the original video image at the destination. Lossless compression algorithms permit complete restoration of the original image; lossy algorithms achieve greater compression by dropping some bits that will not affect human perception (much). Lossy algorithms like.mp3 and. mp4, at most, permit the decoder to restore *almost* the original sound or image.

Now, the information is ready to be sent wirelessly. To do that, an electromagnetic signal of an appropriate frequency must be modulated. *Modulation* signifies that its properties of a *carrier* are changed to represent values of the modulating signal—the *baseband* signal; in other words modulation impresses intelligence on the carrier. It is necessary to increase the frequency of a baseband signal to a higher frequency carrier suitable for transmission on the assigned channel or channels.

Several types of modulation exist. *Amplitude modulation* (AM) varies the amplitude of the carrier so that a binary 1 is represented by a stronger signal, and a binary 0 is represented by a weaker signal. *Frequency modulation* (FM) shifts the carrier frequency a little upward to represent a 1, and a little downward to represent a 0. *Phase modulation* (PM) shifts the phase of the carrier signal. In practical modern wireless applications, the modulation scheme is usually more complicated, Quadrature Amplitude Modulation (QAM) is common. In that scheme, two carrier signals are interlaced. The values of the modulating signal can change combinations of amplitude, phase, and frequency of the carrier signal. The values of different bit positions in a four-bit binary word are determined by the combination of phase angle and amplitude. Depending on the necessary signal-to-noise ratio and transmitter and receiver power and sensitivity, 16, 64, 256, or higher permutations are possible. More sophisticated modulation schemes can encode more data in a given bandwidth but they require a better quality link (a higher signal-to-noise ratio).

In the ubiquitous 802.11 Wi-Fi standard, the coding scheme is OFDM, while a common modulation scheme is 64QAM. Orthogonal Frequency Division Multiplexing (OFDM) is a coding scheme that splits a signal up and multiplexes it—sends small sequences on different paths—on different, but closely spaced frequencies selected so they do not interfere with each other (that is the "orthogonal" part). QAM is a common modulation scheme explained earlier in this section.

No modulation technique can reshape a carrier waveform with finer granularity then the waveform itself has. As the rate of information represented in the modulating baseband signal increases, the carrier frequency must increase at least proportionally. One cannot modulate a 1.6 MHz carrier with a 5 MHz signal, but one can easily modulate a 2.4 GHz carrier with that signal. The baseband signal varies faster than the carrier signal in the first case, and the lower frequency carrier signal cannot capture all the changes in the baseband signal.

Coding and modulation take place on the transmitter end. At the receiver end, the RF signal must be demodulated, to extract the information signal from the carrier, and the information stream decoded, to reinstate the sequence of analog or digital values that can be processed further and, for a video stream, rendered on a display. Modulation is the core function of any RF transmitter. Demodulation is the core function of any RF receiver. Codecs perform the coding and decoding function.

Encoding more complex streams of information, such as full-motion video, usually follows a standard developed by standards committees such as the Institute of Electrical and Electronic Engineers (IEEE) or the Internet Engineering Task Force (IETF). In the United States, for example, the NTSC standard for analog color television, and the ATSC standard for digital television predominate. IETF Request for Comments 6184 defines the standard for grouping bits into packets under the H.264 standard, which was, itself, developed by the International Telecommunications Union (ITU).

Multiple transformations of an information stream often are necessary. In drone imagery applications, the camera encodes the images it captures, then the transmitter may decode it and re-encode it in another scheme which then modulates the RF carrier. There may another codec on the receiver, and yet another on a device that sends it over longer distances as a microwave signal or inserts it into the cellphone infrastructure. At the other end it must be demodulated, decoded, and re-encoded so that it can eventually rendered on a video display.

The bits representing information almost always are grouped into *packets*, permitting the communication protocols to manage packets comprising a discrete number of bits rather than an entire bitstream of indeterminate length. Each packet has a predetermined format that includes meta information such as the length of the packet, the type of information it contains, who sent it, and its destination.

Figure 3.2 shows a schematic of the video processing systems on a microdrone and its associated photographer's console (PHOTOCON). The microdrone camera appears at the upper left. It provides data representing the image impinging on its sensor for quantization, digitization, and compression, sequentially shown from left to right at the top of the microdrone diagram. A codec typically performs these functions. The codec may be integrated into the camera or provided separately in software or hardware.

The codec creates a baseband signal, which the modulator, appearing on the right of the microdrone diagram, combines with the carrier signal from the RF exciter and outputs to an RF amplifier, shown in the lower right of the microdrone diagram. The resulting modulated RF signal is transmitted through the downlink antenna, shown at the top middle of the diagram.

The codec also provides a video signal to an onboard memory card, as depicted at the bottom of the microdrone diagram.

The PHOTOCON, shown on the right-hand side of the diagram, receives the downlinked video signal on its antenna, shown at the top middle of the PHOTOCON diagram, so that it can be processed by the receiver shown at the upper left of the PHOTOCON and then decoded through a codec matching the one on the microdrone. This codec performs any necessary decompression and may also perform format conversions if the data structures needed by the video display and output devices are different from those provided by the microdrone. After the necessary codec transformations are complete, the decoded signal is sent to the video display screen for the photog and recorded to a memory card and transmitted to a destination through the Internet or

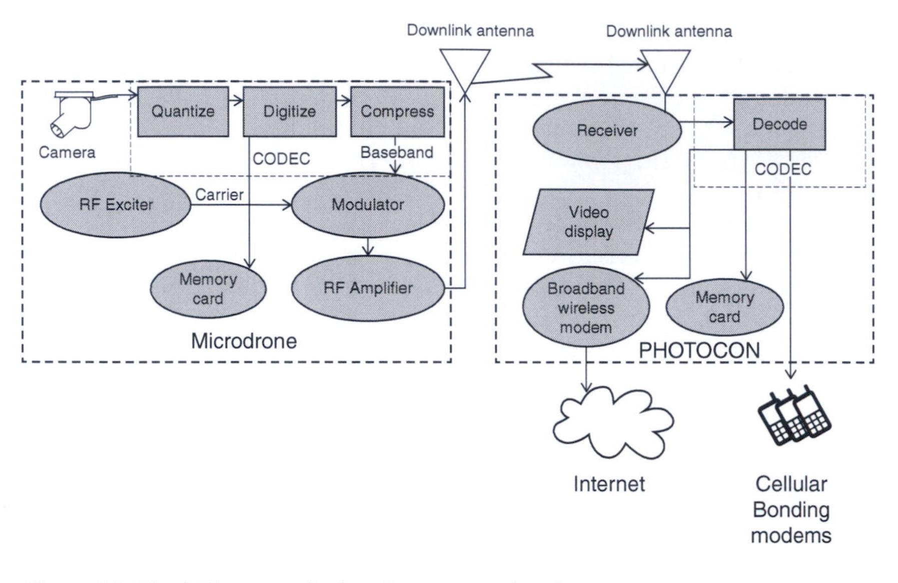

Figure 3.2 Block diagram of microdrone control and power systems

via cellular bonding modems. A typical Internet connection would involve the use of a broadband wireless modem, shown at the lower right of the diagram.

Hardware configurations vary, depending on the particular model of microdrone and its mission equipment and on the ground configuration enabling the photog to control the camera and its gimbal and to transmit the video to its destination. The functions depicted, however, must be performed by any system.

3.9.2 Antennas

Transmitting DROP commands to drones and sending drone-captured imagery to the ground require that the drone and the ground station must be able to acquire each other's signals. Each uses an antenna to transmit and receive the necessary RF signals. For a radio antenna to work efficiently either to transmit or receive a radio signal, its size has to be comparable to wavelength—sometimes a fraction like one-half or one-quarter, but a big fraction. Antenna sizes must be larger for lower frequencies. The wavelength of VLF signals is 6–62 miles. The antennas for the Navy's VLF submarine communication facilities occupy many square miles.

AM broadcast antennas must be the size of the antenna towers scattered about the landscape. Antennas for HF frequencies like the 80 meter amateur band (3.5 MHz) typically are a little over 100 ft long—half of the wavelength of eighty meters. The wavelength of a 2GHz signal is 15 centimeters (5.9 inches). That is small enough for a microdrone to carry comfortably.

Moreover, smaller antenna sizes permit multiple antenna elements, which permit the antenna to be directional, and to be rotated. This enables the receiving antenna or transmitting antenna or both to be pointed at each other.

Aircraft cannot carry big radio antennas, because of their weight penalty and their aerodynamic drag. A microdrone must have an antenna for the control link and another for the downlink, but it must be a small one. Airborne video transmitters, in order to reduce RF power requirements and thus weight and electrical power needs, typically rely on directional antennas at both ends.

Directional antennas have the property that they concentrate signal strength in a particular direction. Directional transmitting antennas concentrate a signal in one direction, resulting in signal strength at the receive site that is equivalent to that from more transmitter power feeding a non-directional antenna. Directional receiving antennas allow receivers to handle weaker signals from the direction in which they are pointed.

Directional antennas thus are excellent tools for locating the source of a signal. Changing the azimuth away from the direction from which a signal is transmitted, even a little, reduces signal level. Indeed, it is exactly this phenomenon that is used to orient consumer satellite television antennas so that they point toward the satellite. A human being can align a directional antenna by looking at a signal-level meter as he moves the antenna, stabilizing it in the position that results in the highest reading on the meter. The process can be automated by writing a computer program that does the same thing by detecting signal level from the antenna in receive mode while it commands the electric motors attached to the antenna shift the azimuth back and forth to maximize signal strength.

And of course, before the signal is first acquired, the same process can be used in a gross way to acquire an image initially. If the antenna system is unclear where the signal from a ground station is coming from, it rotates the antenna through 360° until it acquires the signal, and then makes finer adjustments to point the antenna toward the transmitting station exactly. The system needs to know the precise frequency to listen on in order to be able to receive the desired signal at all.

Many different types of directional antennas exist. *Yagis* are a paradigm. They comprise several antenna elements (the elements are the bars that stick out from the boom) in a row. Only one of these elements (the *driven element*) receives an electrical signal from the transmitter or feeds a received signal to the receiver. The others either passively reflect or concentrate the signal from the driven element. The antenna elements get progressively smaller from back to front. The longer elements behind the driven element reflect the signal, while the smaller elements in front of the driven element concentrate (direct) it.

One of the advantages of operating at microwave frequencies is that they have a very small wavelength, making directional antennas with many more elements physically practicable. More elements mean larger gain and more directionality. Smaller wavelengths also make parabolic antennas—another design for directional antennas—practicable. Parabolic antennas, to achieve the same gain as a Yagi antenna for the same frequency, must have diameters that are multiples of the bandwidth. In other words, a parabolic antenna will be much bigger than a Yagi antenna for the same frequency.

Yagi antennas are common for many microwave applications. Parabolic dishes are more common for satellite communications. Their size, however, makes them unattractive for airborne applications because aerodynamic drag is proportional to cross sectional area, and a parabolic antenna is much bigger than a Yagi antenna for the same frequency.

3.9.3 Polarization

Yagi antennas (actually, all antennas) can be oriented horizontally (the elements stick out to the side of the boom) or vertically (the elements stick out up-and-down from the boom). If the elements are arranged horizontally, the resulting radio signal has *horizontal polarization*. If they are arranged vertically, the resulting signal has *vertical polarization*. *Circular polarization* involves a rotating signal, sometime vertical; sometimes horizontal. Polarization does not affect where the signal goes but rather how the receiving antenna needs to be oriented to receive it most strongly; for best reception, the two antennas should be aligned. Which type of polarization is preferable depends on space considerations. An HF antenna would be awkward if it were vertically constructed. A whip antenna on a police car would be cumbersome unless it were oriented vertically. Circular polarization eliminates the need for antenna alignment.

3.9.4 Multipath problem

The *multipath* phenomenon plagues all mobile wireless systems, from cellphone systems to control links for drones. It is especially acute for digitally encoded signals in the high VHF, UHF, and microwave bands. Ghosts in over-the-air analog television reception result from multiple signal paths. They are usually correctable without much difficulty, because the transmitting and receiving antennas and the obstacles that create another signal path by reflection are fixed in position.

Here's the problem: a mobile receiver collects multiple versions of the same signal, one directly from the transmitter, and others reflected off the ground, buildings, and other vehicles. The reflected signals are almost certainly going to be out of phase with the one received directly from the transmitter. That can be disastrous for digitally encoded information that must be received in real time, because the phase shift easily can cause a one in reflected signal to overlap a 0 in the direct signal, thus distorting the content of the information. For digitally encoded signals, this is equivalent to the stream of bits being shifted to the right or left by an arbitrary number of bits. Suppose a particular bit position in a part of a video signal is a 0. The phase-shifted reflected signal may show a 1 in this position. The result is that the information is garbled. So all mobile wireless systems must have some mechanism for suppressing or ignoring the reflected signals.

Reflection of higher frequency and lower wavelength signals by small objects makes the multipath problem more intense at higher frequencies. When both receiver and transmitter are stationary, multipath interference sometimes can be reduced by using directional antennas. But that is more difficult when either the transmitter or receiver is moving because the origin of the reflected signals changes from moment to moment.

One popular and relatively low cost way to mitigate the multipath problem is *diversity transmission*, in which the same information is transmitted over two different frequencies on different paths at the same time. The different signal paths result from using two antennas spaced at least one wavelength apart. *Diversity reception* is more common. It uses two receiving antennas spaced similarly, receive the signal over two different paths. For the small wavelengths in the 2.4 and 5.7 GHz bands, the spacing criterion allows use of two antennas mounted to the same device.

3.10 Control systems

It is not enough for drone and DROP to be connected by a control link. The drone must do something with the control signals received from the DROP. Drones carry out their tasks with automatic control systems performing the same function as any servomechanism.

3.10.1 Servomechanisms

A *servomechanism* is a source of mechanical force—it could be an electric motor or a hydraulic actuator—that is combined with a feedback circuit, so that an output value, such as position or rotational velocity, can be compared with a command input, causing the power source to cease operating with the commanded value is reached, and to operate again if the output deviates from the commanded value. The system for filling a toilet tank after a flush without the tank's overflowing is a very simple example. It is entirely autonomous. When the toilet's flush handle is activated, the water stored in the tank flushes the toilet. The drop in water level causes a float to drop, which opens a valve from the water supply. The water pressure provides power to deliver water to the tank. As the water enters, the float rises. The mechanical connection between the float and the valve closes the valve progressively until it is fully closed when the water in the tank reaches the desired level.

In the aviation context, a servo motor on a model aircraft receives a command to deflect the ailerons to induce a roll to the left. It moves the ailerons until they have reached the commanded deflection angle, until the rate of movement of the aileron matches the degree of movement on the stick, or until the aircraft reaches the commanded roll rate.

Engineering an automatic control system involves, not so much the design of the motor; almost any small electric motor capable of delivering the necessary torque and having acceptable weight and power consumption will do. The challenges are in designing the feedback algorithm and matching it with appropriate sensors.

3.10.2 Flight control subsystems

Automatic control systems are central to aviation automation. Similar in function to servomechanisms, which are themselves small automatic control systems, they have the capacity to sense the state of a target system such as an air vehicle and automatically to take appropriate action to maintain that state—to make control inputs in the case of an air vehicle. In general, they require sensors to measure indications of state, processors to compare that state with the nominal or desired state and generate *error signals*, and other processors to determine what action to take to restore the desired state. The system functions required for autonomous hover by a microdrone are a good example.

The desired state is a hover, defined as no movement over the ground in any direction, and no change in height above the ground. A *sensor subsystem* determines the actual state of the microdrone so that it can be compared with the desired state. The sensor subsystem includes a magnetometer compass (also known as a *fluxgate compass*), an altimeter, accelerometers, and GPS navigation components. Not all of these are necessary for every function. For example, a sufficiently accurate GPS is enough for navigation, because it can determine the microdrone's position in three-dimensional space. GPS cannot, however, determine vehicle orientation. If no GPS signal is available, a combination of compass, altimeter, and accelerometers can determine orientation and deviations from the original state, but cannot determine if the original state included movement. Accelerometers can only measure changes in velocity; not constant velocity.

The *error subsystem* translates both the desired state and the actual state into geospatial coordinates (it could also use other units of measurement) so that they can be compared on an apples-to-apples basis. The accelerometer package, itself, or the deviation subsystem must integrate raw accelerometer inputs twice to get position along the axis that particular accelerometer measures, and then transform those values into positions on standard coordinates, and height. Either the altimeter subsystem itself, or the deviation subsystem, must translate barometric altitude into position on the height coordinate.

The error subsystem outputs error values along each axis to a separate *command subsystem*—the flight control system (FCS). The command subsystem is preprogrammed with parameters indicating how much force should be applied along each vehicle axis to restore it to the desired state. In the case of a multi-copter, such as a quadcopter, the only forces available are those generated by the thrust of the rotors. The vehicle can be made to pitch forward by causing the aft rotor to generate more thrust, relative to the forward rotor. It can be made to pitch backward by generating more thrust on the forward rotor, relative to the aft rotor, to roll left by generating more thrust on the right rotor relative to the left rotor, and to roll right by generating more thrust on the left rotor relative to the right rotor. Thrust differentials may be necessary to maintain a hover in the presence of wind, their magnitude depending on the direction of the wind and its velocity.

A quadcopter can be made to yaw to the right by adjusting rotor RPM so that there is a net clockwise torque on the vehicle and to yaw left by making a similar adjustment to generate net counterclockwise torque. Almost all multi-copters have an even number of motors and rotors, half of which rotate clockwise, and half of which rotate counterclockwise. As the motors spin their rotors counterclockwise, they necessarily generate an equal and opposite clockwise torque on their booms, which translates into clockwise torque on the fuselage. The clockwise-rotating rotors similarly generate a counterclockwise torque on the fuselage. Speeding up the motors for the counterclockwise-rotating rotors generates more clockwise torque, and if the motors on the clockwise-rotating rotors remain the same, the vehicle yaws to the right. Yaw is thus controlled by altering the balance of torque from clockwise and counterclockwise rotors.

Of course, when RPM adjustments are made to generate yaw, the FCS also must maintain the total thrust, or the vehicle will climb or descend and roll or pitch. So the FCS must be smart enough to generate the necessary torque differential while also maintaining the appropriate total thrust and forward and aft and left and right differentials.

Once the desired thrust and torque values are generated by the FCS, they must be converted into desired motor RPM for each motor, based on preprogrammed values representing the characteristics of each motor and rotor subsystem to relate RPM to the resulting thrust. The thrust-to-RPM calculation may be accomplished in the FCS or in the *power control subsystem*—the power control board (PCB); the dividing line is essentially arbitrary. The power control subsystem then uses preprogrammed values to relate RPM to the voltage to be applied to each motor. Alternatively, the motors may contain servomechanisms that measure their own RPM and report it to the power control subsystem so that it can adjust the voltage applied to that motor to bring the RPM to the desired value.

Even this high-level description of the automatic control system necessary for autonomous hover makes it clear that a large number of stored values must be retrieved, compared with other values generated by sensors to determine and apply the appropriate control inputs. A fundamental design benchmark determines the computer processing power necessary to control a vehicle, based on the expected frequency and magnitude of deviations from the desired state. That benchmark shapes the design of every subsystem. The *Nyquist Criterion* says that the sensor subsystem must sample the position along each of the three axes at least twice as frequently as the position is expected to change. The processors performing integrations necessary to translate acceleration values into position values must keep up with the sensor outputs.

The deviation subsystem must be able to keep up with the inputs it receives from the GPS and inertial measurement sensors, or the outputs that it passes along to the command subsystem will lag the actual state of the aircraft. Delays in generating control inputs to restore vehicle state easily can result in dynamic instability of the vehicle and uncontrollability. Similarly, the power subsystems and the motor and rotor combinations must respond quickly enough so that they do not introduce additional delays.

Vehicle size and configuration have important implications for the design of automatic control systems; different dynamics require different control system sensitivities. A blimp reacts quite differently from a microdrone. If a blimp's position is disturbed by a wind gust, the deviation from the position will develop much more slowly than will a microdrone disturbed by the same wind gust. Thus the sampling rate by the sensors and the processing speed of everything else must be considerably greater for microdrones than for blimps. Lags in control responses never can be reduced to 0; even the best pilot takes several fractions of a second to respond to his sensory inputs. Early in their flight training, most pilots, especially those learning to fly helicopters, have difficulty responding quickly and delicately enough not to trigger instability in the vehicle. The combination of subsystems comprising an automatic control package must contain good models of vehicle dynamics.

Any digital processor requires a finite amount of time to perform an information storage, retrieval, or computation step. This is known in advance when semiconductor systems are designed and tested. The semiconductor chips selected for the components of each automatic control subsystem must be selected to have sufficient processing power so that their delays do not result in instability. Similarly, memory chips require finite amounts of time to store or produce the contents of a memory address. Most real-world automatic control systems have *feedback loops* to minimize instability that otherwise would result from cumulative delays in subsystem responses.

Usually, given any particular state of technology, faster processing comes at the expense of more weight, volume, and power consumption. That is why miniaturization and computation advances play such a major role in the envelope of possibilities for microdrone performance.

3.10.3 Flight control sensors

The sensor subsystems in drones comprise magnetometers, accelerometers, GPS receivers, and inertial navigation systems. Similar systems are used on manned aircraft to perform similar functions. GPS navigation necessitates acquisition of satellite signals; inertial navigation does not.

The following subsections explain each type of sensor and its function.

3.10.3.1 Magnetometers

A fluxgate magnetometer, also known as a fluxgate compass, detects the orientation of the Earth's magnetic field by measuring the difference between the currents on two coils of wire wrapped around a magnetizable core. One coil is energized by an AC signal, which alternately magnetizes and demagnetizes the core and induces a current in the second coil. When the Earth's magnetic field impinges on the core it changes its level of magnetism and alters the current in the second coil. The orientation of the device causes the effect of the Earth's magnetic field to be stronger or weaker, thus enabling associated logic to determine the magnetic heading of the device with respect to the Earth's magnetic field. Miniaturized magnetometers are regularly used to drive a magnetic heading indicator on aircraft. Their size and weight suit them well for microdrones with limited payload capacity. A small compass magnetometer sensor module could be bought on Amazon in October 2015, for less than $10.

3.10.3.2 Accelerometers and inertial guidance systems

Accelerometers are devices that measure acceleration along a particular axis of the device. Small ones used for vehicle control typically comprise a piezo-electric crystal or a capacitor. Piezo-electric materials generate an electrical charge when a force is applied to them. Capacitors change their electric properties, e.g. their reactance to alternating current, or the length of time they hold a static charge when a force is applied. A weight attached to either applies pressure to it when it is put in motion, proportional to the acceleration.

By sensing the resulting voltage of a piezo-electric crystal or the changed electrical resistance of a capacitor, an appropriately designed circuit can measure acceleration along the designated axis. Three such devices, properly oriented to each other, can measure acceleration along each of three orthogonal axes. Logic that accurately integrates the acceleration inputs can determine velocity in any direction. By integrating again, they can determine position in space. Newtonian mechanics defines acceleration as the derivative of velocity with respect to time, and velocity as the derivative of change in spatial position with respect to time. Integral calculus permits determination of position by double integration of the values for acceleration along each of three axes. The inertial guidance system knows where it is. Metaphorically, it draws a map of the speed, distance, and direction traveled.

The availability of GPS systems has made inertial guidance systems less necessary, but they still are useful when a GPS signal is unobtainable, to improve precision in location determination, or to determine vehicle orientation. Inertial guidance systems system can be designed to measure arbitrarily small increments in movement and thus can be much more accurate than the best GPS receiver working with the most advanced satellites. IMUs on multiple-warhead ICBMs, for example, maintain extremely precise position data at the end of a 30 minute flight, despite enormous acceleration during the launch phase.

Small accelerometer packages, available for $5–$10, are common on consumer electronic devices such as laptop computers and cellphones. They are there to detect a mishap such as dropping the device, so that internal subsystems can respond so as to minimize damage; by turning off or spinning down hard drives, for example. Drones use accelerometers to detect direction and magnitude of movement and to reinforce other data indicating position. With an appropriate accelerometer subsystem, a microdrone does not need a GPS signal to know where it is, relative to its position when it is powered up. It calculates its position, direction of travel, and speed, based on inertial measurement of its movements.

3.10.3.3 GPS

The Global Positioning System (GPS) allows a GPS receiver to determine its position in space and to calculate its direction and speed of movement by receiving time-coded signals from at least four specially designed GPS satellites in orbit. By calculating the time it takes a signal from a particular satellite to reach the receiver, the receiver can calculate its distance from that satellite. By knowing its distance from each satellite, it uses trigonometry to calculate its position, a process known as *trilateration*. Applying Newtonian equations of motion and differential and integral calculus, it can determine speed and direction of the receiver's movement. GPS is a mainstay of aircraft navigation, and is scheduled to replace most ground-based aerial navigation facilities by 2020. Satellite-based augmentation systems (SBAS) like WAAS in the United States, improve GPS accuracy by sending updates on satellite signal errors from ground stations. WAAS GPS systems designed for aviation achieve 50 ft lateral and 13 ft vertical accuracy required for Category I precision landing approaches. Typical WAAS accuracy for North America is approximately 3 ft.

3.10.3.4 The cockpit on the ground

The drone not only must contain appropriate control systems to allow remote control of the vehicle; the DROPCON must contain an appropriate interface between the DROP and the control link.

Designing the visual interface for the DROP is a considerable challenge. Because he is not aboard the aircraft, he cannot feel, see, and hear what a pilot feels, sees, and hears. To supplement what he can see from keeping the drone within his line of sight, he must rely on a visual display that combines imagery captured by a drone-mounted camera with digital values representing the configuration, attitude, height, speed, and direction of flight. Chapter 4, exploring human factors, details the design advances necessary to provide good perceptual cues to the DROP.

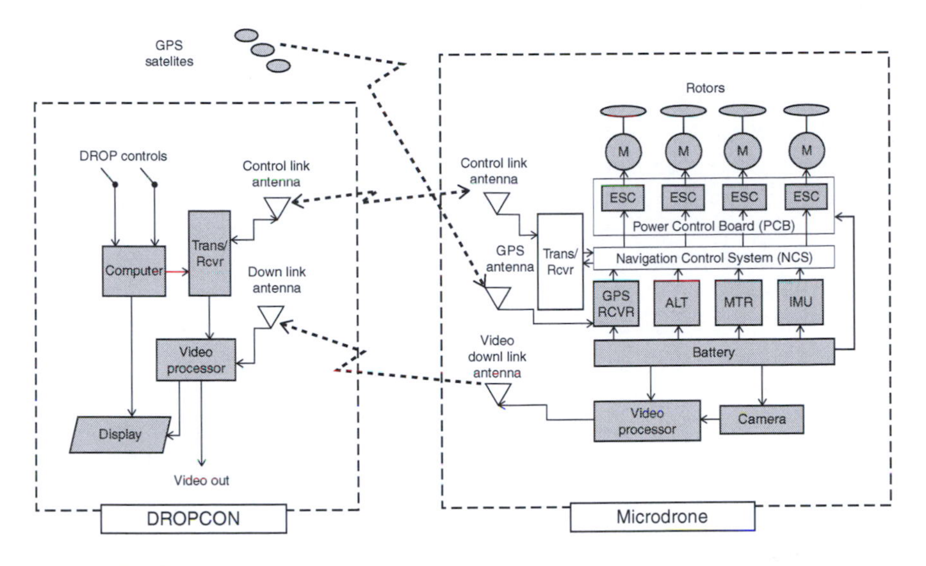

Figure 3.3 Block diagram of microdrone imaging system

Figure 3.3 shows a schematic of a microdrone and its DROPCON. At the top of the microdrone appear its rotors and motors (M). The ESCs that control motor current are embedded in the power control board (PCB). The battery, shown in the lower middle, supplies power through the the PCB to the motors, and also supplies electrical energy to the other electronic components. The GPS antenna, shown in the middle left, and the GPS receiver (GPS RCVR), the left-most of the squares in the middle, acquires signals from GPS satellites, processes them, and sends latitude and longitude data to the navigation control system (NCS), shown just below the PCB. The navigation sensors—altimeter (ALT), magnetometer (MTR), and inertial measurement unit (IMU)—are shown to the right of the GPS receiver. They feed signals to the NCS. The control link antenna, shown at the upper left, is connected to the receiver, which feeds control signals from DROPCON to the NCS.

The NCS processes all of its sensor and control inputs and sends the requisite commands for current to each motor to the PCB. It also sends telemetry data to the transmitter from which it is sent through the control-link antenna back to the DROPCON.

The image capture subsystem is conceptually separate. It comprises a camera, shown at the bottom right, which feeds images to the video processor, which in turn, transmits them through the video downlink antenna.

The DROPCON, depicted on the left side of the figure, accepts control inputs from the DROP through the collective and cyclic sticks and various switches and buttons shown of the DROPCON. The DROPCON has a certain amount of processing power, provided by the computer shown in the left middle, and has a display—either a built-in LCD display or an attached tablet computer, shown at the lower middle (Display). The computer sends control signals through the transmitter to the control link antenna through an RF link depicted by the lightning bolt to the microdrone.

The DROPCON receives telemetry from the microdrone's NCS through the same antenna receiver integrated with the DROPCON transmitter. The computer processes telemetry for display on the video display. The video downlink signal is received through a separate antenna, processed through a separate video processor, from which it can be rendered on the display and sent elsewhere.

The functions depicted in the figure are ones that must be performed by any microdrone carrying an image sensor and its associated DROPCON. Various models of microdrone organize the components performing these functions differently. For example, the Phantom 3 integrates the image processing function with the navigation control and telemetry functions for transmission to the DROPCON.

3.10.3.5 Autonomy

Autonomy must be distinguished from remote control. Sometimes, they can be hard to differentiate. Fly-by-wire systems and autopilots are spreading downward (in terms of aircraft cost, size, and complexity) throughout the airplane and helicopter industries. These arguably are a combination of remote control and autonomous systems. The elevator is not as remote from the pilot in an airplane as a DROP is from his drone, but he cannot touch the elevator, so there is some degree of remote control. Traditionally, the pilot's stick or yoke, which he can touch, was connected mechanically to the elevator by means of a wire or rod. Increasingly, however, particularly in heavier and more complex aircraft, the yoke or stick is not connected to the elevator directly at all. Instead, it sends electrical signals to servomechanisms that apply forces to the elevator.

The servomechanism may have very little autonomy, simply doing precisely what the pilot commands, or it might have quite a lot. Some of the safety features built into ONE Aviation's new Eclipse 550 executive jet include automatic auto throttle advance to maximum power settings when the aircraft approaches a stall and reduction if flaps are extended above maximum-flap-extension speed.[1]

3.11 Mission-oriented subsystems

Airborne imaging systems must have means for pointing the mission sensor (usually a camera) at the subject matter. In simple systems, the airplane or helicopter pilot or drone operator simply points the aircraft toward the target. That still leaves the problem, however, of camera elevation. If the pilot or DROP points the nose of the aircraft up or down, the vehicle will climb or descend.

Imaging mission systems of any sophistication, therefore, have camera stabilization and gimballing systems. Stabilization systems use elastic bushings and accelerometers to isolate the camera from aircraft vibration. Gimballing systems allow a camera operator to direct the camera lens up or down (elevation) or left or right (azimuth). It also is desirable to add a third axis of control, so that the camera can be tilted to compensate when the aircraft rolls.

Gimbals can be automated to combine accelerometers and servo motors automatically to isolate the camera from aircraft movement, for example, to keep it pointed in the same direction when the aircraft turns, climbs, or descends. Higher levels of automation can lock the camera onto a subject and keep it pointed there regardless of aircraft movement.

3.11.1 Cameras

The typical microdrone mission sensor is a video camera. Digital video cameras, just like film cameras, use a lens to focus an image onto something to record the image. In digital cameras, sensors replace film. The sensors comprise one-to-three light-sensitive semiconductor chips at the back of the camera. Two types of light-sensitive chips are in wide use: charge-coupled devices (CCDs) and complementary metal-oxide semiconductors (CMOS). The principles are the same for both types. Higher-end cameras have three of these photosensors, one for each primary color, while lower end cameras alternate pixels specializing in each color. A three-chip camera uses a prism to separate primary colors and send each to the corresponding sensor.

Each pixel (also called a *photosite*) generates an electrical charge proportional to the amount of light falling on it. The CCD in a consumer video camera is about a half-inch wide and contains 300,000–500,000 pixels. The layer of photo-sensitive pixels transfers its pattern of charges to a back layer, which digitizes analog value samples and sends the digital pattern to the camera circuitry while the top layer gets ready to acquire a new image. The rate at which images thus are captured is the same as the frame-rate for the video mode being used—24 frames per second (fps) for movies and 29.97 fps for TV. Either of these rates is more than fast enough to give a viewer the illusion of motion, but higher rates improve quality and editing options. Thomas Edison discovered that a sequence of images presented faster than about 12 per second is perceived as motion.

In a film camera, the image is focused on a frame of film where its patterns of light and dark and color chemically transform silver halide crystals (known as *grains*) arranged in three layers, separated by filters, one for each color. A mechanical drive moves the reel of film so that it stops under the lens briefly as a mechanical shutter opens for each frame to be captured. The film is subsequently developed chemically to fix the image pattern as light and dark grains, stained appropriately to represent colors.

The maximum resolution of film is determined by the number of silver-halide grains per square inch of film; the maximum resolution of a CCD or

CMOS is determined by the number of pixels. Limitations on resolution are perceived as *graininess* in a film image, but the term is less used in the digital video community.

Although the exact dimensions of different quality categories is the subject of debate, some rules of thumb can be articulated. To qualify as SD or standard definition, the sensor must have at least 480 rows of pixels vertically, and 720 columns of pixels horizontally. HD or high definition requires a minimum of 720 rows and 1280 columns. 4K, or UHD (ultra high definition), requires 2160 rows and 3840 columns. Consumer television screens and other video displays have progressively moved from SD to HD and UHD, as practical and affordable display sizes have increased from 12 inches or so on the diagonal, common in the 1960s, to 84 inches or more for home theater screens. At that size, a consumer would be able to see every raster line and every pixel on an SD image.

A CCD or CMOS has a circuit in the form of a matrix in the back layer that represents the lightness and darkness of each pixel on the sensor as a pattern of ones and zeros (binary digits or "bits"). Resolution is affected further by the number of these binary digits available to represent the lightness or darkness of each pixel, sometimes known as the *bit depth* of the sensor. For example, a depth of 8 bits can represent 256 colors, while a depth of 12 bits can represent 4096 colors.

A timing pulse causes the electric matrix to dump its pattern of binary digits at the frame rate (a process known as *clocking*) to upstream circuitry, which handles the resulting stream of binary digits representing the sequence of frames.

The number of bits generated per frame capture is enormous, and thus the circuitry must be capable of handling very high bit rates to capture everything and send it along for downstream processing. For example, a Sony HDR-AS10 head camera has 16.8 megapixels on its CMOS sensor, resulting in 2 effective megapixels. The extra megapixels on the sensor are used to establish a black baseline from which brightness can be measured. At a bit depth of 8 bits, sufficient to represent 256 levels of brightness, each of the 2 million pixels generates 8 x 30 bits per second—the bit depth times the frame rate. 240 bits per second per pixel, times 2 million pixels means a bit rate of 480 million bits per second—480 megabits, or 60 megabytes per second.

Most video cameras perform onboard compression to reduce the bit rate that must be handled by downstream circuitry and logic. H.264, also known as MPEG-4 Part 10, Advanced Video Coding (MPEG-4 AVC), is a popular compression standard for full-motion video. It offers high-quality HD-level results at half the bit rate of the MPEG-2 standard. § 3.9.1.3 explains compression in more detail. Compression is more common for recording and less common for wired streaming, in, for example, a television studio.

Resolution alone, however, does not determine picture quality. Two cameras, each with HD resolution, can produce significantly different quality video. That is so for two reasons. First, bigger cameras have bigger sensors. The 1 inch sensor in a Sony AX100 handycam has 4.2 times the surface area of the 1/2.3 sensor in a GoPro 3+ and 6.6 times the surface area of the 1/3 sensor in an iPhone 5. Larger sensor size for a given resolution makes it possible for each photosite to be bigger than it can be on a smaller sensor. Bigger photosites capture more light. That means that in low light conditions, the signal-to-noise ratio will be higher and thus the perceived quality of the image sharper.

The focal length of a lens (about 55 mm is the standard for 35 mm film cameras) determines the angle of view of the image imparted to the sensor. Bigger sensors require greater focal-length lenses for the same angle of view obtainable on a smaller sensor. Longer focal length generally requires a bigger lens. Stock GoPros have 6 element aspherical glass lenses. Sony AX100s have 17 element glass lenses, with 24–120mm (35 mm equivalent) optical zoom capability, unlike the GoPro lenses.

More elements add to weight and size, but they also allow for reduction in chromatic and optical aberrations. Aspherical element shape permits more correction with fewer elements. Chromatic aberrations produce fringes of color (usually pink) at boundaries between light and dark areas of the image. Optical aberrations include flare, which makes bright images look hazy, and other distortions such as straight lines in the subject being curved toward the edges of the captured image—as is noticeable in GoPro imagery. Curvature of lines results from use of short-focal length wide angle lenses, but the anomaly can be corrected with more sophisticated lens design.

Zoom capability require more elements than are necessary for the same image quality in a fixed focal length lens, known as a *prime* lens.

As always, there is a trade-off. Higher quality lenses with zoom capability weigh more and cost more than prime lenses with the same aperture. Aperture refers to the size of the lens opening, determining how much light the lens can capture. Bigger sensors require bigger camera bodies, which weigh more, and they consume more electrical power. In other words, bigger sensors necessitate bigger drones to carry them.

3.11.2 Other sensors

Drones can carry other mission-oriented sensors, aside from video cameras. Infrared sensors capture light and heat outside the visible spectrum. Radar and sonar can measure height above the ground precisely, allowing that data to be combined with position information determined by the drone's position sensors to conduct topological surveys of the ground. Radiation and chemical detection devices can detect hazards.

3.11.3 Digital signal processing

Transferring images, whether, photographic, infrared, or other, from the drone mounted camera to the ground and from there to the customer requires multiple steps—the same steps required to transfer DROP control inputs to the drone, explained in § 3.4.2: quantization, digitization, compression, and modulation of an RF carrier. Each of these steps consumes computer processor cycles and requires memory to buffer the bits being processed at that step. Processor speed and memory access times determine how long each step takes. Specialized *digital signal processing* (DSP) chips perform better than CPUs, which theoretically could handle the same data and algorithms.

In any event, the time that elapses for each step to be performed is greater than 0. The sum of the times introduces delay or *latency*. Latency does not matter much—unless it causes buffers to overflow—if the imagery is simply being recorded for later processing, as might be the case with aerial surveying. But if it is meant to be consumed live, as is the case with much newsgathering, other aerial video applications, and law enforcement, latency is undesirable, especially when reporter is interacting with the imagery. Latency less than the duration of one frame is undetectable; skilled reporters and producers can handle latency of a few seconds or more, although it may annoy audiences. In the online gaming world, latency is frustrating, because even quick player reflexes come into play too late to achieve the intended result in the gameplay.

The same progress in digital hardware technology that has made drones possible in the first place will continue, increasing processor speed, reducing size and power consumption of processor and memory chips, and generally driving the price lower. This means that smaller and more affordable packages will be capable of increasingly high-quality video imagery with low latency and longer range.

3.11.4 Image processing

Digital processing of images is necessary to make effective use of imagery captured by drones. It also is crucial to the development of sophisticated collision avoidance systems, considered in § 3.12. Three kinds of image processing are relevant to productive drone use. The first, known as *multispectral imagery*, processes still or video images to enhance representation a particular colors or heat signatures. This is useful in agricultural crop surveys, construction site surveys, and railroad, pipeline, and powerline surveys. The second is combining images so that they provide geo-referenced landscapes or mosaics. The third analyzes the fine details of images so as to detect particular kinds of objects and to compute their trajectories.

Multispectral imagery, pioneered for use in satellite-based earth imagery, uses sensors that capture narrow bands of the optical and infrared spectrum, permitting the resulting images to be processed so as to highlight features that emit or reflect radiation in particular spectral bands. Much of the commercial multispectral hardware and software is derived from systems developed for the US National Geospatial Intelligence Agency.

Multispectral imaging is useful in agricultural surveys, because vegetation reflects not only the blue, green, and red colors perceivable by the human eye, but also near-infrared and mid-infrared light. Healthy vegetation is highly reflective in the near infrared spectrum. Overlaying the near infrared light intensity on a traditional core photograph highlights healthy vegetation, and contrasts it with less healthy vegetation that appears less green in the composite image. Soil reflects mid-infrared radiation according to its moisture content. So overlaying mid-infrared light intensity onto traditional core shots, highlight the moisture content of soil.

Multispectral imaging is a combination of photography and spectroscopy. Spectroscopy analyzes different wavelengths according to their intensity. Some multispectral imagery is captured by single cameras with appropriate filters to collect red, green, and near infrared light. Other systems use multi camera arrays, with each camera specializing in a particular spectral range.

Adequate cameras have large enough sensors to capture light across the requisite parts of the spectrum with enough sensitivity and minimal noise from interference between adjacent pixels. Processing algorithms must provide adequate color depth to permit discrimination among the different spectral intensities. Multispectral computation involves more than rendering of the different primary colors into a faithful representation of what a human observer would see. The algorithms must know what spectral regions, such as near-infrared or mid-infrared, to emphasize and how to combine them with primary colors. Multispectral cameras and systems can be tuned to highlight particular types of plant life, either specific crops, or different species of weeds.

The resulting portrayal of a crop field or a construction site thus reveals conditions of interest. For example, an ordinary image of a crop may indicate a generally green color, while a multispectral image reveals that certain parts of the crop are growing well, while other parts are under stress from too much heat, insect infestation, or not enough water. A multispectral portrayal of the same field, tuned to emphasize mid-infrared light, shows varying moisture content.

Construction site surveys can similarly be constructed to highlight moisture content or different types of dirt and rock. Powerline, pipeline, and railroad surveys can emphasize foliage overgrowing the infrastructure.

Designing software for multispectral analysis and performing the necessary computation is the least technically sophisticated of the three types of image processing.

Combining a series of images taken of the same general geographic area to produce a panoramic view requires knowing the geographic coordinates representing the boundaries of each frame in the series. Then the software must be capable of using this geometric tagging to combine the images so as to eliminate overlap and otherwise to align the images so that those representing adjacent part of the area are aligned with each other with the overlap cropped out. Depending on whether real-time review of the consolidated panorama is necessary to determine navigation, speed of processing is not particularly important. The drone can fly the requisite flight paths to collect imagery of the desired in a series of passes over the field, and the resulting image data can be processed after the flight.

Object detection is vital to the functioning of collision avoidance systems. The video processing must be able to determine, in real time, when a shape in a series of frames represents an object of interest—another aircraft, or a stationary obstruction. It must be able to determine whether the shape is moving, its past trajectory and projected future trajectory. Then the programs must also compute the relative position of drone and target object. Those data then can be fed the other subroutines that decide whether the drone needs to take evasive action and, if so, the flight path that it must follow.

The principal challenges are accurate object recognition and sufficient speed of computation to give information in close to real time. Some degree of latency is inevitable, but the software must know what the latency is it compensate for it. If too much latency results from the data analysis, inputs to the collision avoidance subsystem will be so far behind reality that avoidance will not be attempted until after the collision occurs.

Integrity of the signals is essential. False positives are as harmful as false negatives. False-negatives may result in a collision, or at least the need for a last-minute pilot or DROP intervention; false-positives increase aircrew workload and generally dull crew reactions to reports of potentially conflicting aircraft. False positives also may, if they feed automated collision avoidance subsystems, lead to erratic flight paths that actually increase the collision risk.

Imaging and imaging processing R&D frontiers involve improving the processing times for sophisticated pattern recognition, enhancing the ability of software to disaggregate different parts of the spectrum and to recognize shapes relevant to collision avoidance and to predict their trajectories.

3.12 Collision avoidance

The principal limitation causing the FAA not to permit beyond-line-of-sight (BLOS) operations is the absence of reliable collision-avoidance systems at acceptable weight and cost. Even the most capable navigation and control system can only put a drone at a point in space desired by the DROP or commanded by its autonomous navigation system; doing that may run the drone into another aircraft or try to fly it through a tree or a building or smack it into a hill or a wind turbine tower. When the DROP maintains line of sight with the drone, he also can see obstacles and conflicting traffic and can avoid these obstacles by intervening manually with appropriate control inputs.

When the drone is out of sight, this is not possible.

One strategy, as § 3.10.3.4 suggests, is to try to extend the DROP's line of sight by replicating the visual information an onboard pilot would have. Model aircraft hobbyists call this *first person view* (FPV). Then line of sight moves with the drone, and a regulatory line of sight requirement imposes no particular limit on range, any more than see-and-avoid limited the range of manned aircraft. But that probably is not a cost-effective way to attack the problem, as § 3.10.3.4 explains. It likely always would have limitations derived from a lack of three-dimensional perspective in even the most sophisticated cockpit on the ground. The more attractive technology development to enable safe BLOS uses automated target detection and collision avoidance, or automatic airspace access control.

Conceptually, four approaches exist to collision avoidance, beyond unaided (or video-enhanced) line of sight:

1. Ground-based radar systems track aircraft flying in proximity to each other and ground-based operators communicate potential collision risks to the pilots involved, who alter their flight paths accordingly. The radar equipment and operators may be airborne instead of ground-based as in air-force and navy AWACS systems.
2. Aircraft engage in peer-to-peer communication and exchange data about their positions and flight paths. This is the core approach for current TCAS and ADS-B traffic separation.
3. Aircraft use onboard systems to detect other aircraft (and other obstacles) that do not depend on data transmitted from the target. These generally are referred to as *passive*, even when they rely on active transmission of radar or laser pulses from the aircraft seeking protection.
4. The air traffic control system assigns aircraft exclusive rights to occupy a block of airspace, one aircraft at a time.

The current air traffic control system in the United States uses a mixture of all of these, backed up in the end by see-and-avoid. Aircraft on IFR flight plans and under positive control in controlled airspace are assigned particular blocks of airspace to which other aircraft do not receive clearances until it has been vacated by the previous occupants. The assignments are made by VHF voice radio communication. Ground-based ATC radar tracks IFR flights, making use of altitude and other data transmitted by aircraft transponders. Onboard TCAS uses data transmitted by the transponders on targets and, increasingly, ADS-B out transmitters. The availability of ADS-B out signals is reinforced by ground-based TIS transmitters, which rebroadcast ADS-B out data received from aircraft within range of the TIS facility.

Safe operation of drones BLOS requires adapting these basic approaches, automating the collision avoidance function to a greater degree, and automating access control for low-level, high-density drone flight as would result from package delivery.

3.12.1 Finding the targets

A typical sensor and image processing system begins with images captured by any sensor such as a camera, radar, or laser detector. Radar and laser detectors provide data on the range of the target; camera-based imagery does not, requiring range to be inferred from other image characteristics such as size and its increase or decrease with time. Smaller wavelengths associated with light provide greater detail. Longer wavelengths associated with microwave signals used for radar provide less detail. Radar and laser sensors consume power proportional to the strength of the pulses they send to be reflected by targets. Camera-based systems are passive and thus consume less power, but they cannot see in the dark.

Some sensor systems, such as phased array radar, use multiple detection beams, each for a different part of the field of view. Alternatively, as in conventional ATC radar, the antenna is rotated, and the timing of that rotation determines where in the field of view a target reflects the signal. Typical laser-based systems use the same principal as a rotating radar antenna, but the laser beam is deflected by a tiny mirror that provides imagery for a circular plane by progressively aiming the laser beam through all the points comprising the plane (each point comprising a pixel in the resulting image).

The sensor delivers a series of still images organized into a matrix of pixels (e.g. in the case of pertinent Australian research, 1024 x 768 pixel images captured by an ordinary camera at a rate of 7.5 frames per second[2]) to the processor. In the processor, morphological filtering is used to select clusters of pixels that differ from the background and seem to be related to each other. Dynamic programming is used to reduce noise, such as random bright or dark spots smaller than features of the object are likely to be. Measures of quality include the rate of false alarms, the times of first detection, and of consistent detection.[3]

False alarms can be reduced by adding a processing stage that concentrates on images that have a low rate of image movement and a high rate of image expansion.[4] Testing of such systems in the mid-2000s showed that the systems could detect potential collision threats 35–40 percent sooner than human observers, even with heavy cloud clutter.[5]

When the sensor is mounted on a moving platform such as a drone, its image processing logic must compensate for movement of the drone, through inertial measurement data or GPS data or a combination of the two.

Identifying stationary targets, such as terrain features, buildings, towers, or objects is no different in principle from identifying other aircraft. The difficulty is that ground clutter presents more of a challenge for target discrimination than typical cloud cover or light precipitation.

3.12.2 Avoiding collisions

Merely identifying nearby aircraft is not enough, of course, the system must predict the respective paths of the drone and, for moving threats such as other aircraft, the paths of the threats. Conceptually, that is not complicated; it simply involves using trigonometry and transformation of coordinate systems to match up the velocity vectors of the objects on collision course.

The computational load is considerable, however, because of the quantity of data generated by the sensors at any useful sampling rate. The objects are moving, after all, and so their positions change considerably every second. Only recently has computing power been available to handle the data load at acceptable weight and cost.

Identifying collision paths may finish the automation task if its goal is only to alert a human pilot or operator to take evasive action. If the system itself is expected to take evasive action, however, it must have programmed into it evasive strategies—it must decide whether to go right or left, climb or descend, when it finds itself on a head-on collision course. The challenge can be eased somewhat by setting up rules of deviation based on direction of travel—northbound vehicles descend and turn right, and southbound vehicles climb and turn left. Such rules are already built in to the basic rules of the road for maritime and aviation activities.

In addition, it must know the limitation of the flight vehicle on which it is installed. What is the maximum turn rate and radius? What is maximum rate of climb and descent? It must factor those into its trigonometric calculation so that it computes—and is able to execute in time—a course of action that avoids the obstacle.

Collision avoidance algorithms have been around for a long time—since TCAS was mandated for large transport aircraft in 1995—but engineers still wrestle with how to design the algorithms and program the software so that the system does not *increase* the likelihood of a collision if one pilot follows ATC instructions while the other follows TCAS advice, or if considerable latency exists in the control link for a drone.

For short-range drone flight, it may be perfectly adequate to give the DROP traffic advisories and let him decide how to alter his flight path to avoid a collision. Such a system could work just like TCAS and ADS-B in manned aircraft. As the range increases, however, latency in the control link becomes a reality, and autonomous response calculated and implemented onboard is more likely to be necessary.

Improvements in image processing are necessary for these systems to be reliable and useful. A much simplified example helps explain the process. Suppose a matrix or raster of pixels is defined to represent the field of view of a camera. Each pixel signifies a white or black area of the image depending on whether it is a 0 (for white) or a 1 (for black). The first 1080 pixels represent the first line of the raster, the next 1080 pixels the second line, and so on. The image processing program scans the bits representing the pixels sequentially. When the pixel value flips from 0 to 1, software makes a note of the position in a particular line where this occurs. If the bit value flips from 0 to 1 in the same position of the next and subsequent lines, the algorithm recognizes that this is the edge of an object. If the next few positions in the same lines also are ones the object takes on a rectangular shape, its width defined by how many sequential pixels in each line are ones. The software can represent the rectangle by numerical values representing its top, its bottom, its left edge, and its right edge.

If the flipped pixels representing the edge are displaced in subsequent lines, the object may be a circle, in which case it can be represented by vertical and horizontal values representing its center, and another value representing its radius.

Such a description of the object can be matched against a variety of templates, say, one for a square, a second for a vertically elongated rectangle, a third for a horizontally elongated rectangle, and another representing a circle. Depending on which template the object matches, the algorithm may take different actions. A similar template-matching process can be used to recognize other objects, such as helicopters, fixed wing aircraft, and birds of different species. As the image processing becomes more realistic in this fashion, the computations must deal with gray areas as opposed to pure black and white, colors and object orientations not square with the visual frame. The number of processor steps to do this is quite large.

The mathematics required to represent the objects and the templates also is sophisticated. It is not difficult to detect a large stationary obstacle such as a brick wall. But when the obstacle is smaller and is moving, like a vehicle or another aircraft, the image processing algorithms must do multiple computations to determine shape, size, color, and direction and rate of motion. This requires computations of complex transform equations, each application of which requires multiple computer processor cycles.

3.12.3 Migration of automotive technologies

The rapid development and introduction in the consumer market of automated driving features for automobiles will help accelerate technology development for drones. Autonomous automobile operation involves reliable sensors and collision avoidance software, and the risks of system malfunction obviously are enormous. Consumer acceptance seems to be substantial, and the scale of automobile sales covers development costs that otherwise might be an obstacle for development purely for drone flight. In theory, drivers are supposed to be attentive to the automated operation of their vehicles and remain ready to take over if there is a malfunction. The likelihood, however, is that they would relax and be slow to take corrective action.

The motor vehicle industry obviously is serious about developing technologies for autonomous driving. Already, many models of luxury automobiles have self-parking and imminent-collision energy management systems. This level of automation is spreading to lower end, mass market models. The capability of specially equipped test vehicles to navigate ordinary roadways successfully without driver intervention has been demonstrated by Tesla, Google, Mercedes-Benz, Lexus, and others. The Mercedes F105, and similar models from Audi, BMW, Google are in demonstration phases on city streets and will be available commercially within five years or so. The same navigation and collision avoidance technologies are suitable for drones.

Self-driving automobiles and trucks benefit from advantages that make it more likely that they will be widely commercialized before the full array of drones:

- The vehicles are designed to be operated by persons with a little training and relatively low levels of skill. The stands in sharp contrast to manned aircraft, which must be operated by pilots with much more extensive training, experience, knowledge, and skills. It is, thus, far easier for an autonomous system to improve driving, as compared with flying. The public will accept automation of the functions of a sleepy truck driver and of a bumbling automobile driver sooner than it will accept automation to replace the popular image of an airline captain.
- The collision avoidance challenge is inherently simpler for ground vehicles, because it deals with only two dimensions, as compared with three dimensions for a drone. Moreover, ground vehicles operate on fixed guideways—driveways, streets, and expressways. Aircraft operate on essentially arbitrary routes selected directly between origin and destination. The necessary infrastructure for self-driving ground vehicles will develop sooner, because much of it is already built.
- The regulatory agency for ground vehicles, NHTSA, is enthusiastic about autonomy and is pushing the regulated entities to move faster. In contrast, the FAA, the regulatory agency for drones, is reluctant, sluggish, and is pulling industry back.

It is clear that motor vehicle engineers have made substantial progress in designing and proving the effectiveness of the requisite mix of sensors and navigation algorithms that permit autonomous automobiles and trucks to follow the road, adjust speed with the flow of traffic, react instantly to threatened accident, and to avoid stationary objects such as roadsigns.

They are improving their ability to detect pedestrians, predict their likely behavior, and adjust the vehicle's behavior accordingly.

These developments in the ground vehicle industry have profound implications for drones. For one thing, large markets for automobiles and trucks permit development cost to be spread, so that manufactures are willing to invest far more than aircraft manufacturers. Manufacturers often offer the features in their high-priced flagship vehicles initially and then allow them to 'trickle-down' to lower-priced models as costs fall. As that occurs, improvements in sensors' capability at a lower cost and weight will migrate into the aviation world. Once upon a time, migration was the other way, as automatic control system technologies were developed first for aircraft, and then decades later migrated into the automotive world.

There are, however, significant barriers yet to be overcome. 2014 and 2015 Mercedes-Benz tests and demonstrations involved a struggle to avoid the $140,000 cost for sensors used in earlier tests, and algorithm designers continued to try to replicate the kind of multidimensional judgment that an ordinary automobile driver makes when he decides whether a pedestrian is like to continue into traffic or a step back on the curb. As one engineer explained, it is not simply a matter of detecting the presence of the person and determining his velocity and acceleration vectors; it is also a matter of assessing his socio-economic characteristics, his facial expression, and a holistic perception of his prior movements: is he in a hurry? What are his nonverbal cues as he looks left and right at the traffic? Robots and their algorithms cannot do a very good job of perceiving or understanding these things.

3.12.4 Deployment on microdrones

Basic collision avoidance capability is already available. The DJI Phantom 4 has built-in collision avoidance. The drone has sonar sensors that constantly scan for obstacles within a field of approximately 60° either side of the direction of flight. The vehicle can be programmed to stop when it comes within a DROP-specified distance of the obstacle, to fly over it or to fly around it. Tests performed by the authors show that the system can detect foliage and porous objects such as chain-link fences, and react appropriately.

Further R&D, however, is necessary to take the software to the next level: to enable it to determine when sensor data indicates a flying object and then to compute its trajectory and to figure out how to avoid it. The same R&D challenge confronts engineers of self-driving cars, which must be able to identify pedestrians and project their likely actions.

3.13 Neighborhood Access for MicroDrones (NAMID)

Growing interest in delivering packages by microdrones creates the need to manage microdrone congestion over city and neighborhood streets to avoid the risk of collisions between microdrones trying to access neighboring premises. The NAMID system described in this section is very similar to the system proposed by Amazon in two white papers released in August 2015; although the following subsections provide more detail on the architecture and its traffic-separation rules. NASA's UTM project is developing concepts similar to those described for NAMID.[6] It uses a combination of external surveillance by radar, satellites, and cell towers, and peer-to-peer telemetry exchange to track drone positions. Then it relies on onboard collision avoidance and geofencing to control navigation. Internet-linked databases of weather and wind, airspace constraints, and three-dimensional maps of terrain and human-made structures. Human oversight and flight planning is involved, but not moment-to-moment vehicle control by DROPs.

3.13.1 Overview

NAMID utilizes a network of allowable routes from any point to any other point within, say, five miles, developed from detailed data about ground features, already available in Google maps and a wide range of mapping competitors. Basic algorithms in the system do not simply draw straight line between origin and destination pairs; they figure out how to use the existing infrastructure of streets, sidewalks, and expressways. They update their routing strategies based on minute-to-minute information about congestion, construction, and street closures, features already in databases associated with consumer-level GPS navigation systems.

Layered on top of this basic infrastructure map are traffic separation rules. A microdrone tasked by its DROP to fly to a particular destination would broadcast its intentions. Other nearby microdrones would respond with a data block disclosing their position, their route of flight, and their intentions. These autonomous communications exchanges would reference the routes defined in the infrastructure map.

If a particular segment of the route desired by the first microdrone is occupied, it would wait or seek another route. A first-come-first-served rule of thumb is built in. A microdrone delivering a package to the apartment complex at 930 Evanston Street, would request access to the segment between the intersection of Bode Road/ Evanston Street and the driveway into the complex. If that segment is already occupied by another microdrone, it would hold at the intersection until the segment is clear. Battery charge would limit wait time. That means that drones designed for this system must have much larger battery capacity than is available on microdrones currently on the market.

3.13.2 Implementation

Each microdrone is equipped with a transceiver capable of communicating with NAMID, constantly receiving inputs about geographic position from onboard GPS, magnetometer, altimeter, and accelerometers. When a microdrone plans to enter a NAMID neighborhood, it would broadcast an inquiry message, similar in design to the message broadcast by a wireless network equipped computer wishing to connect to a Wi-Fi network, a cellphone seeking a handoff to a new cell tower, or an office computer connecting for the first time to a wired LAN.

The NAMID protocol would cause any other NAMID station within range to acknowledge receipt of the inquiry and a handshake (establishment of a communications link between the two) would result. The new microdrone would then be connected to that NAMID network.

Thereafter, each NAMID node in a network would broadcast position and destination in packets. All nodes would process these packets and determine which microdrones were on the same route.

The network architecture is peer-to-peer; no ground stations would be required for collision avoidance.

3.13.3 Block definition

Implementation of the concept requires defining blocks of airspace that can be assigned exclusively to a single microdrone. The blocks are similar to the blocks represented by base leg, final approach, and the runway itself utilized by ATC for visual approaches at controlled airfields, or blocks implicitly defined by horizontal and vertical separation requirements for IFR approaches controlled by TRACON. They also resemble the blocks used by railroad dispatchers under the *direct traffic control* or *track warrant control* methods of train dispatching.

For NAMID, the blocks could be particular segments between two intersections, or pathways to individual residences in a housing complex. GPS permits blocks to be arbitrarily defined, but they must be known to every vehicle participating in the system.

The blocks can be fixed, or they can move with the vehicle, as in some advanced forms of railroad dispatching and in IFR approach-control operations.

Capacity depends upon the size of the blocks; larger blocks mean less capacity, while smaller ones mean greater capacity. A block would have a vertical dimension as well as a horizontal one. So, for example, one block in a neighborhood might be the airspace between 50 and 90 ft over Carol Lane between its intersection with Green Bay Road and the intersection of Park Place; while another block might be the airspace between 110 ft and 150 ft over the same segment of road. Grid systems in the form of maps already exist for neighborhoods, but NAMID also must have altitude separation.

Some blocks would be defined on an *ad hoc* basis, such as those accommodating descents for landing at a particular destination such as the front porch of a residence. Entry into an arrival block from a predefined enroute block would be indicated by transmission of special kind of announcement message alerting the other microdrones in the vicinity to expect pathway coordinates for the landing profile. Once they receive it, the landing segment block would be like any other block in terms of the exclusive right of the first entrant to occupy it.

3.13.4 Peer-to-peer, or ground infrastructure?

A basic design decision is whether NAMID would be entirely peer-to-peer, or whether it would rely on ground-based infrastructure to some extent. The peer-to-peer system would put all the intelligence and data systems on board the microdrones. An infrastructure-based system would place some of it, the more demanding database and computational activities, in ground-based facilities. The peer-to-peer approach requires onboard memory and computation of sufficient power to handle the volume of data and to execute the navigation algorithms fast enough. Onboard databases also would require frequent updates. The means for updating routing tables in Internet routers could be a model. And infrastructure-based NAMID would require that ground-based antennas be placed within range of every neighborhood in the system—a requirement that would drive up cost and encounter political opposition, unless the antennas were co-located with cell-system antennas or unless NAMID used the cell-system infrastructure itself for communications. The actual ground-based databases and computation facilities could be located anywhere in the world.

Peer-to-peer NAMID would draw on some of the technologies implemented in TCAS, but it would have considerably broader functionality. TCAS does not have route information. NAMID must have it. The NAMI infrastructure also resembles a more finally grained NextGen system, in that aircraft will determine their own route, while ground-based logic determines assignment of requested routes in order to assure traffic separation.

3.13.5 "Track warrant" based?

Another design decision is whether access to a block would depend upon possession of a key. That is the case in the now largely abandoned *token ring* system of controlling access to local area network segments, or the track warrant system used by railroad dispatchers. There is also the possibility that it would work like CDMA (collision detection, multiple access): the protocol used to control access to local area networks on the pervasive Ethernet, as another. When a computer connected to an Ethernet network is ready to send a packet, it listens to the network to see if a packet is already being transmitted by another computer. If the segment is thus occupied, it waits before transmitting its own packet.

To apply this principle to NAMID, microdrones flying in NAMID would constantly broadcast their locations, thus allowing other drones to know when a block is occupied. They also would listen to other drones broadcasting their locations. If another drone is already in a block, a second drone would not enter.

Once a block is free, a drone could enter it and thereafter would have exclusive authority to remain in the block, perhaps limited by time. Time limits for block occupancy might be fixed, or they might depend on the size of the block, being defined, for example, by a certain direction multiplied by the length of the block. The latter approach would give flexibility for blocks of varying sizes, which in turn would depend upon capacity needs for that particular portion of the network. A largely rural area with one farmhouse per square mile would have blocks a mile long. Limiting access to that block by only one drone at a time would not impose significant cost; the likelihood of other drones waiting for access is small. On the other hand, recalling that only one drone can occupy a block at one time, the blocks in a residential apartment complex would need to be quite small, probably the size of segments of sidewalk extending from the projected sidewall of one building to its opposite side wall.

3.13.6 Holding

If a block is occupied when a microdrone wishes to enter it, its onboard NAMID system would cause it to hold at the nearest block boundary. Because microdrones would be rotary wing in configuration, holding would not take up much space; it would involve hovering at a particular height above the ground. The risk of collisions between multiple microdrones holding would be managed by deeming a block still occupied as long as a microdrone is holding at its far end. For example, a drone's route calls for it to traverse Block A and Block B. Block A is unoccupied when the drone arrives at its boundary. The drone enters and has exclusive rights to Block A until it moves beyond Block A's boundary into Block B. Block B is occupied by another drone. The first drone must hold inside Block A until Block B is cleared. While the drone holds, Block A is still occupied, and no other drone may enter.

3.13.7 Emergencies

The system would have to be designed to handle three types of emergencies appropriately. One emergency is loss of propulsive power. This type of emergency is well accommodated by the basic design of NAMID. The microdrone losing power would already have exclusive rights to be in a particular block and therefore would represent no risk with other microdrones—as long as it stays in the block. It would, of course, represent a potential hazard to persons or property below it. Moreover, a number of malfunctions by different drones could make the network unusable in relatively short order, because multiple blocks would be unavailable.

A second type of emergency—lost contact between drone and NAMID—is more challenging, because there is no assurance that the microdrone losing contact with NAMID would stay in its block, or even know its block's boundaries after the failure. Lost communications protocols in the FAA's instrument flight rules could be used as a rough model, clearing the airspace between the drone's last known position and its destination. Another possibility for dealing with this emergency is to clear all of the blocks between the point of last contact and that microdrone's launching point. Either would depend, of course, on the system knowing origin and destination for each microdrone in the system. That could be accomplished by defining the initial announcement message to include fields similar to the origin and destination IP address fields in an Internet Protocol packet. Like an IP packet, a NAMID packet would include the drone's origin and destination.

Impending battery exhaustion is a third type of emergency. This emergency could be handled in the same way as lost contact: the system would clear all the blocks from the point at which the low battery alert occurs.

The problem with the block-clearing emergency procedures is that the blocks to be cleared might presently be occupied by other microdrones. They could be programmed to sidestep to the next block, but it might be occupied as well. Design would have to accommodate the resulting cascading effect of an emergency, as one block after another is cleared, pushing drones into other blocks that might be occupied.

Implementation of the proposed emergency procedures would be more feasible in less congested NAMID airspace.

Higher-level blocks would have larger lateral boundaries and be reserved for the enroute portions of sorties, while lower-level blocks would be smaller to accommodate arrivals and departures and transition from the enroute blocks.

3.13.8 Drone swarms research

Research on controlling swarms of drones is highly pertinent to making NAMID a realistic possibility. The same needs for inter-drone communication and control systems that respond to the data received from other microdrones in the swarm are similar to the challenges of developing a logic for NAMID.

3.13.9 Economics of implementation

The foundational principle of the NAMID concept is that drone collision-avoidance communications would be peer-to-peer; no new ground-based data or communications infrastructure would be necessary. Practical implementation of such a system would depend upon the deployment of communication services on top or alongside of cellphone data, and other broadband wireless networks. A ground-based system, depending on central or regional servers would be economically sound only where the commercial demand for delivery services is sufficient to support the investment. A certain population density is necessary to provide revenue to support the capital cost of ground facilities. If only one drone per week makes a delivery to a particular neighborhood, it would be hugely wasteful to invest the resources necessary to map that neighborhood and provide traffic separation from traffic that doesn't exist.

But that is for only a ground facility proximate to that neighborhood. Every part of the United States is already mapped. The marginal investment to extend the existing database and mapping algorithms to a new geographic area is quite low. And for the peer-to-peer version of NAMID, the marginal cost of a microdrone in a new neighborhood for the first time is zero; the vehicle or vehicles wishing to go there already are equipped with the necessary NAMID subsystems.

So the ultimate investment decision depends, not on whether a particular neighborhood has enough package-delivery density, but whether the nation as a whole does. It is reasonable to assume that each drone delivery service will have its own drones and can equip them with compatible NAMID systems.

3.14 Protecting RF security with encryption

Security of control links, telemetry, and video feed can be enhanced by encrypting the signals. Low-cost encryption schemes have been available at the consumer level since the early 1990s. Encryption is used by almost everyone interacting with the Internet. Web browsers use it to protect passwords and usernames used in log-in transactions; cell phone providers use it to protect the credentials used by subscriber hardware to acquire access to their networks.

They rely on an *asymmetric public key* system to encrypt one time keys for a *symmetric* system that is used to encrypt and decrypt the message contents. *Message*, in this discussion, refers to data structures, the contents of which represent control signals, telemetry about aircraft orientation, position, flight path, fuel or battery state, or bitstreams comprising video images. Either type of cryptology can be accomplished by software or by specialized hardware. Specialized hardware is much faster.

Encryption does not completely protect against interception or sabotage; sophisticated state-sponsored operatives can often compromise cryptography, not by brute-force cryptanalysis, but by human and technical means of compromising the encryption scheme. An IT professional can be bribed or blackmailed to reveal secret keys; malware can be installed to reveal keys or signals before they are encrypted; interception of Wi-Fi signals behind the encryption can reveal keys or details that make cryptanalysis easier. Moreover, encryption, no matter how strong, cannot entirely protect against sabotage; the saboteur always simply can jam the frequencies being used. Nevertheless, encryption protects against casual hackers, saboteurs, and cyberthieves.

Both asymmetric and symmetric encryption rely on *keys* that determine how a message is transformed to encrypt it. A simple example, not of any practical use, would transform each alphanumeric character in a text message by substituting the character two places further along in the array of characters. Using this key, *EMS* would be encrypted as *GOU*. Two characters forward of *E* is *G*; Two characters forward of *M* is *O;* two characters forward of *S* is *U*. A recipient of the message *GOU* can apply the same key to recreate the original message (known as *plain text* in the cryptography field) *EMS*.

This example illustrates *single-key* or symmetric encryption, because both sender and receiver must have the same key. Actual symmetric encryption works by taking blocks or streams bits in the signal and transposing them through a complex scheme of rotation, each step of which involves the key. A recipient who has the key easily can reverse the process and restore the original signal. The difficulty of breaking the encryption scheme is determined by the length of the key.

Exchanging the single key between parties to symmetric encryption is its central disadvantage. It might be stolen en route without either of them realizing it. Securing the means of transit increases the complexity of the logistics.

Asymmetric encryption is attractive because it eliminates these difficulties. Asymmetric encryption does not require both parties to an encrypted communications to have the same key; instead, they use different keys. Each has his own *private key* and an associated *public key*. The private key is known only to one party to the communication; the associated public key is known to the world to be associated with its owner. Linking public keys to their owners is the responsibility of *certificate authorities* maintained by government agencies or trusted private entities such as Microsoft or Apple.

Asymmetric encryption offers two capabilities: communication security, and communicator authentication. Communication security occurs when the sender encrypts a message with the receiver's public key. Only someone possessing the receiver's private key can decrypt it.

Communicator authentication is achieved when the sender encrypts a bit stream comprising his *digital signature* with his private key. The sender is authentic only if that message can be decrypted with the sender's public key.

The underlying algorithms rely on complex polynomial computations with prime numbers for both encryption and decryption. Even in their intended use, they are computationally intensive. The amount of computation required of someone not possessing the private key is so great that brute-force cryptanalysis is infeasible. As with symmetric encryption, the strength of the scheme depends on the length of the keys. Brute-force cracking of an algorithm with a 128-bit key would take multiple supercomputers more than a billion years.

Asymmetric encryption is much slower than symmetric encryption and decryption. Longer keys make it even slower. So most encryption systems use the two types of encryption in combination. Asymmetric encryption is used only to encrypt the single key for symmetric encryption and to authenticate sender and receiver. The encrypted single key is used to encrypt and decrypt the message. Using the two together offers the benefits of simplified key exchange and the speed of symmetric encryption.

Even the most efficient encryption systems, however, slow down the processing of signals. A transaction such as moving one character from DROPCON to transmitter now requires multiple computer processor cycles. On the receiving end, moving it from receiver to vehicle control logic subsystem and then decrypting it requires multiple processor cycles. To achieve the same speed of communication, both ground and airborne systems must have faster hardware and software.

To be traded off against this cost of encryption is the risk reduction that it buys. In the context of drone operations, the strongest candidate for encryption is the control link. There is no reason that anyone aside from the DROP needs access to the control link, and corruption of it can have disastrous consequences.

Telemetry is a weaker candidate for encryption. Other aircraft, air traffic control, and privacy and civil liberties advocates, benefit from access to telemetry showing that drone position and flight path. On the other hand, a DROP's competitors may obtain proprietary business information if they have access to telemetry.

Whether the imagery downlink should be encrypted presents a more difficult set of questions. As this section explains, encryption and decryption slows down signal transfer at both the transmitter and receiver. Imposing encryption on a real-time, live, high resolution video signal likely will require upgrading hardware and software to increase speed. More bandwidth should not be required, because the encrypted signal is no more complex or information intensive then its unencrypted version. As § 3.9.1.1 explains, Shannon's law says that the bandwidth required for a signal depends on the signal's information richness and the speed at which information is transferred—the bit rate. Key-length makes no difference, except for the encryption and decryption processes, which occur before the encrypted signal enters the RF channel.

Interception of unencrypted high-value imagery enables misappropriation and copyright infringement. Interception of unencrypted confidential imagery compromises confidentiality.

3.15 Safe design: reliability engineering and fault analysis

Both types of drone are even more highly automated than manned aircraft, and manned aircraft automation is intensifying. Increasingly, the center of gravity in airworthiness certification will shift away from mostly mechanical systems to mostly electronic systems. Automation involves advanced software. Experience teaches that it is difficult to get bug-free software, particularly in early versions of complex programs.

Software fails, not because something wears out or breaks, but because of a programmer mistake or because some kind of unanticipated interaction occurs in the computer code. Physical inspection cannot determine functionality and reliability.

Good programming practice has two people working on the same section of code. One person creates the code, while the other programmer checks the code. Good practice also requires commenting on code so others can understand its purpose and logic. Still, problems often arise when multiple people work on different parts of a complex computer program, which is constructed as largely independent objects or subroutines. As the program executes, subroutines are typically loaded in and out of memory by the operating system. Predicting all of the permutations of states that might exist among different program elements is almost impossible. Accordingly, a state can occur that was not anticipated by systems analysts or programmers. When this happens, the program may stop executing, exhibit anomalous behavior, or produce incorrect values.

Hardware faults are less likely, because semiconductor chips are electrical, not mechanical, but they do occur, especially when temperature, moisture, voltage, or current levels exceed specifications. Hardware faults can often be detected by visual inspection; software usually must be executed to find its flaws. The interaction of hardware faults with the software is particularly difficult to predict and program around.

Aviation safety depends on its technologies being reliable. Since the earliest days of aviation regulations, regulators have insisted that designers and manufacturers of aircraft demonstrate their airworthiness. That, in turn, has spawned a new technology known as *reliability engineering* or *safety engineering*. Analytically, reliability engineering requires: (1) inventorying every fault that can occur in every aircraft component, (2) quantifying the probability of that fault occurring, and (3) assessing the risk of failure.

An example would be the failure of a pitch link on a helicopter rotor blade. The probability of this occurring depends on the design of the link and the properties from which it is made. The consequences would be catastrophic. Asymmetric lift as between the two rotor blades would probably cause the rotor blade to separate from the rotor hub.

In a microdrone a fault might occur in the power supply to one rotor because the soldered connection of one of the motor leads to the power distribution board has failed, resulting in an open circuit to that motor. The probability of that occurring is relatively high because wire connections in which the only strength is provided by the solder itself are brittle and weak. The consequences would not likely be catastrophic in a multirotor design, because the thrust of the rotors still in operation can be ramped up to ensure stable flight or at least a controlled landing.

A capacitor on an integrated semiconductor circuit board might fail, rendering a micro-drone's GPS navigation system inoperative. The consequences of an inoperative GPS navigation system depends upon how else the drone could navigate in that particular flight regime.

The three-step process is known as *fault analysis,* which also recognizes that multiple faults can occur at more or less the same time. Rigorous fault analysis must consider all of the possible permutations of faults.

The results can be quantified by use of a *fault tree*, in which generally accepted probability analysis multiplies and adds the probabilities to determine the joint probability of various combinations of multiple faults.

When faults have been identified, their probabilities estimated, and their consequences assessed, designers and regulators decide what should be done to reduce the risk of failure. One possibility is to redesign the failing component to reduce the probability of failure occurring. Depending on the way in which the failure such as that of the pitch link occurred, the component link could be redesigned to be made of stronger material, to be larger in dimension, or to attach the link to the pitch horn of the blade or to the upper swashplate in a different way.

In the case of the broken solder connection, assembly procedures could be modified to require that the wire be mechanically connected before is soldered, as by wrapping the wire around or hooking it through a terminal, or twisting two wires together before the connection is soldered.

If redesign is not likely to be cost-effective, redundancy is another corrective action. Each rotor blade could be equipped with two pitch links, either strong enough to adjust the pitch of the rotor blade throughout its operating range. Two power connections for each leg of the electrical circuit could be provided for each motor.

Another mitigating strategy is to revise component specifications so as to narrow operating limits in terms of speed, temperature, or turbulence.

Increasing automation means that more of the critical aircraft systems are implemented by computer software rather than mechanical structures, assemblies, connections, and movements. Faults in software are far more likely to be due to mistakes in coding logic than to physical failure. In the world of computer programming, fault analysis and mitigation is known as *debugging*. The more complex the program, the more difficult it is to debug. A fault may manifest itself in the overall failure of the system of which it is a part. A program may cease execution or produce wrong values, but isolating exactly what fault caused it is challenging. Advancing the frontier of reliability engineering for aircraft requires better techniques for fault analysis of computer software and automating them.

In many cases, the best solution is some means of indicating failure of a system or of a component, allowing a backup system to take over, or alerting the pilot or DROP so that he can use his training, human instincts, and ingenuity to take appropriate corrective action, one not programmed into any of the systems in advance.

For example, when the engine of single-engine helicopter fails, it is less important for the pilot to know whether it was a tooth on the bevel gear in the main transmission that failed than it is for him to know that the propulsion system, as a whole, has failed. To make sure he knows that immediately, helicopters are equipped with both an annunciator light and an aural alarm for low rotor RPM, an immediate manifestation of an engine failure.

Engineering science permits the designers of physical components to determine their strength and other properties and thus to establish the conditions under which they will break, bend, or suffer fatigue processes likely to resolve into an eventual fatigue fractures.

But things often behave differently under real-world conditions than theory predicts. Data on actual behavior is essential for good failure analysis, and it often is unavailable in sufficient quantities to complete fault analysis feasible before an aircraft enters operation.

Usually, full fault analysis is not possible until after an aircraft system is in service for many months or years. Before that, averages of test results can be used, but averages such as mean time between failure (MTBF) are not enough. Failures often exhibit wide deviations around the average, and a particular fault may have such catastrophic consequences that it would be insufficiently protective of safety to focus the average circumstances under which it will occur, rather than conditions that might cause it to occur at the 10 percent, 5 percent or 1 percent probability level.

Chapter 9 considers options for collecting operational data on autonomous system function to enlarge the store of data necessary for statistically robust failure analysis.

Notes

1 Matt Thurber, Eclipse 550, Aviation International News, (Oct. 2014), p. 58 (reviewing capabilities of the latest version of light twin-engine jet).
2 Ibid. at 2852.
3 Ryan Carnie, *et al.*, *Image Processing Algorithms for UAV "Sense and Avoid,"* Proceedings of the 2006 IEEE International Conference on Robotics and Automation (Orlando, Florida, May 2006), [hereinafter "Australian studies"].
4 Ibid. at 2852.
5 Ibid. at 2853.
6 NASA, *UTM: Notional Scenario*, www.utm.arc.nasa.gov/index.shtml.

4 Human factors and labor markets

4.1 Introduction

Human capital is essential to any economic enterprise, including drone operations. An enterprise starting a new drone business or an existing enterprise establishing a new line of business needs the right kind of personnel. Chapter 11 explores the full range of issues to be considered when starting a drone operation. This chapter focuses only on the human capital aspects. Other chapters also consider effective use of human resources. This includes chapter 10, which considers the relative advantages or disadvantages of independent contractor status as compared to employee status.

This chapter considers both human factors and human relations. The two are distinct conceptually, but they are related. Together, they connect machines with human organizations. The human factors end of that link addresses the interface and interaction between a human operator like a pilot or a DROP and the hardware and software he operates. Human relations addresses the opposite end of the link: the interaction between the individual human being and the organization in which he is embedded. Because little drone—or other aviation—activity involves individuals operating completely on their own, one must evaluate the interaction between business organizations and the tools that they use. That can only be done by considering both ends of the link. The link between organizational management and the drone itself thus comprises both human factors and human relations.

Understanding human factors involves more than understanding how to promote safety. Labor (economists' old-fashioned term for human capital) is one of the three traditional factors of production. Just like physical capital, such as aircraft or hangar facilities, labor must be well suited to the production activity and managed so that it produces the desired output efficiently. Thus, the human capital subject implicates a full range of determinants of business success, not just guidelines on how to avoid crashes. The scope of this chapter is defined accordingly.

The chapter begins by reviewing the interaction between aircraft and the human beings that fly them. It then moves to consider DROP qualifications and training, some required by regulation and others imposed by business judgment. Any analysis of drone use must consider the qualities, availability, and cost of human capital, including education and training costs. Personality attributes and technical skills are among the criteria that should shape DROP selection criteria. Innate behavioral characteristics reinforced by effective training leads to sound aeronautical decision making (ADM).

The early sections emphasize qualification criteria and selection of DROPs. They begin with the longstanding criteria for manned aircraft pilots and explain why those criteria may be only partially suited for DROPs. While drones do not have pilots on board, they do have human operators. Accordingly, effective and safe drone flight depends on the qualities of the human operators.

Building on this analysis, this chapter explores organization design and management practices that result in good performance by human resources, and concludes with a section on the level of rule compliance under various approaches to human resource selection and management. It explains fundamental differences between human relations in work organizations that stress internal career development and those that assume employees will use employers as stepping stones to better jobs. It stresses the frequent divergence between employee selection, recruitment and development for management jobs, and corresponding practices for pilot jobs.

In traditional aviation enterprises, the human capital available to fly aircraft is limited by regulatory requirements for pilot and mechanic certification. In any market-oriented economy, the selection, training, and management of DROPs will result from a combination of governmental regulatory requirements and private enterprise decision making. Understanding human factors—the qualities, skills, experience, and attitudes that make pilots safe and productive—inform every aspect of pilot selection, training, and qualification. The same thing should be true of DROPs. Understanding the differences between DROPs and pilots should shape evolving DROP training, experience, knowledge, and certification requirements.

This chapter emphasizes selection, training, performance, and management of operating personnel, including, not only DROPs, but also mechanics and avionics technicians. That does not exhaust the human factors subject, however. Management and support personnel are important to the safety and success of drone operations, as well.

Economists distinguish between internal and external labor markets. External labor markets involve the interaction between potential employees and potential employers. Internal labor markets involve the interaction among employees of the same enterprise and between management and employees. This chapter covers both types of labor markets: external labor markets in its analysis of hiring criteria, and internal labor markets in its assessment of management philosophy and style. Training considerations implicate both types of market, because training requirements may be imposed as a condition of hire, or training may occur after a DROP is on the payroll.

One important question, probed more fully in chapters 10 and 11, relates to assumptions about direct labor costs embedded in a DROP's business plan. Some business plans, especially those for microdrones, assume labor costs substantially less than what they would be for an operation using manned aircraft. This might be because the compensation for DROPs would be much less than that for pilots. Or it might be because the DROPs are assumed to have higher productivity because their duty time is more flexible, or because they fly more, or both. For macrodrones, similar assumptions might be made, although it is more likely that labor costs would be higher for macrodrones because the complexity of the control systems and collision avoidance systems would require more highly trained DROPs. Questions of compensation are mainly reserved for chapters 10 and 11.

4.2 DROPs

This section recognizes that significant differences exist between the human factors concerns for microdrone DROPs and the concerns associated with macrodrone DROPs. Microdrone DROPs, for the most part, operate their drones within their line of sight. Macrodrone DROPs, for the most part, do not. The assessment of a DROP's ability to control the aircraft successfully, to get it to its destination, and to avoid other aircraft depends on quite different variables, depending on which kind of drone the DROP is flying. Flying a macrodrone is much more similar to flying a manned aircraft. Not only are the skills different for microdrones, so are the interfaces between airman and aircraft.

4.2.1 Human factors at the DROPCON

The aviation community appreciates the role that human factors play in aviation safety. Two of the seventeen chapters in the FAA's Pilot's Handbook of Aeronautical Knowledge[1] address human factors. The FAA's Aviation Instructor's Handbook[2] begins with a chapter on "Human Behavior" and concludes with a chapter on "Risk Management." A 2010, 750-page book collects and analyzes empirical and theoretical data on human factors in aviation.[3] Safe flying depends not only on machines performing according to good design; it inherently involves interaction between human beings and flight machines. The literature identifies the following four main clusters of issues:

- Flight crew performance of essential piloting functions, considering the necessary combination of kinesthetic and cognitive activities, the role of information processing by the crew, including the ability to multitask, set priorities and avoid distractions; the importance of situational awareness and the impact of workload
- The impact of aircraft design on crew performance, including the design of human interfaces

- The effect of automation on flight crew engagement and proficiency
- The influence of organizational cultures and management styles on flight crew behavior and performance.

Flying a drone is like many other activities in which physical movements must be combined with mental processing. This includes tennis, golf, playing the piano or guitar, swimming, and skiing, to name just a few. Control inputs by DROP must be smooth, reasonably precise, and well-timed. Many recurring phases of flight require carefully developed instincts to couple accurate sensory perception with movements.

Some emergency procedures require almost instantaneous and correct action by the pilot to avoid disaster.

DROPs learn to do these things, just like one learns to hit a tennis ball or to ski, by much repetition. In the right kind of teaching environment, they have available to them an instructor who points out their errors and coaches them on performing the maneuver correctly. Eventually, the appropriate inputs become automatic—a part of muscle memory, and the DROP need not think about it; he just reacts instantly as soon as he recognizes the problem and, better yet, anticipates a problem and takes corrective action before it occurs.

A DROP must practice enough that he detects any deviation from the intended attitude or flight path and makes appropriate control inputs right away. Experience flying helicopters and airplanes is relevant, although the size of the control movements and the control forces are quite different; one usually operates the controls on a microdrone DROPCON with one or two fingers, while a pilot wraps at least part of his hand around the stick in an airplane or helicopter. Conversely, the need to detect deviations quickly and to make small and correct control inputs is the same.

The amount of time required to learn to fly a microdrone successfully appears, from all the evidence, to be much shorter than the amount of time required to learn to fly a helicopter or airplane. One important reason is the autonomous modes of flight built in to most microdrones. Without these features, most multicopter microdrones would be uncontrollable, regardless of DROP skill.

Automatic physical reactions to flight phenomena, however much practiced, are not enough; DROPs also must think before they act, and the need for action requires them to think quickly.

A microdrone operator is gathering breaking news. His photog tells him that he should try to get an angle for a shot that will force him to fly further away, close to or beyond the drone's range. Should he attempt it, or should he refuse, perhaps considering relocation of his position to get the desired shot. Pressing his luck and depending on the reliability of the autonomous return-to-home feature in the microdrone may result in a crash.

Or, suppose that experience has taught a microdrone DROP that nothing is ever wrong with his microdrone when he pre-flights it. So he abandons that routine set of steps that now seem like a waste of time. On a particular occasion, something had happened to the microdrone to disrupt the calibration of its GPS system or its magnetometer. Or, perhaps, repeated vibration has dislodged the wire from the control board to the power distribution board. He takes off quickly, and realizes the drone has become uncontrollable. It flies away at top speed while the DROP fruitlessly seeks to regain control.

Avoiding these mishaps falls into the realm of what aviation regulators and instructors call *aeronautical decision making* (*ADM*). Practicing responses to certain hazards can be helpful, such as refusing to fly beyond the specific range of a microdrone, but ultimately, safe flight depends on good judgment.

Processing a flood of information quickly and accurately is an essential aviator skill.

Multitasking is an important skill to facilitate information processing. At the most basic level, a DROP must be able to coordinate control inputs to change direction of flight simultaneously with climbing, descending, or coming to a hover. He must fly the drone while coordinating with other personnel such as visual observers and photogs.

Part of multitasking is knowing what tasks have priority. This is particularly important in handling emergencies. Priority setting in emergencies is encapsulated in the popular aviation slogan: aviate, navigate, communicate. The first priority is to *aviate*: to keep the aircraft under control, and, for example, if the GPS link is lost, to act quickly to re-attain manual control.

After performing these aviate steps, the next priority is to *navigate*: to position the drone so that it is headed in the desired direction and to establish an appropriate rate of descent. The margin for error in accomplishing the navigate priority may be unforgiving; if the drone deviates too much from the assigned path, it may hit an obstruction. A DROP must make a quick decision whether to land immediately or trigger a return-to-home function, once a GPS lock is obtained, depending on his assessment of obstacles.

The third priority is to *communicate*: to report the problem to observers and, if ATC communication is involved, to air traffic control and to declare the DROP's intentions.

Situational awareness is also essential. The most basic forms of situational awareness are to know where you are and which way is north. Beyond that is an appreciation of the position of other traffic. Situational awareness also includes a sense of where nearby obstacles are. One does not want to start a left turn only to smack the drone into the side of a building.

Other aspects of situational awareness include a perception of what the drone is doing. Is it climbing? Descending? Turning? Is a microdrone about to exhaust its battery charge?

Loss of situational awareness by a microdrone DROP primarily involves losing sight of the microdrone or being unable to determine its orientation and thus which way it will move if it is commanded to go forward. Otherwise, he knows where he is and where the drone is and where it is going simply by taking in his surroundings and watching the drone. Many other possibilities exist for loss of situational awareness by a macrodrone DROP. His vehicle will be out of sight much of the time, and therefore he cannot know where it is, simply by watching its position relative to its surroundings. Instead, he must rely on instrumentation similar to that available to a pilot in the cockpit and on video images significantly more limited than those available to a pilot.

Skills in performing discrete tasks and maintaining situational awareness are not enough; DROPs must manage workloads that comprise multiple tasks. DROPs, photogs, and visual observers should share duties. Good practice for drone flight leaves the DROP to fly the drone, while a separate photog or systems operator manipulates the camera and other sensors.

Automated capabilities reduce aircrew workloads. A DROP should make effective use of automated capabilities such as automatic takeoff and landing, automatic orbiting, and automatic tracking of a defined path.

To manage workloads, maintain situational awareness, and multitask successfully, all while setting appropriate priorities, DROPs must avoid distractions in some phases of flight, such as those for which the FAA requires a sterile cockpit. A sterile cockpit prohibits communications or activities by the flight crew not directly pertinent to the particular flight operations of the moment.

Mitigating distractions by adherence to a sterile cockpit rule is more difficult for a microdrone DROP than for a macrodrone DROP. A macrodrone DROP is far more likely to be performing his functions in a specialized operations center, insulated from outside forces. Unlike an airplane or helicopter pilot, he is subject to being interrupted by his supervisor or by other DROPS flying different aircraft, but he nevertheless enjoys a measure of isolation resembling that of a cockpit.

A microdrone DROP operates in a far different environment. He is likely to be outside, near other operating personnel—reporters, law enforcement and fire-suppression professionals in ENG applications; other first responders in public safety applications; and engineers and construction workers in construction-monitoring and surveying applications. He must be able to concentrate on flying his drone safely, ignoring the distractions.

A further consideration is boredom. Bored DROPs let their attention wander, and they are less likely to notice automation malfunctions immediately and to react quickly and appropriately. Boring flight profiles can cause atrophy of manual flying skills for maneuvers not part of the routine.

Too many things happen quickly in microdrone missions for DROP boredom to be much of a problem. The opportunity to get bored is inherently limited by having to fly back every 20 minutes or so to swap batteries. Still, undue reliance on advanced automation, such as automated takeoff and landing and automatic orbiting can induce boredom just like undue reliance on automation in manned aircraft. During longer macrodrone flights, the temptation may be strong for a DROP to engage in non-flight-related activities while automated systems fly the mission.

4.2.2 Interface design

Aircraft design has always depended on human factors. Within the larger question of the interaction between aircraft design and human control is the question of interface design. That question looms large for drone operations because DROPs—especially macrodrone DROPs—depend almost entirely on mentally processing what they see on mechanical and electronic interfaces rather than on the physical experience of being in a cockpit.

Instrumentation and interface requirements depend on whether a manned aircraft is equipped to fly under Instrument Flight Rules (IFR) or only under Visual Flight Rules (VFR). The conditions for VFR vary somewhat, depending on altitude and type of airspace, but the most basic rule is that VFR requires visibility of at least 3 miles, a ceiling of at least 1,000 ft, and that the aircraft remain 1,000 ft above, 500 ft below, and 2,000 ft horizontally from any clouds. VFR operations do not require ATC clearances except for landing and taking off at airports with control towers.

An aircraft may be operated IFR under in almost any meteorological conditions, but the aircraft must have an ATC clearance, must adhere precisely to altitudes and courses assigned by ATC, and must follow published procedures for landing and taking off. A pilot or DROP flying IFR needs instrumentation and interfaces that permit him to monitor his compliance with ATC assigned altitudes and courses and to control the aircraft without outside references.

Microdrones, for the foreseeable future, will not be flown IFR. Macrodrones, to realize their potential, will.

In VFR conditions, collision avoidance depends upon the see-and-avoid rule: aircrews must be vigilant in watching for other aircraft so that they can see them and alter their course of flight as necessary to avoid conflicting aircraft.

Flying VFR, a microdrone DROP needs even less instrumentation than the pilot of a very simple VFR airplane. A pilot needs airspeed, altitude, turn-and-bank, magnetic heading, engine speed, oil pressure, oil temperature, and fuel quantity information. But the risks that this basic instrumentation allows a pilot to avoid are different from the risks a microdrone DROP must avoid. A multirotor microdrone is not going to stall if it goes too slowly; accordingly knowledge of airspeed is less relevant to a microdrone DROP than to a pilot. Similarly, because a multirotor microdrone cannot stall, it cannot spin. Furthermore there is really no such thing as crossed controls—applying aileron in one direction and rudder in the other—which increases the risk of a spin in an airplane. So a microdrone DROP does not need a turn and bank indicator. Altitude is important for both pilots and DROPs, mainly to ensure compliance with various airspace restrictions defined in terms of altitude. For a microdrone DROP, however, height above the ground is a more useful than the altitude above mean sea level displayed to pilots.

Microdrone DROPs also need information that is irrelevant to a pilot: strength of the control-link signal and video downlink signals, and information about remaining battery charge on both the drone and the DROPCON.

The kind of information a macrodrone DROP needs to execute an IFR flight plan is the same that a pilot of a manned aircraft needs. The interface for macrodrone DROPs must present the same information provided to an onboard pilot. Whether the interface presents it in exactly the same way as cockpit interfaces do depends mainly on whether cockpit interfaces present the information effectively. If they do, the same interfaces are appropriate for macrodrone DROPs. If they do not, the same design improvements that allow more effective presentation for DROPs also would improve presentation to pilots. In other words macrodrone interface design will influence manned aircraft interface design.

An airspeed readout can be made easier or more difficult to read and understand, depending on the interface. Coupling a number with a moving ribbon indicating airspeed or altitude along with different colors or pointers representing limits is an example of interface improvement. The DROP can get a holistic sense of trends, while mentally processing the number.

Placement of displays is an important part of improving the interface. Instrument pilots are taught to scan attitude, direction, airspeed, and altitude instruments, rather than fixing attention on only one. If those instruments are placed randomly around the instrument panel, the scan is much more difficult. If they are clustered together, right in front of the pilot, the scan is much easier.

Good interface design is essential for drones. The FAA reported one example in which the sequence of commands to adjust the lights on the Predator was almost identical to the sequence that shut down the engine.[4]

Sound design is about much more than merely placing values for flight parameters so that they are easy to find and easy to read. It also involves designing switches and annunciator lights so that their operation is intuitive. It would, for example, be undesirable to lay out a row of switches so that some of them are on when they are up and others are off when they are up. Equally undesirable are annunciator lights all of the same color, some of which indicate safe conditions, such robust video downlinks, and others of which indicate malfunctions, such as loss of GPS lock or loss of the control link.

Sequences of layers and operating control presentation must be intuitive. They are not for some microdrone software; it is difficult to remember what icons a DROP must press to change altitude or range limits, or to prescribe the profile the drone should fly when the DROP commands return-to-home. Commonly used or emergency functions must be available near the top, or on every layer.

The fact the DROPs are not in the cockpit and thus cannot see and avoid other aircraft as well means that DROP interfaces must compensate for this deficiency in perception.

Many popular microdrones, including the DJI Phantoms, downlink telemetry information that can be displayed on a screen overlay, including height, distance, speed, heading, and battery charge remaining. This is sufficient to support flight within the DROP's line of sight, but it is not nearly enough for flights beyond line of sight (BLOS).

The see-and-avoid principle determines the most basic interface requirements for drones. When aircraft fly in IFR conditions, see and avoid is replaced by adherence to an IFR flight plan, and the macrodrone DROP needs essentially the same inputs that an onboard pilot flying under IFR needs. Already, more sophisticated aircraft have autopilots that can automatically execute IFR flight plans from takeoff to landing, although the current FARs impose some limitations on autopilot-coupled takeoffs and landings.

Over the long term, the number of drone flights in congested areas will overwhelm the conventional system for managing air traffic. The need for traffic separation in such an environment will accelerate an existing trend for IFR traffic management. More responsibility for traffic separation will be decentralized, so that the system relies more on peer-to-peer data exchange between aircraft and less on air-to-ground voice communications.

As this occurs, manual or autopilot control of the aircraft to adhere to an IFR flight plan will be replaced by automated flight planning and coordination and automated responses to traffic alerts. When automated sense and avoid is implemented, the DROP responsibilities will become radically different—aimed at monitoring the automation rather than flying the drone. Eventually, this will occur with macrodrones and some microdrones—those equipped to operate in a NAMID-like environment. Macrodrones are not likely to fly very widely until automated sense and avoid is available. Traffic separation displays for pilots now mostly rely on the pilot to take action in response to potential traffic conflicts he sees displayed on a screen. Fully automated sense and avoid systems, likely to be necessary for high volumes of drone traffic in congested areas do not require this type of display.

Until such a level of automation is achieved, see and avoid will continue to be the default rule. In order to see and to avoid, a DROP, like a pilot, must be able to see other traffic. This is not a problem for microdrone DROPs, because they can see not only their drones but also other aircraft near it.

For macrodrone DROPs to have visual information comparable to what they would have if they were in the cockpit, the ground station layout must have a multiplicity of high definition video screens that provide real-time imagery captured from multiple cameras on the drone. The images on the screens would replicate a field of view roughly 270° laterally and 90° up and down.

The biggest concern is whether, even with high-definition cameras on the drone and wraparound video displays at the console, the DROP will be able to see as much as he could if he were in the cockpit. Ensuring adequate field of view is possible with the multiplicity of cameras and video screens or with the virtual reality approach. Zoom lenses theoretically would give him greater visual acuity, but it takes longer to zoom a lens than it does to swivel your head and focus on a particular object.

Providing depth perception, however, is a challenge without an obvious solution. Depth perception is important in spotting other traffic. For example, the FAA's standards for pilot medical certificates require depth perception, even in pilots with serious visual deficiencies in one eye.[5] Depth perception is not possible with even the best two-dimensional image produced by the best monocular camera. 3-D video and medical technology might address the depth perception problem.[6] Lens focal length also is important; iPhone images show things further away than they are. Many sideview mirrors in automobiles do likewise.

The complexity of two cameras, however, shooting each angle with just the right aiming points and fields of view would enormously complicate the sensors on the drone and would be collateral to mission equipment. They would be necessary only for control and traffic separation. So this would represent weight and power consumption unnecessary on a manned helicopter.

Empirical work is necessary to evaluate the adequacy of visual information for a macrodrone DROP to see and avoid other aircraft. Here is how useful data can be collected:

> Take a manned helicopter and equip it with the same cameras that engineers are contemplating for a macrodrone to perform a similar mission. The helicopter would have ADS-B Out and ADS-B In equipment to send and receive telemetry about the position of all similarly aircraft in the vicinity. Pairs of pilots would participate. Pilot number one in each pair would start out as the DROP, seated at a fully equipped DROP console. Pilot number two would fly the helicopter. An audio visual recording with a finely grained timetrack would be made of the DROP and the pilot throughout each test run.

A target aircraft would fly flight paths unknown to the DROP and the helicopter pilot. The route would be designed to put the target aircraft at varying distances, relative altitudes, and bearings from the helicopter. For half of each test run, the target would have its ADS-B Out transmitter on and for the other half of the test its ADS-B Out would be off. The target aircraft would be equipped with mapping software that records its route, as would the helicopter.

When either the DROP or the onboard pilot sees the target aircraft, he would say the usual, "Traffic, one o'clock, left to right," and so on.

After a series of tests, exploring the various permutations of traffic position and ADS-B generated traffic advisories, the two pilots would switch places and repeat the exercise. This is necessary to correct for any differences in visual acuity and physiological profiles between the two.

To get a larger sample size, the process would be repeated with 10 or so pairs of pilots.

Alternatively, data on control movements could be collected automatically for both pilot and DROP. These data should permit a statistically valid evaluation of differences in traffic detection from the cockpit and the DROP station.

4.2.3 Monitoring system health

For highly autonomous drones, understanding the aircraft systems, particularly the navigation systems is as important learning how the aircraft behaves and manipulate the controls. Being able to fly the aircraft by hand competently and safely remains important, but far more energy must be invested in how to use the control and navigation subsystems. Nevertheless, DROPs incur risks when they rely unduly on automation. They may fail to monitor automated flight regimes. They may be tempted by distractions. They may lose their proficiency in manual control.

4.3 DROP qualifications

DROPs must have the right physical capabilities, personality attributes, knowledge and intellectual curiosity, and attitudes toward regulatory compliance. Those recruiting them must understand how their investments in training motivates them with regard to job selection and progression.

4.3.1 Physical capabilities

To perform as a pilot or a DROP, one must have the necessary physical capabilities. The armed services and the FAA have demanding physical selection requirements for flight crews. Candidates to be army helicopter pilots, for example, must pass the Alternate Flight Aptitude Selection Test (AFAST) or Selection Instrument for Flight Training (SIFT) which have criteria for vision, hearing, kinesthetic sensitivity, hand-eye coordination, strength, and dexterity. The FAA has requirements on similar subjects with increasing stringency as one advances from the third, through the second, to the first class medical requirements. A fundamental question for drone operations is whether these traditional requirements are suitable for DROPs.

Whether existing FAA medical certification requirements are appropriate for DROPs is an open question. In an operation requiring a single DROP, the risks associated with a heart attack, stroke, seizure, or a diabetic attack might be serious. In a two-crewmember operation, the risk would be much less, especially if the second member of the crew has had some training in basic control of the vehicle.

How much risk is associated with a medical emergency involving the DROP also depends upon the presence of others in the operating area—DROPs are not by themselves in a cockpit which no one can reach to provide aid. DROPs typically are, for microdrones, in or near crowds of people, and for macrodrones, in a control center with other crewmembers and supervisors. In any event, they are on the ground and therefore easier to reach than if they were flying around in the air.

Medical qualification requirements, whether imposed by regulators or by drone operators themselves, should proceed from empirically based analysis of:

- How a particular medically related episode would affect safety of flight, according to how much it would interfere with the crewmember's performance of essential tasks
- How likely such an episode is, given objectively measurable physical attributes and conditions.

Growing sensitivity in the society at large to irrational disability discrimination, and the large number of disabled veterans from the Iraq and Afghanistan conflicts invite attention to criteria that unnecessarily exclude those with physical or mental disabilities. By extension, it informs the larger questions of medical qualification criteria.

As the American with Disabilities Act requires, the question should be, not "what is your disability?" The only relevant question is, "Can you perform essential functions of the job?" The important point for drone operators is, not how they can avoid violating the ADA; the point is that they can take advantage of the talent pool represented by disabled applicants. Some aspects of the ADA are good conceptual models for sound personnel decisions.

In the drone context the room for disabled applicants should be broader than for manned aircraft. Physical and motor skills are important for DROPs, just as they are for pilots, but different skills are required. Most DROPCONs do not have pedals, for example, and thus it is much more likely that a lower leg amputee or paraplegic could perform all of the physical functions of a DROP without modifying the requirements or installing any assistive devices, than it is that the same person could fly an airplane or helicopter safely.

Moreover, persons with disabilities who have successfully conditioned themselves to perform well despite the disability are likely to have higher levels of motivation and focus than the average applicant.

4.3.2 Personality attributes

Physical strength and dexterity is not enough. DROPs must have personalities that induce them to use their physical capacities effectively and to channel them into relevant skill development. They must not have personality defects or psychoses that make them unsafe. This was made abundantly clear in the notorious example of Andreas Lubitz, who deliberately crashed a German Wings airline transport to commit suicide, killing 144 passengers and five other crew members in the process.

A 2006 study conducted for the army evaluated pre-enlistment test instruments and recommended changes to evaluate personality attributes and skills that had been validated as predictors of success in flight training.[7] Some attention to the army study is appropriate, because personality attributes are important to effective operation of any aircraft.

The army study focused on attributes that pilots of high-performance manned aircraft should have. Many of these are irrelevant to DROP qualifications, especially for microdrone operation. But some of the factors identified as having predictive effect, like general reasoning and spatial ability and achievement orientation are worth considering.

In July 2012, based in part on the study, the Defense Department replaced the longstanding Alternate Flight Aptitude Selection Test (AFAST) with the Selection Instrument for Flight Training (SIFT).

4.3.3 Motivation to be a DROP

An important unresolved question is whether DROPs will find their jobs as satisfying as pilots do. Focus groups of USAF DROPs conducted by GAO revealed significant levels of job dissatisfaction by DROPS, although much of the dissatisfaction related to military requirements that disrupted schedules and to forced reassignment of pilots to be DROPS. Low job satisfaction is corrosive to the labor market. If DROPs are unhappy, the word will spread, and it will be more difficult to recruit DROPs. "In fiscal year 2013, the Air Force recruited 110 new RPA pilots, missing its goal of 179 pilots by around 39 percent,"[8] the GAO said.

Romance has always infected aviation. Many young men and women have been drawn to it, hanging around local airports from an early age to watch airplanes and helicopters. Leonardo da Vinci might have been speaking to them when he wrote:

> For once you have tasted flight, you will walk the earth with your eyes turned skywards, for there you have been and there you will long return.

Some people who might make good DROPs may reason that, while unmanned flight makes for an exciting revolution in aviation history, it also means missing out on the romance of aviation. "The smell of jet fuel, a screaming turbine, and mighty Gs felt down to your core, won't ever be replaced with the air-conditioned DROP seat," as one young pilot said.

The authors undertook informal opinion research to test young people's motivation to become DROPs. All of the respondents were selected to reduce the likelihood that they have prior connections with the aviation industry and preconceptions about flying. The same question was posed to all respondents:

> You have received a job offer that is so attractive you can't pass it up. The job gives you the option either of flying drones from an air-conditioned room, or flying an airplane or helicopter as a pilot. Which would you choose, and why?

A group of 31 respondents were second-year law students at Chicago-Kent College of Law, ranging in age from mid-20s to mid-40s, most in their upper 20s. Sixteen of them said they would prefer to be DROPs; fifteen said they would prefer to be airplane or helicopter pilots. Those who chose DROP emphasized safety, security, comfort, and family responsibility. Those who chose pilot emphasized adventure and being "cool" in the eyes of others.[9]

Four other respondents included two actors, a professional computer systems analyst with a computer engineering degree from Carnegie Mellon, and a New York talent agent.[10] All four are in their mid to late twenties. The talent agent chose DROP on the grounds of greater comfort. The computer engineer chose pilot because it would involve more adventure. Both actors chose pilot for the same reason.

Interviewees for magazine articles by the authors often said things like, "No! I'm not going to add drones to my offerings; I'm a pilot."[11] A classmate of co-author Perritt's research assistant told the research assistant that he had little interest in drones; "I'm a *pilot,*" he said.

Dalton Thompson, a 19-year old private pilot and college student who aspires to a career as an airline pilot, says this about drones:

> Becoming a professional drone operator does not sound appealing in comparison to actually being in the airplane. Many pilots will tell you that the flight deck is the best office in the world. That is what flying is all about, enjoying the world around you. Taking that away and putting a pilot in an office eliminates the passion.

He supports the advent of drones, however, and would consider flying them as a sideline:

> Although less fun, I would consider being a drone operator on the side [while being] an airline or helicopter pilot. If you think about it, there will be a growing need for drone pilots, so why not help out the cause?[12]

The GAO/Air Force DROP study shows aviators sometimes make worse DROPs than those without experience flying manned aircraft. Some of this may be due to different skill requirements. Some of it undoubtedly is attributable to a perception that there is romance in flying in the cockpit, but not in operating aircraft from the ground.

Both possibilities are important to understand more thoroughly in specifying qualification requirements, especially those focused on personality traits.

The "Aviation knowledge, interest, and experience in aviation," trait identified as a significant predictor of success in the army study, may not mean a romantic attachment to manned aircraft, past, present, and future. It may mean something different, something present in video-game enthusiasts.

A Pew Charitable Trust survey showed that the most popular video game genres involve auto racing and sports.[13] Players of those types of games are likely to be drawn by the excitement of developing and showing off skills in fine motor movements that translate into competitive success—an urge obviously pertinent to commercial drone operation.

4.3.4 Knowledge and intellectual curiosity

Selection criteria for pilots emphasize relevant knowledge. It is difficult to fly an airplane if one does not know what the aileron does, or to fly a helicopter if one does not understand the function of the tail rotor. Accordingly FAA pilot certification rules require training in and testing of certain elements of basic aviation knowledge:

- Basic aerodynamics and principles of flight
- Meteorology
- Safe and efficient operation of aircraft
- Aircraft mechanical, propulsion, hydraulic, electrical, and electronic systems
- Tools of air navigation
- Aeronautical decision making
- Regulatory requirements

Similar knowledge requirements, tailored to the environment within which microdrones or macrodrones fly, are appropriate qualifications for DROPs.

The ideal candidate for almost any job takes a genuine interest in job content. She is eager to learn on her own about the history, technology, economics, and alternative ways of performing the customer's mission and their assigned jobs to support it. Candidates should be eager to express creative ideas about how their jobs can be done better, and the organizational culture must be one that invites such suggestions. Often, deficiencies in operating practices will be apparent to operating personnel on the front lines long before they are apparent to higher levels of supervision or operations analysts.

4.3.5 Flight experience

Obviously, experience flying drones for a hobby is desirable for candidates to be commercial DROPs.

The evidence is mixed on whether significant experience piloting manned aircraft is necessary to be a successful DROP. The answer is likely to be "yes" for macrodrone DROPs and "maybe" for microdrone DROPs.

The GAO study of USAF DROP satisfaction indicates that pilots may prefer to fly manned aircraft rather than drones, however, and that suggests caution in focusing DROP recruiting too narrowly on active pilots.

On the other hand, experience in actual flying, or having at least taken flight lessons, is likely to translate well into mastery of the requisite knowledge and skills for performance of DROP duties.

Pilots have developed hand-eye coordination that permits them to make fine control movements to correct for small deviations from desired aircraft orientation, altitude, and course. They have been trained and tested on long-distance navigation, Federal Aviation Regulations, weather, and handling emergencies. All of this knowledge and these skills are helpful for safe drone flight. The weather and navigation knowledge is more pertinent to macrodrone operations than microdrone operations. Macrodrones, like helicopters and airplanes, are more likely to fly beyond line of sight, from one weather system to another and to require use of long-distance navigation techniques. Microdrones flying within line of sight or otherwise within close proximity to the DROP experience the same weather the DROP himself experiences and do not require long-distance navigation.

On the other hand, as chapter 8 points out, the overlap between pilot knowledge and skills and those required for DROPs is incomplete. Safe operation of a both types of drone requires at least basic knowledge of RF control links and video links; safe operation of airplanes and helicopters does not. Moreover, in most airplane and helicopter control systems, aerodynamic forces on control surfaces and rotor blades are transmitted back to the pilot's hands and feet. Sensing those forces and responding appropriately is an important part of pilot skill. Drone control systems do not do this. The onboard electronics sense forces and movements and automatically correct for certain deviations from the desired orientation and flight path. The DROP feels nothing against his fingertips, except possibly the resistance of spring-loaded joysticks.

Flying a microdrone is more like playing a videogame. Video game experience may be as relevant to DROP competence as manned aircraft experience. The hand-eye coordination for drone flying is quite similar to the hand-eye coordination required to play a videogame, for example, aiming a weapon and directing fire in the game *Call of Duty*. The operator senses movement of the target with his eyes, and then makes very fine movements on joysticks he works with his fingers to make the target do what he wants. The physical movements required on a DROPCON are indistinguishable from those required on video game joysticks, and quite different from those required on the yoke, cyclic stick or pedals of an airplane or helicopter. There, full arm or leg movements are required as well as finger movements. Moreover, the magnitude of movements by an airplane or helicopter is coarser than those for flying a drone—feet or miles rather than inches or fractions of inches.

Situational awareness is also similar as between videogames and microdrones. In *Call of Duty*, the player must be quick at detecting constantly changing threats and avoiding them. The same thing is true for DROPs in avoiding obstacles and potentially conflicting traffic. Airplane and helicopter pilots scan much larger regions of sky and deal with traffic conflicts that are likely to take minutes to materialize rather than seconds or fractions of seconds.

Sound is an important cue for an airplane or helicopter pilot. One can hear noise intensify as speed increases. One can tell when engine RPM is increasing or decreasing by changes in sound. Kinesthetic sense is available in manned aircraft but not in drones; a pilot can sense when the aircraft pitches up or down, when it yaws away from correct trim, or is rolling into or out of a turn. These inputs are not available to DROPs.

4.3.6 Attitudes toward regulatory compliance

Boldness is a virtue for DROPs as well as pilots, but rebelliousness is not. For a relatively trivial example, different sequences in performing preflight checks often reduce the time required and result in equal or greater safety. "We've always done it that way," or "everyone does it that way," passes up an easy opportunity for improvement based on suggestions by those who use the checklists.

At the same time, work organizations must be hierarchical, to some degree. When more than one person is involved in performing the work, someone must have the authority to choose among options and make other decisions. For the organization to be effective, other employees must be willing to conform their conduct to a decision. Creative enthusiasm and the instinct to ask, "why?" must be balanced by a willingness to follow orders, whether the boss is looking over an employee's shoulder at the moment or not.

DROP selection, training, and supervision strategies must deal with a major difference between pilots and other DROP candidates. Few of the forces that socialize aviators to a culture of compliance operate with respect to DROPs. There is no popular literature spanning 100 years about the romance of drone flying, as there is celebrating the romance of flying airplanes and helicopters. DROPs do not attend DROP flight schools and participate in hangar talk—there are no DROP schools, yet.

Because DROPs do not participate in a training infrastructure, they have no occasion to learn about FAA regulations. There is no reason they would have had any awareness or contact with the FAA, unlike airplane or helicopter pilots, who have been steeped in the role of the FAA.

No microdrone "community" exists in anything like the same sense that the traditional aviation community exists. Barriers to entry are minimal. Anyone who really wants to can scrounge up a few hundred dollars to buy a Phantom on Amazon, and he can take it out of the box and begin flying it, however awkwardly, as soon as it arrives on his doorstep. There is nothing special about being a member of a club that is so easy to get into.

No infrastructure exists for training or socialization. Most people who want to fly an airplane or helicopter think, "I need lessons." Most people who buy Phantom microdrone on Amazon believe they can fly it without any formal instruction. Few already participate in model aircraft clubs, and they are unlikely to join. Model airplane operators traditionally fly their aircraft in conjunction with other members of clubs organized at the local level. Through the clubs, they exchange design ideas, compete to show off their proficiency in flying, and generally help one another. Most DROPs are not active in such organizations, because they are more interested in aerial photography than in building vehicles to see how well they fly. Moreover, the attractiveness of clubs and other casual, recreational social attachment appears to be diminishing. As Robert Putnam's popular book, *Bowling Alone*,[14] explains, Americans are less and less likely to join organizations than they were half-century ago.

It does not occur to a DROP that he's doing anything wrong when he flies his microdrone; the likelihood of a police officer reacting adversely to his activity is no more likely, he thinks, than that a police officer would object to his playing football with a buddy. And, he certainly has no fear that an FAA inspector will drop by; he probably doesn't know what an FAA inspector is.

Over time, as commercial drone operations become more common and DROPs become more professional, a community will develop. It may or may not be closely integrated with the existing aviator community, but that does not matter; it will be a community, capable of effecting the same kinds of socialization that aviation traditionally accomplished. The goal, of course, is to make sure that the new community reinforces safety and rule compliance rather than undermining them.

Few work organizations want outlaws to act on their behalf. Accordingly, a willingness to obey rules, whether it are part of the duties imposed by aviation law and regulations or whether they are imposed by employers and reinforced by a sense of community, is a natural part of the selection process.

Once an operator selects suitable DROP candidates, it must understand how it will socialize them to an appropriate safety culture.

4.3.7 Earning back the training investment

Getting jobs that allow them to earn back what they have invested in training is an important factor in pilot assessment of job opportunities—and for becoming a pilot in the first place. Whether this factor operates as strongly for DROPs depends on how much training costs and on who pays for it. What appear to be dramatically different costs of creating a competent pilot compared with the costs of creating a competent DROP will influence recruitment and compensation.

In 2014, the USAF estimated that it costs $65,000 to train a drone pilot and $557,000 to train a manned aircraft pilot (the average between $679,552 to train fighter or bomber pilots and $434,418 to train transport pilots).[15] These dramatically different cost figures are misleading, however. Most of the difference in the USAF figures results from fuel costs for manned aircraft training. Moreover, much higher capital and labor productivity for large manned aircraft, compared with that for drones, makes it economically rational to spend more money to train the personnel who fly aircraft that are more productive. The inherent disposability of drones—even more expensive ones—compared with manned aircraft carrying troops or tens of millions of dollars' worth of equipment—justifies greater investment in training pilots.

But training cost differences are rational. Flying a drone is different from flying a helicopter or airplane, as many preceding sections of this chapter explain. The USAF study puts it this way:

> [T]hese aircraft are not "flown" in the traditional sense of the word. Only one of the aircraft reviewed (Predator) has a pilot/operator interface that

> could be considered similar to a manned aircraft. For the other UA, control of the aircraft by the GCS pilot/operator is accomplished indirectly through the use of menu selections, dedicated knobs, or preprogrammed routes. These aircraft are not flown but "commanded." This is a paradigm shift that must be understood if appropriate decisions are to be made regarding pilot/operator qualifications, display requirements, and critical human factors issues to be addressed.[16]

Most observers assume that adequate DROP training should cost much less than an equivalent level of pilot training. In any event, training microdrone DROPs will certainly cost less than training macrodrone DROPs, because the risks of the two types of operation differ greatly, and because microdrone DROPs need less understanding than macrodrone DROPs of weather, long-distance navigation, and flying the aircraft with reference to instruments only.

Significant differences in training costs have several implications. First, to the extent that applicants for pilot or DROP positions are expected to have completed training before they apply, training costs represent a significant economic barrier to entry to the more expensive professional airman tracks. A young person contemplating a career as a pilot who finds out that her training may cost as much as $100,000 is less likely to pursue her aspiration than someone interested in flying drones who estimates training costs at $10,000 or less. A pilot who invests $100,000 will demand more compensation—or not enter the field at all unless she thinks she can earn compensation commensurate with training costs.

Few airplane and helicopter operators are willing to hire pilot aspirants with no training at all and to train them after they are on the job. The cost of doing so would drastically alter their business models. Conversely, drone operators, especially microdrone operators, will be more willing to pay for the more modest costs training after DROPs are on the job.

The combined effect of these phenomena drastically enlarges the pool of potential DROPs.

4.4 Mechanics and electronics technicians

Human factors and labor market considerations extend beyond flight personnel. Commercial drone operators must have access to personnel capable of maintaining the drones, troubleshooting problems, and fixing drone subsystems.

When an operator is already flying airplanes or helicopters for commercial customers, it has a support infrastructure in place that can be adapted relatively easily to a new line of drone business. Receptionists, crew schedulers, and customer liaison personnel who handle airplanes or helicopters easily can do the same thing for drone crews and customers. The requisite skills are the same, and little additional training is required.

Maintenance on manned aircraft must be performed by licensed A&P (airframe and power plant) mechanics.[17] Certain mandatory inspections must be performed by an A&P mechanic who also has IA (inspection authority) certification.[18] Airplane and helicopter mechanics must have detailed knowledge of the aircraft subsystems on which they work: airframes, power plants, communications, electrical power, automated control systems, and navigation systems. Many aircraft and powerplant mechanics subcontract work on avionics and major overhauls of propellers and power plants. Indeed, basic A&P mechanics are not authorized to perform major repairs or alterations of propellers or instruments.[19]

An A&P mechanic may not perform work on a particular model or component unless he has previously performed work on the same model or component under the supervision of another A&P experienced with the work involved.[20]

Depending on the size of an operator's fleet, it may be able to keep a mechanic busy full-time, in which case the operator may employ one or more mechanics. Operators with smaller fleets are more likely to contract out maintenance or hire mechanics on a part-time basis.

Drones, of whatever size, are sufficiently different in their details from manned aircraft that a good helicopter mechanic is not likely to be qualified to go to work on a drone immediately without further study and training. His existing skills and knowledge set may be a better fit for a macrodrone, which is more likely to have drivetrains and propulsion systems similar to those of its manned counterparts. Microdrones, on the other hand, resemble manned aircraft less, in almost every respect. Their principles of operation, in terms of basic aerodynamic, thermodynamic, electrical circuit theory, radio wave propagation, and Newtonian physics are the same, but not much is the same at lower levels of implementation.

The skillset for mechanics working on manned aircraft centers on understanding aircraft structures, material such as aluminum and steel, and how different kinds of fasteners tie everything together. Airplane and helicopter mechanics must understand how to troubleshoot, repair, and a assure prescribed tolerances of mechanical, hydraulic, and electrical systems for moving control surfaces, flaps, and landing gear. Skillsets for drone mechanics are different. They need to know less about the construction and functioning of vehicles over 10 ft long and wide, and more about microelectronics and RF data communication. It is not difficult for a good auto mechanic to become a good aircraft mechanic, after appropriate training and certification. An electronics hobbyist, someone who writes computer code for fun, or someone proficient with model airplanes, is a better candidate to be a good drone mechanic. A clear understanding of Ohm's Law—$I = E/R$—is more valuable than thorough mastery of the basic equation of Newtonian mechanics: $F = ma$.

- $F = ma$ represents the relationship between the force exerted on a physical object, its mass, and acceleration, signifying that a larger force produces greater acceleration, and that a body with a larger mass requires greater force to produce the same acceleration.

- I = E/R represents the relationship among current, voltage, and resistance in an electrical circuit, signifying that a higher voltage produces a larger current in a circuit of fixed resistance.

Accordingly, when drones come into an operator's inventory, mechanics must go through training similar to what they would go through on a new model of aircraft or helicopter. Many manufactures and final assembly vendors offer training programs, and enrolling in such programs as an efficient way to gain familiarity with particular unmanned aircraft systems.

An operation, such as a startup enterprise, that does not already have a maintenance infrastructure, must decide whether to build maintenance capability through licensed airplane and helicopter mechanics, either by hiring them or contracting with them, or to seek personnel with different basic qualifications.

The most useful skillset for microdrone maintainers is different from the skillset of a manned aircraft mechanic. Microdrone maintainers will spend little time on airframe components, beyond repairing an occasional broken bracket, a stripped screw, or replacing a broken rotor blade or carbon fiber boom. They will spend most of their time troubleshooting and fixing the control and navigation subsystems, which resemble those in wide use on model aircraft built by or sold to hobbyists. Accordingly, knowledge and experience with model aircraft maybe more useful than experience with manned aircraft. Model aircraft experience must be backed up, however, by solid theoretical knowledge. Merely knowing how to dive into a model aircraft blog and swap war stories is insufficient; the maintainer needs to know how and why something works, including electronic components, computers, and the tightly integrated electronics and electric power distribution systems for propulsion.

Increasingly, user programmable microdrones are entering the market at relatively low prices. The DJI Matrice 100 and 3Drobotics' Solo are examples. DJI offers a programming language it calls SDK for developers of DJI-specific mission applications. 3Drobotics promotes multiplatform autopilot hardware and open-source software. Operators taking advantage of these features need human resources that can program sensors, gimbals, and other mission equipment, and set parameters for automated flight control over a wider range of possibilities than is available in basic microdrone models. Such human resources must have good computer programming skills, the ability to adapt to new programming languages, and facility in molding computer code to mission requirements. In the jargon of the computer programming community, a desirable person will have system analyst skills as well as coder skills.

4.5 Finding, recruiting, and managing managers, sales personnel, and strategists

An entrepreneur who knows little beyond how to fly a drone well is unlikely to be able to establish a successful drone business. He can have extensive knowledge of the detailed capabilities of different models in the constantly changing market. He may know how to program new navigational features into drone software. He may be able to fly the drone manually in close quarters to get just the right video shot. But unless he has reasonable capability in the other aspects of starting and running an enterprise, considered in chapter 11, success will prove elusive. Most successful small businesses succeed because they find a way to recruit the full range of talent needed for a viable business; rarely is that full range found in one person. As chapter 11 explains, a drone enterprise must be attentive to finance, marketing, operations management, purchasing, and strategy. The most desirable marketing resource will already understand the industry segments the drone firm intends to target as customers and, better yet, some contacts in those industries. The financial resource will have a strong understanding of financial accounting, business plans, capital structures, and means for raising capital. Better yet, she will have contacts in the venture-capital and investment banking communities.

Procurement and operations management functions will benefit from the kind of knowledge typically possessed by drone enthusiasts, model airplane enthusiasts, and aviation specialists, but also need an appreciation of the economics of efficient aviation operations: especially programming human and physical assets so as to maximize utilization of capital of human and physical capital—a subject probed more deeply in chapter 10.

Formulating a good strategy requires a combination of boldness and vision on the one hand, and cautious attention to reality on the other. If the entrepreneur is bold, he needs a team member who can balance his boldness with appreciation of risk. If the entrepreneur is risk-averse, he needs a team member who is bolder. Vision and imagination and important for any enterprise; the business environment is always changing and successful enterprises are willing to adapt and to take advantage of new technologies and market shifts. This is especially important in a new industry, like drone operations, where no clear rules of thumb for success yet exist, and undue adherence to long-standing practices and assumptions imported from other industries can be dangerous.

4.6 Specialized training and other post-high-school education

A high-school diploma is the minimum educational requirement because anyone without one is unlikely to possess the requisite ability to communicate and to have the self-discipline to conform to a work environment. Education beyond high school level is appropriate for operating more sophisticated equipment because it involves greater levels of specialized knowledge or makes it likely that the applicant has the breadth of knowledge and analytical capability to take larger responsibilities. Enough formal education or other theoretical training is necessary for a DROP to be able to adapt to new drone technologies and to move up the ladder from microdrones to macrodrones, when this is expected. It also facilitates a DROP's learning electronics as well as purely flying functions.

Whether a college degree is desirable depends on whether the operator wants a worker who performs the same, relatively narrow function for the duration of his employment, or whether the operator intends for a successful applicant to have non-flying duties, either immediately or as part of a long-term career path.

The different practices of the US armed forces illustrate the difference. The Air Force, Navy, and Marine Corps commission all pilots as officers, which requires a college degree. They justify this because aviation officers in those services are eligible to take command positions as they gain seniority. In any business, a college degree similarly helps ensure the breadth of knowledge and the flexibility of mind to manage and lead an enterprise.

The army, on the other hand, allows high school graduates without college degrees to become pilots as warrant officers. Warrant officers serve their entire careers in flying assignments and are not eligible for command outside aviation units. For someone who spends his whole career flying, a college degree is less relevant.

Formal education requirements also serve as a shortcut: if Harvard Business School vouches for a candidate's business strategy acumen, the hiring enterprise need not expend the effort to evaluate on-the-job training unique to the applicant. If a candidate for chief technology officer has a masters' degree from MIT's electrical engineering program, MIT has already tested the candidate's qualifications.

Purely technical training has its limits. Education necessarily involves theory and context. An aeronautical engineer or a pilot is not educated if all she understands is the airflow through a particular model of Rolls-Royce turbine engine, or the mechanism for changing the pitch on the blades of a particular model of helicopter. She is educated only if she also understands the dynamics and aerodynamics of rotary wing aircraft and propulsion systems. That way she has a foundation to evaluate innovations in propulsion systems and rotorcraft design and to come up with some of her own. Also, and this may be of more immediate importance, she had a basis for understanding how a particular design introduces risks.

An entrepreneur is not well educated in new business ventures merely because he knows how to construct a financial statement and can make a compelling pitch to investors. He needs to understand basic microeconomics theory, such as the idea of diminishing returns and product substitution. He needs to understand that capital can be embedded in an enterprise in wide variety of ways, and to be familiar with the legal and accounting mechanisms for implementing some of them.

Likewise, someone who operates at the interface between business and its customers or between business and government can do so more effectively if she has been exposed to the world's store of knowledge about how societies embrace or reject disruptions to tradition, and how social and economic forces in the society at large produce particular political institutions and drives their dynamics. Effective functioning in most positions over the long term requires understanding that challenges that confront his business are not unique or unprecedented. Similar challenges have been confronted by other enterprises from which the drone business can learn.

And anyone whose job assignment requires written communication must know now to translate knowledge into well-organized, clear memoranda and reports. Anecdotal evidence from the armed services context suggests that college graduates (all officers) are, on the whole, competent in that activity, and that non-college graduates (most enlisted personnel) have more trouble, no matter how great their skill in their specialties, no matter how motivated they are, and no matter how smart they are. It is hard to get a college degree without having to write many papers that have reasonable analytical structure. Some of the resulting analytical and communications skills sticks.

Formal education provides context for real world action. Getting a degree also demonstrates the capacity to stick through a program that involves many subjects and experiences of no apparent immediate relevance. Studying aerodynamics, meteorology, and the FARs is one thing, when one is enraptured with the romance of becoming a pilot. Soldiering on through a required curriculum that contains elements the utility of which may not be apparent to the post-adolescent student but create broader capacity is a different badge of character.

4.7 Screening and recruiting

4.7.1 Job postings

In applying and extending the following examples of job postings, operators face a tradeoff: they do not want to receive applications that they will not consider, but they also do not want to exclude potential candidates by articulating unnecessarily stringent requirements. Helicopter and airplane pilot job postings regularly articulate minimum flight hour requirements that the operators do not actually apply. They never see potentially interesting applicants who may have something special to offer, but who take the published minima at face value. DROP recruiters should avoid this error. Accordingly, the proposed criteria do not include minimum flight hours.

4.7.1.1 Microdrone DROP—minimum requirements

- FAA airman certificate
- Model aircraft or microdrone flying experience
- Manned aircraft experience a plus
- Videogame experience a plus
- Basic knowledge of radio-control electronics and computer programming
- Job experience in related field
- High school diploma
- 18+ years of age

4.7.1.2 Macrodrone DROP—minimum requirements

- FAA airman certificate
- Manned aircraft or macrodrone experience
- Radio-control electronics experience
- Mission related experience, e.g. ENG, utility, firefighting
- Post-high-school education or training in relevant fields, including appropriate theoretical foundation instruction
- 18+ years of age

4.7.1.3 Mechanics and electronics technicians

- Experience troubleshooting and repairing model aircraft, helicopters, or airplanes
- Training in aerodynamics, mechanics, electronics, or computer hardware and software
- 16+ years of age

4.7.1.4 Entry level managers

- Experience, paid or unpaid, in organizing work and performing basic marketing, operations management, and finance functions
- College degree
- Demonstrated activities or interest in aviation or technology
- 21+ years of age

4.7.2 Interpretation of posted requirements

The purpose of any job posting is to attract applications to enable further steps in the recruiting process. Operators should post jobs in the usual places that traditional aviation jobs are posted: on operator websites, in career services offices of universities with aviation programs, on traditional aviation pilot job matching services like JS firm, indeed.com, and pilothiring.com, and on general services such as monster.com. In addition, operators should consider posting notices on job matching services that cater to electronics and computer enthusiasts. Some of the pool of potential DROP applicants includes those who envision careers in computers as well as those who envision careers as pilots. A number of specialized web-based services focused on matching pilots and DROPs with commercial drone opportunities are also beginning to appear, and they are obvious channels for advertising DROP jobs as well.

The specific requirements in a job posting perform two functions: they eliminate applicants that the operator will not seriously consider; and they eliminate applicants who will not be interested in the job, if it is offered. The items enumerated in the proposed job posting deliberately state requirements in general terms. Instructions should make it clear that applicants must explain how they meet each requirement. Although DROPs need not be good essayists, they should be articulate enough to communicate effectively about aviation matters. Applications should include, not only a resume, but also a cover letter that links items on the resume to the stated job requirements.

The requirement for an airman certificate is stated generally, because regulatory requirements for DROPs are evolving in the United States and elsewhere, and any required airman certificate is almost certain to vary depending on the size and type of drone to be flown. If an operator knows a traditional pilot certificate is required, as for commercial operations under a section 333 exemption, and as it is likely for macrodrone DROPs, the job posting should state that instead of the more general requirement. When the sUAS operator certificate proposal in the NPRM is finalized, the job posting should state the requirement specifically as to a job that involves flying microdrones only.

The proposed training requirement ensures that candidates have some theoretical framework for understanding what they are doing rather than being prisoners of rudimentary trial and error. Similarly, some degree of training is necessary, to ensure that candidates' abilities are not limited to a particular model of drone, but rather allow adaptability as technology changes.

The experience requirements reflect the analysis earlier in this section. The high-school, specialized training, and further education requirement should be interpreted in accordance with the considerations presented in § 4.6.

The proposed requirements for mechanics and electronics technicians differ significantly from those proposed for DROPs. Many of the best qualified candidates will be youngsters who have steeped themselves in tinkering with model aircraft, computers, and radios. The criteria should not exclude a radio amateur, for example, or a go-to teenager in a model airplane club who can fix problems that frustrate everyone else. The proposed age requirement is lower because many high school students can do what the operator needs. The minimum age requirement of 16 is intended to make sure that the candidate qualifies for a driver's license; otherwise, the logistics of getting to and from assignments are too complicated.

Job postings for managerial and administrative personnel can imitate postings for similar aviation positions more generally.

Operators must also be as clear as possible on whether they are recruiting personnel whose duties will never extend beyond direct drone operations, or personnel whose capabilities extend to performance of administrative, maintenance, and marketing functions. Depending on which of these is the case, operators can include an additional criterion such as:

- Must be willing to participate in vehicle troubleshooting and maintenance, to make marketing contacts with customers and potential customers, and to assist with administrative matters.

They also need to distinguish between a workforce that has the capacity to grow into managerial and executive roles that determine the strategic future of the enterprise, and a workforce that comprises journeymen in a single professional activity. Accordingly, they can include or exclude the following:

- Must have the capacity to grow into management responsibilities.

Operators must understand when they need someone who is seasoned by several years of relevant experience, and when a youngster who can be more completely molded by the organization is more desirable.

4.7.3 Interviews and skills evaluation

The next step after the operator receives and screens applications is an interview, preferably in person—although telephone interviews can be a preliminary step. The interview evaluates:

- Whether the applicant shows up on time and is resourceful enough to find the location of the interview, as opposed to getting lost
- Whether the applicant is dressed appropriately
- Whether the applicant observes professional courtesies such as shaking hands and maintaining eye contact
- Whether the applicant would be credible and professional in contact with a customer and
- Whether the applicant would be enjoyable to work with on a daily basis.

The interviewer should ask the applicant to explain how he meets the posted criteria. His answers will flesh out the specifics of training, education, and experience.

Then the interviewer should give the applicant some realistic operational scenarios and ask him to talk through how he would handle them. The scenarios should involve compliance with restrictions such as height limitations, line of sight, avoidance of overflight of people, and handling confrontations with police or bystanders.

Then, the applicant should demonstrate his ability to fly the drone he will be operating—unless the operator has decided to provide post-hire training. This is a more straightforward part of the evaluation process than it would be for a helicopter or airplane pilot because of minimal cost of flight time and the modest amount of time required to take a microdrone outside to a pre-selected space and have the candidate fly it for 20 minutes, performing prescribed maneuvers. For macrodrones, the flying demonstration part of the evaluation more closely resembles an airplane or helicopter demonstration flight, and therefore might be deferred until later in the process.

Mechanic and electronics technician candidates should be given scenarios and practical exercises involving hypothetical types of drone malfunction.

Managerial and administrative candidates with nonflying duties should be given relevant scenarios, and need not demonstrate practical skills, although role-playing exercises may be helpful.

The intensity of the post-application evaluation process will depend on the level of responsibility successful candidates will exercise and the expected duration of the relationship. An electronics technician hired merely to be on call for six months does not need as much scrutiny as a candidate to be the principal DROP.

4.8 Training and certification

Certification and training requirements will be a combination of FAA regulation, implementation of the regulatory requirements by private entities such as existing flight schools and new drone training and certification organizations, individual practice and self-certification, and internal training by operators after DROPs are hired.

Drone operators cannot fully control what licensing requirements will be imposed on DROPS. Ultimately, the minimum licensing requirements will be imposed by the FAA and equivalent regulators in other countries. But they can decide what level of licensing is necessary and work through political and regulatory channels to advocate it.

Some of the possibilities discussed in the section for certification will be preempted by evolving airman certification requirements in the U.S, and elsewhere. Such regulations are unlikely, however, to prescribe government-operated training and testing for all categories of DROPs, and it is likely that self-certification, and only general training and experience requirements will be prescribed for some categories of microdrone DROPs.

Many other skills-oriented governmentally prescribed training and licensing programs exist, for lawyers and physicians, for cosmetologists and professional engineers, and for commercial and amateur radio operators. All of them require some kind of governmentally prescribed test as the final step before licensing. Getting a license to drive an automobile or truck is an experience almost everyone has had. There are no particular requirements for training; only a simple knowledge test and checkride, usually one conducted by a state employee.

It's entirely up to the candidate whether she wants to attend a driving school before taking the tests. Many private driving schools exist, and some are integrated into public high school education.

No licenses are required to operate recreational boats. Boat-handling schools exist, especially for sailing, and many other short training programs focus on specific safety issues, but anyone can go into a boat dealer, buy a boat, and sail it or drive it away. To the extent the recreational boaters need instruction, they get it informally through conversations with other operators where they dock or through a proliferation of boating clubs.

One could argue that the experience of thousands of microdrone owners before the FAA had rules for drone flight shows that little DROP training is necessary. Formal training requirements may not be necessary. Anyone could buy a microdrone and fly it, pretty much as happens now with people who want to fly them for fun. One has never needed a license to fly model airplanes. The same is true for operating many things that can be dangerous: chainsaws, lawnmowers, tractors, and bulldozers.

Even without government licensing requirements, incentives for competent and safe operation exist. Usually, the person most vulnerable in an accident is the operator. Someone who operates a chainsaw doesn't want to cut his own leg off. Someone who operates a bulldozer or tractor doesn't want to roll it over and be crushed.

A variety of resources are available to anyone who wants to learn how to operate a machine with which he is unfamiliar: books, text and video tutorials on the Web, explanations and modest levels of instruction from dealers, friends, and family.

The same thing now happens with microdrones. Someone buys one, reads the documentation that comes with it, maybe buys a book or two, and sits through a selection from the array of tutorials and demonstration videos on YouTube. If the operator confronts some particular roadblock, he is likely to run across someone else who flies a similar microdrone and simply asks.

This works satisfactorily. Most people with ordinary physical skills have had many experiences playing golf or tennis, bowling, or playing volleyball, where it is necessary to learn how to process visual information about the position and direction of movement of an object and how to condition reflexes so as to send it where he wants. Playing video games is almost the same thing, and self-teaching occurs there also. The typical DROP flying a microdrone is no different. One can look through web postings in vain searching for a report that anyone gave up in frustration after trying to fly a microdrone.

Basic operating instructions shape the way in which any product is used. The starting point for becoming competent and safe in piloting an aircraft is to become familiar with its Pilot's Operating Handbook or Flight Manual, required by the FAA for aircraft certificated after 1979. The same thing is necessary for safe drone operation, whether a DROP obtains it on his own or as part of a more structured institutional training activity.

DROP training need not be live at a physical teaching site. Many good video tutorials already are available, and knowledge is widely distributed on how to construct video training materials that require some kind of testing, before the student is allowed to proceed. Homeland Security training modules for emergency response personnel are examples. So is the typical requalification test for flight instructors.

Well-crafted distance learning techniques are an effective way to conduct DROP training, especially for subjects such as regulations, meteorology, aerodynamics, wireless technologies, computer electronics, and maybe flight simulation.

Indeed they may be superior to traditional lectures or face-to-face tutorials backed up by pencil and paper sketches or diagrams drawn on a whiteboard. Distance-learning is in wide use by colleges and universities, including top ones like MIT. As with any other instructional technique, it can be well or poorly constructed. Simply presenting a student with a series of screens displaying text offers few advantages over assigning the student a passage in the textbook.

On the other hand, distance-learning modules can be constructed so that they force student interactivity by presenting discrete pieces of information, illustrating them, and then requiring a student to answer questions with a certain level of correctness before being allowed to proceed to the next module. More sophisticated modules can permit or require the student to interact with animated simulations. NASA, for example, has a number of online programs in aerodynamics and meteorology, apparently aimed at high school students. One of them allows the user to change parameters for an airfoil and see how it changes airflow, lift, and drag.

Dynamic testing algorithms can shape the modules. A student struggling with a certain block of material can be offered additional examples and taken more deeply into it, while being allowed to proceed more quickly through material he has mastered.

Distance-learning, when properly designed, forces students to pace themselves. For example, a multi-day or multi-week course can require a student to complete certain exercises by a specific time on a specific day, after which that exercise disappears or locks out the students. That mitigates one of the major risks of self-paced learning: procrastination. Students are not allowed to act on the rationalization that they can cram all the assignments into the last minutes before the final test.

Along with structure, pacing, and forced interaction, distance learning also offers flexibility and convenience. Students need not travel for ground instruction, nor need they learn at any particular time.

The disadvantages are substantial, however. Good distance learning modules are hard to construct; it is far easier just to copy and paste text the designer grabs from somewhere and organize it into a series of web screens. Designing and constructing effective modules requires significant investment of time and intellectual energy by a good teacher who is knowledgeable in the subject matter. The teacher would have to exert significantly less effort simply to give a series of extemporaneous lectures or tutorials, informed by pencil-on-paper sketches or whiteboard illustrations on an ad-hoc basis as the need arises. And, a major reward of traditional teaching is face-to-face interaction with the students—a quality missing altogether when the same teacher designs distance learning exercises and interacts with students online.

An important goal of traditional education is professional socialization. This arises only from fully dimensioned regular interaction between students and role-model instructors and among the students themselves. As § 4.3.6 explains this might be especially important for DROPs because of the need to bring them into the aviation community. Distance learning is entirely incapable of delivering this aspect of professional education.

Hybrid programs can fill this gap. They can be designed to combine the strengths of distance-learning with the strengths of intensive social interaction. For example, participants in the course can be brought together for an initial "boot camp," where they form social bonds with each other, develop a certain amount of trust, and appreciate one another's idiosyncrasies. Northwestern University's Kellogg School of Business encourages MBA students to participate in a week-long outdoor adventure with discussion and reflection of various leadership issues. Experience with such hybrid programs teaches that students interact more in a distance learning environment if they already know each other. That means that the boot camp should come first, following by distance learning.

Despite the validity of these principles of coursework design, semester-long distance-learning instruction requires far more student time and investment than is practical for DROP training. The basic concepts for effective distance-learning must be miniaturized to suit the particular needs of DROP training and the particular subject matter of drones.

4.8.1 Adaptation of existing certification requirements

The existing framework for pilot training and certification can be adapted to DROPs. The most straightforward way to do that is simply to add a new category of airman: *UAS operator*, as the NPRM proposes.

The content of the requirement is molded to account for the differences between the skillset needed by DROPs and the skills and knowledge needed by pilots.

Airplane and helicopter pilots cannot be licensed, even at the private-pilot level, to fly until they have completed a demanding program of flight training and passed a written examination. A candidate must complete at least 40 hours of training, evenly divided between dual instruction by a certificated flight instructor (CFI) and supervised solo flight. Additionally, she must pass a written examination covering basic aerodynamics; aircraft performance, stability, and control; FAA regulations; navigation; and aeronautical decision making. Most candidates, especially those intending to fly helicopters, take longer than the minimum of 40 hours. Practical skills are tested by an FAA-designated examiner, independent of the CFI who gave the candidate flight instruction.

To fly larger, more complex, aircraft designed to carry larger numbers of people, pilots must have a type rating as well as basic category and class ratings. A type rating requires familiarity with and demonstrated skill in flying a particular aircraft model.

This training and testing regime is manifestly overkill for microdrone pilots. Microdrones are easier to fly then helicopters or airplanes, and their flight envelopes present much lower risks to life and property. Many of the qualification requirements for manned aircraft pilots aim at reducing the likelihood that crew or passengers will be injured or killed in the event of a mishap. That threat is not present for drones.

The rules also are aimed at reducing the risk to persons and property on the ground if the aircraft crashes. Microdrones are capable of doing much less damage if they crash, because of their lighter weight and because they do not have inflammable fuel on board. Microdrones will not be flown cross-country. They will be affected only by local weather, and not by the vicissitudes of weather over a large geographic area. When they experience loss of power they behave differently from either airplanes or helicopters. Existing requirements pertaining to these areas of operation are irrelevant to microdrone operations.

On the other hand, safe operation of microdrones depends on a variety of systems not found in manned aircraft, notably wireless control links and autonomous flight regimes. They also depend on external visual reference. DROP certification should assure DROP knowledge and proficiency with these unique features of microdrones.

A greatly streamlined system for assessing the qualifications of microdrone DROPs is appropriate, one focused on the features of microdrones that differentiate them from manned aircraft.

Macrodrones, on the other hand, present many of the same risks to persons and property on the ground as do manned aircraft. They are likely to weigh much the same, and those with reciprocating or turbine engines will have fuel on board. Nevertheless, the differences in the way they are operated, compared to manned aircraft, necessitate different content for training and testing.

For example, losing the control link is the most serious emergency a drone can suffer, and it is not improbable given the vicissitudes of wireless data communication. Helicopter and airplane pilots do not confront anything analogous to loss of a control link. Macrodrones, even more than microdrones, will have significant degrees of automation and autonomy than smaller, lower-cost helicopters and airplanes. An increasing number of light airplanes have autopilots, but they operate with more human involvement than drone features such as automatic hover, return-to-home, and refusal to fly in certain airspace. Accordingly, familiarity with wireless technologies, avionics in general, and control systems specifically, should be a much more prominent part of macrodrone DROP training than of pilot training.

It might be that certain limited training would be a desirable component of the regulatory regime. But it is hard to enforce a training requirement unless some kind of license is required which can be obtained only after successful completion of the training. The burden of mobilizing an inspection force to check whether every operator of a small drone has a license is unworkable.

The text of present rules, and the NPRM-proposed rules, adapted for drones, follows:

§ 107.73 Aeronautical and systems knowledge

(a) *General.* A person who applies for a microdrone DROP certificate must receive and log ground training from an authorized instructor, or complete a home-study course, on the aeronautical and systems knowledge areas of paragraphs (b) and (c) of this section, [and pass a knowledge test prescribed by the FAA].

(b) *Aeronautical knowledge areas*

(1) Basic aerodynamics and the principles of flight;
(2) Applicable Federal Aviation Regulations that relate to:

[A] DROP privileges, limitations, and flight operations;
[B] Operating microdrones within the National Airspace System;

(3) Safe and efficient operation of microdrones;
(4) Significance and effects of exceeding:

[A] Microdrone performance limitations;
[B] Regulatory restrictions on microdrone flight;

(5) Aeronautical decision making and judgment;
(6) Maneuvers, procedures, and emergency operations appropriate to the microdrone model;
(7) Accident reporting requirements of the National Transportation Safety Board.

(c) *Systems knowledge areas*

(1) Basic theory and operation of wireless control links;
(2) Principles and functions of microdrone systems;

(i) flight controllers
(ii) power distribution
(iii) batteries.

(3) Principles of antennas and propagation

(i) range
(ii) multipath problems
(iii) GPS.

The bracketed language would be included or excluded, depending on whether a knowledge test is required for a particular class of drone.

Simply knowing how something works and what it is supposed to do is not enough; a DROP must actually be able to do it. As with someone seeking a manned-aircraft pilot's license, a microdrone DROP must be able to demonstrate his ability to fly a microdrone within model-specific and regulatory limitations.

The text of a possible adapted rule follows

§107.73A Flight proficiency

(a) *General.* A person who applies for a microdrone DROP certificate must receive and log ground and flight training from an authorized instructor on the areas of operation of this section.
(b) *Areas of operation*

  (1) Preflight preparation;
  (2) Preflight procedures;
  (3) Hovering maneuvers;
  (4) Takeoffs and landings;
  (5) Effective use of autonomous flight regimes;
  (6) Keeping microdrone within mission area;
  (7) Performance maneuvers;
  (8) Ground reference maneuvers;
  (9) Emergency operations, including:

    [A] Loss of control link;
    [B] Loss of visual reference;

  (10) Special operations;
  (11) Postflight procedures.

The 2015 ICAS report[21] also contains recommendations on DROP qualification, including subjects for knowledge tests[22] and tasks for practical tests and training.[23] The ICAO suggestions contain only a few modifications of requirements for manned aircraft pilots. Little is included that is pertinent to multicopters.

One flight test demonstrating minimum proficiency does not assure that a microdrone DROP can consistently meet the requirements. Certain levels of experience also could be required before a DROP is fully licensed to fly a microdrone without supervision.

Microdrones are simpler to fly than manned aircraft. They operate in limited geographic areas, dealing only with localized weather phenomena. DROPs are aided by autonomous capability to perform certain maneuvers. DROPs can learn to fly them safely and consistently in much less time than is required for manned aircraft.

The text of a possible rule, adapted from existing FAA requirements follows:

§107.73B Aeronautical experience

(a) *For a microdrone DROP rating.* A person who applies for a microdrone DROP certificate must log at least 5 hours of flight time as a DROP that consists of at least 5 hours of actual flight time, including flights at three or more locations, at least two with obstacles.

The FAA will face continuing pressure for basic operating rules. Such rules could:

- Forbid operators from tampering with built-in law-abiding features
- Require pre-flight inspection, for example, to assure that the control link is operational and that the GPS and compass are calibrated
- Require basic maintenance, for example, to replace damaged rotors
- Forbid flight with an inadequately charged battery
- Forbid the DROP to fly the drone behind objects that obscure the DROP's view of it.

Regulators should not, however, impose operating or airman qualification rules with anything like the detail and complexity of the existing FARs on those subjects for manned aircraft. Nor should any such supplementary rules be thought of as primarily being enforced by the FAA; instead, they would represent basic standards of conduct to inform state and local enforcement of state and local law, and to provide a foundation for civil litigation evaluating the DROP's adherence to the basic standard of care.

4.8.2 Schools

One can expect a mix of training programs and material offered by drone vendors, third-party face-to-face training institutes, existing flight schools that are grasping the new business opportunity, and traditional aviation and technology publishers. Diversity in training opportunities and materials is entirely workable; it has been the norm for manned aircraft flight training for many decades. A few independent training activities are beginning to emerge, and that phenomenon will intensify, especially if the FAA, as expected, mandates successful completion of a knowledge test as part of the certification process for DROPs.

The FAA will supplement these efforts with its own publications for drones modeled on its popular helicopter and airplane flying handbooks. Both of them well written and comprehensive primers on the content of flight training.

Training can be delivered within a variety of governmental and private frameworks. The norm for pilot training is that a pilot candidate receives training prescribed in detail by the FAA through private instructors, often working through flight schools, but also allowed to work on their own. Two types of flight schools exist: those certified by the FAA under Part 141 of the FARs, and those operated more informally under Part 61. Part 141 schools must prepare detailed syllabi and have them approved by the FAA. The syllabi must articulate goals and objectives for each flight lesson and each component of ground training. When changes are made, the changes must be submitted to and approved by the FAA. Part 141 schools also must have minimum facilities and staff. Part 61 flight schools need not meet any specific requirements; indeed, Part 61 makes no explicit reference to flight school or the way in which flight instruction is given. Typically, however, a Part 61 flight school advertises itself under a business name, has a fixed locus of operations, and offers its students a variety of charts, instructional books, logbooks, devices for making navigation calculations, and other paraphernalia. Part 61 flight schools usually have one to a half-dozen aircraft and have lines of business other than flight instruction, such as aircraft rental and aircraft chartering. Part 61 flight schools offer student-instructor relationships that generally are less formal and structured than the student-instructor relationships at Part 141 flight schools.

In either type of flight school, the student may not take the written knowledge examination until the instructor concludes that he is prepared and endorses the student's logbook to that effect. Once the student passes the knowledge exam, the final set up to receive his pilot's license is the *practical test* (universally referred to as the *flight test* or a *checkride*), administered by an FAA inspector or a designated pilot examiner—a CFI who has specially been designated and accredited by the FAA to give flight tests to pilot candidates. The examiner must be someone other than his usual flight instructor, although she may be an employee of a Part 141 flight school. As with the knowledge test, a candidate may not take a practical test until a CFI has certified that she is ready.

Established flight schools are beginning to offer DROP training as an option within their flight training programs.

Most of the longstanding aviation universities, such as Embry Riddle Aeronautical University, have added DROP training as a major. Other universities with aviation programs, such as the University of North Dakota, Lewis University, and Auburn have done the same.

Another possible approach is to look to manufacturers and distributors to conduct training. It is not unheard of for vendors of other complex systems to offer training packages in conjunction with sale of the systems. Most manufacturers of helicopters offer some kind of factory training.

It could be left to the market to decide when such training should be offered and to would-be operators to decide whether they need it and want to pay for it. Or, it could be required as a precondition of sale. Some states prohibit gun sales unless a purchaser can demonstrate successful completion of a prescribed safety course.

Looking to private organizations for DROP skills certification activities is desirable because it opens up more possibilities for building a community of DROPs in which peer group pressure can reinforce safe practices and skill development. Also, the private sector has greater flexibility in decision-making, compared with governmental agencies.

A number of non-aviation models exist for this. One of the most interesting is the Professional Association of Dive Instructors (PADI). PADI emerged in 1966 because of a perception by its founders that the existing organizations offering training and certification of SCUBA divers were poorly organized and not very effective.[24] It has grown into an elaborate organization that offers diver certification at multiple levels.[25] It is difficult for a diver to rent diving gear unless she can show a certificate of completion of at least the basic course. There is almost no governmental involvement. The market enforces the requirements and is backed up by the possibility of liability and insurance requirements.

There is no reason that a PADI-like organization cannot be erected for DROPs. Similar to PADI, it would offer memberships, recruit instructors, administer training programs and standards for certifying them, and issue instructor certificates. It would pair certified instructors with divers or would-be divers. The organization would develop tests that could be administered directly by instructors or online. It would issue certificates of completion of various levels of instruction and testing.

A second interesting model is the one adopted by the FCC for licensing amateur radio operators. The FCC enters into agreements with Volunteer Examiner Coordinators (VECs), organizations such as the American Radio Relay League,[26] which in turn accredit Volunteer Examiners (VEs), individuals holding amateur licenses who actually administer written tests to candidates for licenses. When the VEC certifies test results, the FCC issues the license.[27]

Other models can be adapted from lifeguard certification, a process in most states administered by the non-governmental Red Cross and backed up by hiring practices for lifeguards.[28] Also helpful is young hunter, motorboat, snowmobile education and testing requirements in states like Illinois. Under most of these programs, the requirement for certification is expressed in statute and enforced by the state Department of Natural Resources, but mostly private instructors certified by the DNR conduct the training itself.[29]

The most promising infrastructure for training, testing, and certification of microdrone and perhaps for macrodrone DROPs as well, would be modeled on PADI, amateur radio operator licensing, Red Cross lifeguard certification, and on hunter, snowmobile, and boating training programs. It would draw upon and strengthen the private center of gravity of manned aircraft pilot licensing, as well.

Preparing DROPs for safe operation of microdrones would be the responsibility of a private association tentatively known as AADI—American Association of Drone Instructors. AADI would be a private nonprofit membership organization organized under the laws of one or more states.

AADI would develop detailed curricula for DROP training, follow the NPRM specifications for DROP knowledge, develop skills standards, and offer training materials in book and online form and practical test standards for DROP skills testing.

AADI would maintain a database of certifications. Once a DROP instructor certifies a DROP candidate, the instructor would submit an online form that automatically would make an entry in the database. AADI would periodically audit, on an essentially random basis, the training and testing activities of its instructors. The database of certified DROPs would be available to the public through AADI's website.

AADI could operate purely in the private sector. PADI and the Red Cross (for its lifeguard certification), for example, do not enjoy any governmental imprimatur. Even though no federal or state rules require SCUBA divers to get PADI certification, they do it anyway. One reason is that diving equipment rental enterprises view PADI certification is an easy way of assuring that the customer is unlikely to have an accident that would result, at least, in the loss of equipment, and might result in litigation and insurance claims.[30]

The widespread interest in drones has also spawned the establishment of several specialized drone training programs that advertise aggressively on the Web. It is not yet clear which of these offer bona fide and well-structured training, and which may turn out to be essentially consumer frauds. Unfortunately, fraud is rampant in Internet-advertised training in all fields.

4.8.3 FAA accreditation

Accreditation of training and testing programs is a possibility. In other words, a training or qualification requirement based on knowledge testing would be satisfied only if the candidate completes a course or a test that has been preapproved by the FAA. To some extent, this is how Part 141 flight schools function. It is not, however, the norm for most aviation testing. For pilot and mechanic certification, the required test instruments are developed by the FAA itself or by contractors working for the FAA. Then they are made available for administration by private entities that meet requirements to ensure test security. The FAA could establish a mechanism to review test content and validation studies deemed necessary and let the market produce as many different test brands as supply-and-demand suggest. The process would be roughly similar to that used by the FDA for new drug approval.

Accreditation of training programs is well understood. The accreditation process for institutions of higher education is comprehensive, robust, yet also varied and flexible. The educational institution takes the initiative to define its goals and to develop a program of courses and other activities to satisfy the goals. It then explains to accreditation institutions how the program meets whatever benchmarks are established by the accreditors. The same process could be applied to DROP training. Designers of the accreditation process could start with the Part 141 approval process and modify it for content and skills appropriate for DROPs as opposed to pilots.

4.8.4 Operator training

Regardless of governmental training requirements for DROPs, drone operators must consider some kind of training program of their own. DROPs should be intimately familiar with the operating details of the vehicles that an operator flies, especially its sensors, electronics, and wireless links. The typical practice of manned aircraft operators for training new pilots is a good model. The new pilot must perform a set certain number of hours in the cockpit with an experienced pilot in an informal flight-instruction/checkride environment. The new pilot may not fly on his own until the instructor pilot believes he can perform typical missions satisfactorily and safely. A certain amount of ground training also is desirable. DROPs, like pilots, must learn about, and demonstrate mastery of, details of the operator's flight planning procedures. These methods often include weather requirements more stringent than those under the FARs, preflight planning and approval, and any peculiarities of aeronautical decision making, such as keeping a company dispatcher as well as ATC and other aircraft informed about the progress of the flight and of any difficulties. It is in these ground-based training interactions, as much as in the instructional cockpit, that the DROP or pilot becomes socialized to the operator's culture relating to safety and otherwise.

4.8.5 Developing a culture of safety

International aviation organizations are placing more and more emphasis on how safety practices get built into the operation of healthy aviation organizations.

That is an important facet of organization design. No matter how safety oriented a pilot's training was, the effect of the training will recede into irrelevance if he flies in an environment that places higher priority on considerations that conflict with safety.

Many people have had the experience of being a freshly trained new hire and having a supervisor or more experienced coworker say almost immediately, "Forget what you learned in the Academy; we're going to teach you how it's done in the real world."

That kind of gap between training and workplace socialization renders the training irrelevant. The designers of effective organizations understand this, and they either manage their work environment so that it reinforces training, or redesign their training so that it matches the realities of their work.

Astute managers know that informal norms and peer-group influences are often more important determinants of workplace behavior than formal rules. This is particularly likely to be true in situations where employees are bound together by a common professional attachment. Physicians working for hospitals, lawyers in a law firm, college faculties, and pilots working for an EMS operator are all examples.

Effective organizations find ways to harmonize their business objectives and their organization of work with these peer-group forces, so that they push in the same direction rather than playing tug-of-war.

Received wisdom dictates that pilots must periodically demonstrate their flying skills. Even recreational and private pilots must undergo biennial (in some cases annual) flight reviews. Commercial operators may not keep their certificates unless they subject their flight crewmembers to annual recurrent training and competency checks. DROPs similarly must be subject to recurrent training and testing.

As automation increases and as drone technology expands and shifts the boundaries of aviation activities, healthy organizations must keep up. Even if they do not wish to be first-adopters, they will be confronted with change, as old system models become obsolete and are replaced by models employing new technologies of aircraft control and presentation of flight data.

They must respond with ongoing training and a willingness to shift the nature of their human resources. As § 4.3.5 explains, someone who was a superlative pilot in a 1960-era airplane may lack the innate physical instincts, the kind of intelligence, and the personality traits that make him a good macrodrone pilot. An awkward twenty-year-old, who might have had an extremely difficult time learning to fly a light helicopter, might be a quick study in learning to fly a microdrone.

The experience that produces good judgment in the world of manned flight may be a poor guide to good judgment in microdrone operations.

4.9 Detection of violations and enforcement

This is not the first time that inexpensive technologies have led to massive noncompliance with rules based on older technologies. Music file sharing is an example.

If licensing requirements match reality, compliance will be higher and coercive enforcement less necessary.

Some kind of licensing regime for DROPs provides an incentive for rule compliance. A DROP fearing the loss or suspension of his certificate is more likely to inform himself about and to comply with FAA rules. Even something as simple as the FCC's procedure for obtaining a restricted radiotelephone operator's permit would have this effect. A similar approach for drone hobbyists might be desirable, although it would be fiercely resisted by the effective model aircraft lobby.

An important compliance risk factor is present for microdrone DROPs as compared to macrodrone DROPs or pilots. Because they face fewer requirements for certification as DROPs, especially if this book's suggestion for law-abiding drones shapes the regulatory philosophy, their ties to aviation are weaker than those of pilots. They will have invested much less in training—it's not unusual for a commercial pilot to have invested $150,000 of his own money to achieve a commercial and instrument rating. As a result, DROPs are less likely to have committed themselves single-mindedly to a long-term career as an airman. If a pilot commits a serious rule violation or has an accident, he may lose his career. If a microdrone DROP commits an equally serious violation or has an accident, he may lose his job, but move on fairly easily to a job in a different field with minimal economic and psychological cost.

Furthermore, if a pilot makes a serious mistake while in the air, he is likely to get killed and to kill his passengers. A microdrone DROP is not going to get killed, whatever happens, and the odds of the microdrone doing something serious on the ground that will haunt the DROP are small.

Finally, more relaxed training means that microdrone DROPs are less likely to internalize a culture of compliance with FAA rules and safe practices than pilots. There may be intellectual understanding of rules and good practices, but pro-safety peer-group forces are likely to be much weaker.

All this means that instinctive, internal adherence to safe practices will be weaker for microdrone DROPs, and likely weaker than for macrodrone DROPs, depending on the intensity of their training.

Part of the enforcement equation can be requirements imposed by vendors of microdrones as preconditions of sale.

The major microdrone manufacturers offer a variety of online tutorials in both textual and video form. Some of them have been produced by the manufactures themselves; others were produced by third parties and then made available by the manufacturers. The quality is mixed, and the organization is random. Most vehicles arrive with quick-set-up instructions, and the purchaser can download more complete operating manuals from manufacturer websites. A significant amount of the material is outdated, referring to earlier versions of software or vehicle configuration. Few of the tutorials are interactive, and few track user progress through them. While the spotty nature of instructional materials may be understandable, given the rapid pace of innovation, and the difficulty of documentation's keeping up with it, the result is that purchasers struggle to become proficient with their microdrones and to master their features.

Just as aviation regulators can prohibit the sale of microdrones that do not have certain limitations built into them—an approach developed in chapter 9, vendors could be required to obtain certain information about microdrone purchasers before completing the sale. Virtually all e-commerce sites require some kind of user credential—if no more than a credit card number—to complete a purchase. Many also require limited amounts of additional information including address, age, and other demographic information.

This familiar framework for forming e-commerce relationships can be adapted to screening potential DROPs. Microdrone regulations can prohibit the sale of microdrones over the Internet unless the vendor obtains certain information from the purchaser. The burden on the purchaser must be modest, or compliance will be low. For example, certification that the purchaser has attended a training program likely would be unworkable, especially if only preapproved training programs qualify. On the other hand, filling out a short questionnaire would not be terribly burdensome, and it would enhance understanding by new DROPs of some of the risks of microdrone flight. It's not unlike the several pages of safety warnings that accompany almost every consumer product now, except that the presentation of the safety information would be such as to ensure that the purchaser has read it. Whether or not the purchaser is a hobbyist or intends to use the drone commercially, would influence the content of the questions.

More demanding training requirements can be enforced by vendors' requiring entry of a certification number from a flight school, university, or other private certifying organization, much as one must enter insurance information to renew a vehicle license plate online in most states.

Notes

1 FAA, *Pilot's Handbook of Aeronautical Knowledge*, FAA-H-8083-25A (2008) (chapter 16—Aeromedical Factors, and chapter 17—Aeronautical Decisionmaking).
2 FAA, *Aviation Instructor's Handbook*, FAA-H-8083-9A (2008).
3 Eduardo Salas & Dan Maurino, eds., *Human Factors in Aviation* (2nd edn 2010) [hereinafter "Human Factors"].
4 2004 FAA Accident Report at 10.
5 www.faa.gov/about/office_org/headquarters_offices/avs/offices/aam/ame/guide/app_process/exam_tech/et/31-34/cl/.
6 Research on laparoscopic surgery using two-dimensional flat-panel displays compared with three-dimensional simulation with da Vinci cameras and polarizing glasses showed significant improvements in performance times and surgeon satisfaction with the 3-D simulation. It also showed, however, that experienced surgeons use monocular cues to compensate for lack of depth perception, including motion parallax by shifting the camera, relative position and size of instruments and anatomical structures, and shading of light and dark. Kazushi Tanaka, *et al.*, *Evaluation of a 3D system based on a high-quality flat screen and polarized glasses for use by surgical assistants during robotic surgery*, 30(1) *Indian J. Urol.* 13 (2014), www.ncbi.nlm.nih.gov/pmc/articles/PMC3897045/.
7 Cheryl Paulin *et al.*, *Review of Aviator Selection*, Technical Report 1183 (US Army Research Institute for the Behavioral and Social Sciences 2006) (Army Project Number 633007A792) [hereinafter "Aviator Selection"].
8 2014 GAO DROP study at 17.
9 Nov. 13, 2014 interaction between co-author Perritt and the students in his Civil Procedure class at Chicago-Kent College of Law.
10 Telephone conversations between the authors and the subjects, Nov. 2014.
11 Mar. 2, 2015 telephone interview between co-author Perritt and commercial helicopter operator.
12 Aug. 19, 2015 email from Dalton Thompson to Eliot O. Sprague.
13 Amanda Lenhart *et al.*, *Teens, Video Games, and Civics* (2008), www.macfound.org/media/article_pdfs/PEW_DML_REPORT_080916.PDF.
14 Robert D. Putnam, *Bowling Alone: The Collapse and Revival of American Community* (2000), Simon & Schuster.
15 2014 GAO DROP Report at 7.
16 FAA Office of Aerospace Medicine, A Summary of Unmanned Aircraft Accident/Incident Data: Human Factors Implications at p. 13 (Report No. DOT/FAA/AM-04/24 2004) [hereinafter "2004 FAA Accident Study"].
17 14 C.F.R. §§ 65.71–65.89 (basic A&P).
18 14 C.F.R. §§ 65.91–65.95 (inspection authorization).
19 14 C.F.R. § 65.81(a).
20 14 C.F.R. § 65.81(a).
21 ICAO *Manual on Remotely Piloted Aircraft Systems*, Doc 10019, AN/507 (2015), www.wyvernltd.com/wp-content/uploads/2015/05/ICAO-10019-RPAS.pdf [hereinafter "2015 ICAO Report"].
22 2015 ICAO Report § 8.4.31.
23 2015 ICAO Report at § 8.4.41.
24 PADI History, PADI, www.padi.com/scuba-diving/about-padi/padi-history/ (last visited on Mar. 30, 2015).
25 PADI Open Water Diver Course, PADI, www.padi.com/Scuba-Diving/padi-courses/course-catalog/open-water-diver/ (last visited on Mar. 30, 2015).
26 www.arrl.org/ (website of principal US association of radio amateurs).
27 47 C.F.R. § 97.9 (describing requirements for amateur radio operator's license); Ibid. § 97.17 (describing application process); Ibid. §§ 97.521–97.525 (prescribing rules for selecting VECs and accrediting VEs).

28 Lifeguarding, American Red Cross, www.redcross.org/take-a-class/program-highlights/lifeguarding (last visited on Mar. 30, 2015).
29 Illinois DNR, *Safety Education*, www.dnr.illinois.gov/safety/Pages/default.aspx (explaining certification programs for young hunters, archers, and snowmobile and boating operators).
30 *Why PADI*, www.padi.com/scuba-diving/about-padi/why-choose-padi/ (explaining the advantages of PADI certification).

5 Emergency procedures

5.1 Introduction

All vehicles malfunction occasionally. Automobiles have flat tires; construction cranes lose the capability to move their booms; motorcycles get stuck in a particular gear. Airplanes lose hydraulic pressure or engines. Helicopters lose their engines or their tail rotor effectiveness. Microdrones lose their control links; macrodrones lose their engines. A particular malfunction may be rare—engineers try to design for that—but the probability of its occurring cannot be reduced to zero.

Some malfunctions, such as the loss of an alternator on a helicopter or an airplane, may pose no immediate hazard to flight; aircraft engines traditionally were designed to run without electrical power. Other malfunctions, such as the loss of the critical engine on a twin-engine airplane, or the loss of an engine in a single-engine helicopter, require immediate and appropriate action by the pilot to avoid a catastrophic accident, likely to destroy the aircraft and kill the pilot and any passengers. Malfunctions that lead to an inability of the aircraft to continue normal flight or a threat to aircraft controllability are classified as *emergencies*.

Drones are no exception, but the likelihood of different kinds of malfunctions is different, as are their consequences. This chapter reviews the kinds of malfunctions that may occur with drones, analyzes their consequences, and specifies the appropriate reaction by a combination of autonomous responses and DROP inputs. As with most other topics related to drones, substantial differences exist between microdrones and macrodrones, and the chapter is organized accordingly. As appropriate, it draws upon received wisdom with respect to analogous malfunctions in manned aircraft.

It begins with a general analysis of the sources of risk associated with different types of malfunctions and different types of aircraft.

5.2 Levels of hazard

Engineering is always driven by cost-benefit trade offs. If heroic measures, such as quadruple redundancy, are built into a design to prevent low probability malfunctions that would have a few consequences, the cost of the vehicle will skyrocket without proportionate gains in safety benefits. Designing emergency response features of microdrones at the level warranted for the design of the space shuttle simply are untenable economically. Emergency responses must be tailored to particular emergencies, the probabilities of their occurrence, and the consequences if they do occur.

5.2.1 Hazards to crew and passengers

The most basic determinant of the degree of hazard resulting from an emergency is whether there are people on board the aircraft when it occurs. A secondary determinant of the level of hazard is whether the emergency occurs when the aircraft is in flight or when it is on the ground.

Determining the appropriate response for an emergency is greatly influenced by whether the aircraft is carrying people. Certain emergencies in manned aircraft threaten the lives of the pilot and the passengers. The magnitude of that threat warrants a high-level of prevention in the design process: careful design of systems, elaborate testing, and system redundancy to mitigate the risk if a malfunction occurs; as well as rigorous pilot training to handle the emergency.

Drones eliminate this risk altogether, because they are unmanned and do not carry passengers. Accordingly, emergency risk management for drones focuses on other hazards, one of which is to mitigate risks to persons or property on the ground.

5.2.2 Kinetic energy

For drones, the accident risk analysis starts with the kinetic energy of the malfunctioning vehicle.

Kinetic energy is determined by the well-known equation from physics:

$$E_k = mv^2/2$$

This signifies that the amount of kinetic energy is one half the mass of the vehicle, multiplied by its velocity squared.

Thus, a heavier vehicle has more kinetic energy than a lighter one, and one moving faster has much more kinetic energy than one moving more slowly.

Physics also teaches that potential energy can be transformed into kinetic energy, and vice versa, and that different forms of kinetic energy can be exchanged for each other. For example, a helicopter in autorotation can transform potential energy in the form of height above the ground into kinetic energy in the form of rotor RPM. It can transform kinetic energy in the form of flight velocity into rotor RPM and can transform kinetic energy in the rotor into reduced kinetic energy in vertical descent rate.

Autorotation is an aerodynamic phenomenon in which the upward flow of air through rotor blades causes the blades to spin and to generate lift despite the absence of torque from the propulsion system. Without power the aircraft descends rapidly, making autorotation possible. The phenomenon was first analyzed in the nineteenth century by the Scottish physicist James Clerk Maxwell, most famous for first formulating a theory of electromagnetic radiation. Maxwell developed the aerodynamic theory of autorotation based on longstanding observations by others who watched maple seeds falling from a tree spin as they fall to the ground.

The fixed pitch blades of most microdrones are incapable of autorotating in this manner. If the motors stop operating, and the drone remains upright, the upward flow of air will spin the blades backwards in a partially stalled state, generating some resistance to the descent rate, but not nearly as much as an autorotating helicopter rotor. Moreover, multirotor drones depend on differential thrust from powered rotors to maintain their orientation. Without it, the vehicle is more likely to tumble than to remain upright, in which case it becomes completely uncontrollable.

The various ways in which a flight vehicle can store kinetic or potential energy influences the amount of kinetic energy that must be absorbed by an impact.

The amount of damage done by a vehicle when it collides with another aircraft or impacts the ground is determined by the total amount of kinetic energy that must be absorbed by the collision or the impact. The relatively short fixed focal length of drone camera lenses means that the drone must be relatively close to the target of its image collection. Typically a slant range of no more than 25 ft at an offset angle of 45° is necessary to capture reasonable detail of smaller objects or persons. That represents a height of about 20 ft off the ground and a horizontal distance of about 20 feet from the subject. Acceptable imagery captured from an orbit at that distance necessitates flying the drone slowly, no more than 2 to 5 knots.

At that speed, the kinetic energy of a small drone like a DJI Phantom or Inspire is quite low. A US Navy "Crash Lethality" study in 2012, drawing on new empirical data generated by the Navy and on studies of automobile accidents and accidents involving baseballs, concluded that UAVs pose a lethal threat to persons on the ground only when they have kinetic energy exceeding 54 foot-pounds. A small drone flying at 5 knots has a kinetic energy of only 3.12 foot-pounds for a Phantom and only 8.29 foot-pounds for an Inspire. A Phantom would have to be flying more than 20 knots at the time of impact to cross the threshold.

For drones whose kinetic energy crosses the threshold, additional mitigating measures such as parachutes or crashbags may be appropriate.

Even modest deformation of the aircraft skin or windshield can absorb all of the energy of a microdrone. On the other hand, a 55 pound drone with a similar velocity would have 564,190 foot-pounds of energy, and a windshield or skin designed to withstand impacts with smaller objects cannot absorb that much energy without suffering penetration. Aluminum absorbs more energy before it fractures under impact than plexiglas.[1] Energy absorption of either is linearly proportional to thickness, for relatively thin structures.[2]

A similar relationship between mass and damage to persons or objects on the ground exists: the heavier the object, the more damage and the greater the hazard.

The centrality of kinetic energy in accident risk analysis justifies drawing a bright line between microdrones and macrodrones and between different weights in each of the two categories. A microdrone weighing 10 pounds or less can do little damage to something else regardless of how it malfunctions. Its rotors, if they are still spinning when the drone hits something, might cause cuts, but they do not have enough mass to cause serious injury. Similarly, total weight is not enough to penetrate a roof, the wall of a structure, or the windshield of an automobile.

In contrast, a macrodrone weighing 500 pounds is likely to kill someone if it falls on her and could easily penetrate an automobile windshield or a window in a structure even if it has low vertical speed. If substantial horizontal speed is involved, the damage would be even greater.

Accordingly, accident prevention measures and emergency procedures should be more elaborate for macrodrones than for microdrones.

5.2.3 Fires, fuel, and lithium polymer batteries

Fire is a risk associated with an accident involving any aircraft carrying fuel. The magnitude of the risk depends on the quantity of fuel, and the probability of a fire is determined by the flammability of the fuel (aviation gasoline is more flammable than jet fuel) and the likelihood of an ignition source coming in contact with the fuel.

Accordingly, significantly different accident risk analysis and emergency procedures are appropriate for drones that carry fuel. More modest emergency procedures are appropriate for microdrones, most of which have electric propulsion systems, compared with macrodrones, most of which will have internal combustion systems requiring an onboard fuel supply.

Li-Po batteries, the most common source of electricity for microdrone propulsion, present risks of their own, however. The worst case involves *thermal runaway*, a situation in which the battery overheats and the overheating releases additional products for combustion from the battery chemistry. Common causes of thermal runaway are improper charging, discharge rates exceeding the current limits of the battery, and penetration of the protective cover of Li-Po cells. Analysis of the level of risk arising from thermal runaway of Li-Po batteries resulting from collisions or crashes begins with assessment of the likelihood that impact can set off a thermal runaway. Underwriters Laboratories' UL 1642 sections 12–15 prescribe crush, impact, and shock tests to quantify the risks of collision or crash-like stress.

There is not much a DROP or autonomous safety system can do to reduce the risk. Risk mitigation depends on original battery design. Design of the battery mounting hardware on the drone must shield the battery from penetration or excessive deceleration in the event of a collision or crash and—of course—must reduce the probability of a collision or crash.

5.3 Examples of emergency procedures for manned aircraft

The most serious emergencies that can confront manned aircraft and their pilots are engine, control, or structural failures. All airplane and helicopter flight manuals must contain detailed procedures for these and other emergencies, and pilots must train for proficiency in executing these procedures.

For example, multi-engine airplane pilots must react promptly and appropriately to an engine failure, particularly failure of the engine that creates the greatest threat to aircraft controllability (the *critical engine*: usually the left engine on a twin-engine aircraft with clockwise-rotating props). The pilot must detect which engine has failed, and within two or three seconds feather the prop on the failed engine, apply substantial (often more than 100 pounds) of rudder force opposite in direction to the failed engine, and roll the airplane toward the functioning engine. If he fails to do this, the airplane will quickly roll to an inverted attitude from which recovery is difficult or impossible within the available altitude.

A single-engine helicopter pilot experiencing an engine failure has about two seconds to lower the collective to flatten the pitch on the main rotor to allow it to enter autorotation. If he fails to do this, rotor RPM will decay to a level from which recovery is impossible.

5.4 Preventing emergencies by good preflight inspections

The best approach to emergencies is to prevent them. That begins with a good preflight inspection. Before every flight DROP should carefully inspect the drone to ensure that:

- All the connections are secure, and that there are no loose wires
- The battery is secure and fully charged
- The rotor blades are undamaged and secure
- The structure is undamaged and all its elements are tightly fastened together
- The center of gravity is within limits.

To the extent necessary, the DROP should refresh his recollection of control inputs:

- Review the function of all switches and knobs on the DROPCON
- Review how to activate all features of the drone
- Review the effects of different flight modes.

After application of electrical power, the DROP should

- Perform a compass and GPS calibration according to the operating manual, or confirm that one has been done for that location recently
- Ensure that light and buzzer patterns are correct for takeoff
- Wait for a *GPS lock* before attempting takeoff, unless the DROP intends to forego most of the autonomous safety features.

After increasing power for takeoff, the DROP should confirm correct operation of the following functions:

- Automatic takeoff
- Automatic landing
- Automatic hover
- Automatic return to home commanded by the DROP
- Automatic return to home resulting from lost link (triggered by turning off the DROPCON).

All of this can be done within a minute or two, at low heights, within a dozen feet or so of the DROP.

5.5 Microdrone emergencies

This section identifies specific microdrone emergencies. For each, it specifies the appropriate response to be taken by a combination of onboard autonomous systems and the DROP to minimize the risk of damage and injury.

Malfunctions in mission sensor or downlink systems do not constitute emergencies.

5.5.1 Lost control link

Because microdrones without autonomous safety features depend upon remote control by a DROP, a lost control link is one of the most serious emergencies that can occur. If the drone continues to fly at significant speed away from the DROP, the control link is unlikely to be recovered. Where the drone is when its battery is exhausted and it falls to the ground is completely unpredictable. Or it may collide with an obstacle or another aircraft before the battery is exhausted. Moreover, unless it is inherently stable because of its aerodynamic characteristics or a combination of those characteristics and its autonomous control systems, it may, simply fall out of the sky, probably tumbling in orientation as it does so.

Dealing with this emergency requires, first, that loss of the control link be detected, and second, that the microdrone have built-in systems to respond to it. A response by the DROP obviously is irrelevant because, by definition, the lost control link means that he cannot communicate with it.

The appropriate emergency procedure, available in all current microdrones except for the smallest nanodrones weighing only a few ounces, is an autonomous return-to-home protocol. The vehicle's autonomous control subsystems detect the loss of the control link, and provide control inputs to put the drone on a flight path so that it returns to its launching point for that particular mission. The return-to-home profile must employ appropriate heights and speeds. Lower height of return means greater likelihood of colliding with an obstacle such as a tree or a building; lower speed means greater likelihood of the return being impeded by wind.

A successful outcome depends on the drone having determined its original launching point as it first obtained GPS lock. DROPs must be careful not to change the launch point by powering the drone down and restarting it at a remote point after the mission first begins, unless they intend it.

5.5.2 Power loss

Because most microdrones are multi-rotor in design, the loss of torque from a single electric motor is not as serious an emergency as the loss of the sole engine in an airplane or helicopter. The thrust from the remaining motors usually allows the vehicle to remain upright if a single motor is lost and to conduct a controlled landing, if not a return to home. This obviously depends on the margin of thrust available when the motor is lost. The margin of thrust depends on total gross weight, including payload.

The emergency procedure should be a return to home, if possible, and an immediate, controlled, landing otherwise.

Continued microdrone fight, including a successful return to home, requires that enough electrical energy remain in the battery to accomplish that. Accordingly, drone flight beyond a certain battery reserve constitutes an emergency. The likelihood of a microdrone flying into its reserve can be reduced by autonomous systems that detect a low battery and automatically trigger return to home while enough energy remains in the battery. But if this feature is not built in, or if it malfunctions, the default emergency procedure of returning to home is not available. Then, the appropriate emergency procedure is to land immediately while enough battery remains to control the landing. Determining the remaining charge in a battery is not foolproof; most charge-remaining detection circuits rely on battery voltage, but voltage is only a rough approximation of how many ampere hours remain.

Most microdrones, unlike most helicopters, have fixed-pitch rotors; rotor thrust is controlled by varying RPM rather than by varying the pitch of the blades. Fixed pitch rotors lack the ability to enter an autorotation. The upflow of air in a rapid, power-off descent may spin the rotors backwards, and generate some aerodynamic resistance, slowing the descent, but not nearly as much as an autorotation. Moreover, a microdrone experiencing a complete power failure on all its motors is likely to tumble, because there is no differential thrust available to oppose random pitch, yaw, and roll forces inevitable in any descent.

Thus no emergency procedure can compensate successfully from a complete loss of power. The only mitigating approach is a parachute or a means of destroying the aircraft, considered in §§ 5.7.1 and 5.7.2.

5.5.3 Loss of visual contact

A loss of visual contact is different from loss of the control link, considered in § 5.5.1. The drone can tell if the control link is lost; it cannot tell if the DROP cannot see it. Loss of visual contact and loss of the control link may be proximate in time if they result from the vehicle's flying behind an obstacle or at too great a range, but the effect on the drone's operation and the available recovery options are different.

Whether loss of visual contact with a microdrone constitutes an emergency depends on the protocol under which it is being flown. Many national regulators distinguish among line-of-sight (LOS), extended line-of-sight (ELOS) and beyond-line-of-sight (BLOS). ELOS operation signifies that FPV, observers, or additional DROPs can see the drone. BLOS signifies that the drone is designed to be flown beyond line of sight of the crew.

In LOS operations, control of the drone's orientation and flight path depend on the DROP being able to see it. The appropriate recovery techniques depend on the reason that visual contact is lost. If it is lost because of range, especially high altitudes, or because the drone flies into a cloud or other meteorological obscuration, the best recovery technique is to trigger return-to-home. The effectiveness of that response depends upon whether return-to-home can be activated easily through a preset switch on the DROPCON. Alternatively, landing at a safe intermediate point may be safer. For example, if drone travels from point A to point B, and the failure occurs ¾ of the way to point B, it would make more sense to finish the mission and land at B instead of returning to A.

If the reason visual contact is lost is because the drone has flown behind a physical obstacle such as a tree or a building, recovery is more challenging. If return-to-home is activated in those circumstances, the drone may attempt to return to home at an altitude and direction that will cause it to collide with the obstacle. One way to avoid this is to reprogram the return-to-home feature, as a part of pre-flight preparation, so that the drone first climbs to a height above any obstacle to the area before it attempts to return. Even then, it may get snagged by tree branches or some other physical object that is attached to whatever interrupted the DROP's visual contact. Wind also, as § 5.5.1 points out, may prevent a successful return-to-home maneuver. Alternatively, the DROP could set up multiple "safe zones," A, B, C, D, causing the drone to fly to the closest zone in the event of a malfunction. That strategy is available only if the drone has the capability to accept multiple landing points.

If the drop has FPV (first person view) available, he may be able to use his FPV imagery to navigate around the obstacle and regain sight of the drone. The difficulty of navigating a rotary wing aircraft by means of FPV should not be underestimated, however. Typically, the field of view available through

FPV is too narrow for the DROP to get good situational awareness; he may be able to see a tree or a building, but not easily figure out where the drone is. FPV goggles, as opposed to screens, give the illusion of a wider field of view, but the actual field depends on the focal length of the lens on the camera. The DROP may be able to program the camera as a part of his preflight so as to mitigate this limitation.

If the DROP reacts quickly, he may be able to fly from behind the obstacle by a rapid sideways movement or gain in height. But, unless something equivalent to the DJI Phantom's *carefree* mode is available and has been activated, the DROP may not know the drone's orientation, and therefore may fly into the obstacle when he applies control inputs to fly sideways or backwards.

In any event, the most prudent course may be to trigger automatic hover to give the DROP a few seconds to figure out the best recovery technique under the circumstances.

The fallback option is to trigger an automatic landing, which may improve the odds of finding the vehicle and minimize the risk of damaging it or inflicting damage or injury to other persons or objects.

5.5.4 Loss of GPS guidance

A GPS lock is necessary for the functioning of most autonomous features. A microdrone may lose its ability to navigate with reference to GPS coordinates for variety of reasons. The requisite number of satellite signals may be lost because of terrain or obstacles. The GPS processing systems may malfunction, or the necessary calibration may not have been accomplished successfully at launch. Whatever the cause, loss of GPS navigation constitutes an emergency, because it impairs the drone's ability to return to home and thus negates the most common emergency procedure discussed in the preceding sections. It also may impair the DROP's ability to control the vehicle visually.

It may not mean, however, a complete loss of controllability if the microdrone also has accelerometers and a magnetometer, as most do.

Because of the almost certain impairment of return to home capability, the most appropriate response is to land immediately in a controlled fashion. The response can be triggered by the DROP when he detects erratic flight, or autonomously when the microdrone's onboard systems detect loss of GPS capability.

5.6 Macrodrone emergencies

Macrodrones resemble manned aircraft much more than microdrones do. Their weight, speed, range, lift, and propulsion systems are like those of manned aircraft. Most of them are fixed-wing in design. Rotary-wing designs are more likely to employ single main rotors than multi-rotors. In most cases, the only significant difference is that they are flown from a ground-based DROP rather than by an onboard pilot.

The emergencies that macrodrones encounter are likely to be the same as those encountered by manned aircraft, with the important addition of lost-control link, which does not pertain to manned aircraft, which do not have control links with a ground operator.

Therefore, the emergency response procedures for macrodrones, except for lost-control links, will be the same as for manned aircraft.

If a rotary-wing macrodrone encounters an engine failure, the procedure is to enter an autorotation, either triggered by an autonomous response by an onboard system, which can easily detect a loss of torque to the main rotor, or by the DROP. An important difference between a macrodrone and a helicopter, however, is that the DROP is unlikely to have as complete a sight picture to allow selection of an appropriate landing spot within the small autorotation range.

Likewise, an engine failure in a single-engine fixed-wing macrodrone requires maintaining control of the aircraft, selection of an airspeed, and attitude that results in the best lift-to-drag ratio and selection of the best place to land. Here also, a DROP will be disadvantaged because of a more limited sight picture.

The response to other emergencies, such as loss of hydraulic control systems, flight into IMC or icing, also will be similar to that in an equivalent manned aircraft, although the DROP almost certainly will have fewer kinesthetic, aural, and visual cues than a pilot as to what is wrong, and what changes in speed, altitude, or attitude will ameliorate the problem.

5.7 Vehicle destruction systems and parachutes

Emergency responses involve autonomous or DROP control of a drone can be supplemented by emergency systems such as parachutes or explosive devices that destroy the drones before it can do unacceptable damage.

5.7.1 Parachutes

The consequences of a drone emergency involving loss of control can be mitigated by equipping the vehicle with a parachute that deploys automatically or in response to the DROP in certain rare circumstances.

The function of a parachute in these circumstances is to limit the kinetic energy when the vehicle reaches the ground.

Throughout the history of aviation, aircrew members often have been equipped with personal parachutes to permit them to eject safely from an aircraft that is certain to crash. Equipping the vehicles themselves with parachutes has been rare, except for space vehicles returning from orbit and lacking maneuverability.

The advent of the Cirrus SR22 general aviation airplane in 2001, however, invited attention to the possibility of emergency parachutes in small aircraft. Originally adopted because of certain control anomalies that otherwise would have prevented airworthiness certification, the parachute system has become the center of a marketing message from the manufacturer. The Cirrus illustrates the practicability of a parachute to mitigate the effects of emergencies likely to be catastrophic without it.

Small parachute systems exist, suitable in size and weight for installation on microdrones the size of the DJI Inspire and larger vehicles like the DJI S1000. Typically, they contain parachute fabric packed in a canister about the size of a small flashlight. The parachute is spring-loaded, and an electrically triggered release of a catch permits the parachute to deploy. Installation must allow the DROP to trigger deployment so that it follows motor shutdown to avoid the parachute becoming entangled in the spinning rotors.

There may be certain drone design types and mission profiles that warrant consideration of this unusual mitigation measure. Heavier drones can do considerable damage if they fall to the ground at terminal velocity, parachutes may be warranted as a safety feature. Parachutes are less appropriate for smaller, electrically powered vehicles because of their weight and the complexity of systems to activate them. "Crashbags"—resembling automobile airbags—may also have a role to play.

5.7.2 Devices to destroy the drone

Almost all rockets intended to propel vehicles into space have onboard systems that permit operators to destroy them if they malfunction on lift-off. The consequences of uncontrolled flight by a vehicle loaded with tons of highly explosive propellants and oxidizers are simply too much to tolerate, despite the cost of vehicle destruction.

The question arises whether similar systems may be appropriate for drones when their control link is lost or other emergencies occur. The answers is almost certainly "no" for microdrones. Their limited weight and speed, and their proximity to a DROP who can see them limit the adverse consequences of emergencies. But the answer may be different for larger macrodrones resembling manned aircraft in weight, speed, and quantity of fuel on board.

Destruction mitigates the consequences of a crash, because it separates a heavy vehicle into pieces with less kinetic energy and causes onboard fuel to be consumed before it comes into contact with other aircraft or persons or property on the ground.

Vehicle destruction can be accomplished by explosives, as on space vehicles, or by non-flammable systems such as the chemical systems used on automobile airbags. Explosive systems are unlikely to be attractive because of the risks they pose to normal operations and because of their weight and complexity.

Chemical systems might be appropriate however. They could separate the wings from the fuselage of a fixed-wing macrodrone, for example, causing an immediate crash into unpopulated terrain rather than continued flight into more populated areas or more congested airspace.

Such an approach to emergency risk management is obviously extreme and should be reserved only for the most hazardous missions intended to be confined to sparsely populated areas. Even then, proponents of such systems should be able to establish why the danger of a crash outweighs the danger of an exploding airborne vehicle.

5.7.3 Emergency control triggers

Loss of battery power undermines most conceivable safety features in electrically powered drones. Almost all fixed wing aircraft are designed so that they are statically and dynamically stable, concepts explained in chapter 3. That means that, if the propulsion system of a fixed wing aircraft, including a fixed wing drone, fails, the aircraft will begin to glide. Even if its flightpath is disturbed, and the aircraft begins to turn, it will continue to glide. The lift will be the same as for the similar angle of attack and speed as in powered flight. No likely combination of environmental forces acting on it will cause it to stall.

The behavior of a multirotor drone with fixed-pitch rotors is dramatically different. As § 5.2.2 explains, these vehicles, unlike helicopters with controllable pitch rotors, are incapable of entering autorotation.

More seriously, control of the attitude of a multirotor depends on differential trust being generated by the different rotors. When the motors are not turning and generating torque to spin the rotors, attitude control is lost. Depending on the vertical placement of the center of gravity relative to the center of lift, the aircraft may become inverted almost immediately. Even if this does not happen, the drone is likely to tumble. After it tumbles, it will descend at its terminal velocity and be entirely uncontrollable.

Because its rotors are not now being spun by the motors, their potential to cause damage by slicing into something is low. The risk is determined entirely by the kinetic energy in the crashing drone, which in turn is a function of its weight times its terminal velocity, squared. As section § 5.2.2 explains, the consequences of a crash of a lightweight microdrone are modest enough to be tolerated. That is why almost all the world's regulatory regimes adjust regulatory requirements according to the weight of the vehicle.

For heavier microdrones or macrodrones, the risk increases. For these vehicles, as § 5.5.1 and § 5.7.2 explains, the only design to mitigate this risk are either an explosive device that will destroy the drone before it hits something, or the deployment of a parachute. In either case, the question is: how can the drone sense that it is in free fall, when the safety device has no electric power?

Two answers are available—conceptually. First, the drone could have a backup battery. The battery would have enough energy to allow the drone to return to home or to land at an appropriate site.

Second, the drone could have a parachute could be triggered as soon as the safety system senses the main power is lost. For example, a parachute could be deployed by a gas canister containing compressed CO_2 or nitrogen—or simply by a spring. Discharge of the canister would inhibited by the application of an electric voltage. When the voltage is lost, the canister will fire, deploying the parachute. Safe design of such a system would need a mechanism to inhibit firing of the canister when electric power is removed on the ground, for example, to recharge or replace the batteries.

Notes

1 Figure 16, www.kazuli.com/UW/4A/ME534/lexan%20VS%20Acrylic4.htm.
2 David Roylance, *Introduction to Fracture Mechanics*, (Jun. 14, 2001), p. 11 ocw.mit.edu/courses/materials-science-and-engineering/3-11-mechanics-of-materials-fall-1999/modules/frac.pdf [hereinafter "Roylance"].

6 Politics

6.1 Introduction

Four phenomena shape drone politics: first, the rapid proliferation of microdrones has surprised almost everyone; second, the general public opposes their wider use; third, a regulatory vacuum persists, despite the FAA's NPRM and section 333 exemptions; and fourth, microdrones are so numerous, so cheap, and so easy to fly, that even a mature regulatory regime would have difficulty channeling their use.

Law is the codification of political will. In a democratic society, political will is rooted in the desires of the people. It is not the product of a scientist, an engineer, and an economist going into a back room and working from theoretical principles to define particular prohibitions and permissions. Instead, democratic institutions channel individual wants and wishes, combine them with the wants and wishes of others, and eventually produce an expression of political will for the entire society. An understanding of drone law requires an understanding of drone politics.

This chapter begins with a deeper exploration of the four phenomena, then reviews major principles of democratic political theory and explains how the manifestation of the four phenomena will demonstrate the theory and produce law. It goes on to identify the major interest groups likely to influence the content of drone regulation and explains how advocates and opponents of wider civilian use of drones will work with interest groups and mold public opinion to get the regulatory outcome they want.

Psychological and sociological phenomena operating on individuals enter the legal equation twice—not only at the outset, as new regulations are written; they also affect the level of compliance with regulations once they are on the books. Compliance is addressed in chapters 4, 7, and 8.

6.2 Major political currents

As the introduction points out, four political currents shape drone politics: first, the rapidity with which microdrones have proliferated has surprised almost everyone; second, the instinctive attitude of the general public is negative;

third, a regulatory vacuum persists, even after the FAA released its notice of proposed rulemaking and granted more than 5,000 section 333 exemptions; and fourth, microdrones are so numerous, so cheap, and so easy to fly, that even a mature regulatory regime will have difficulty channeling their use.

6.2.1 Rapid proliferation of civilian microdrones

The first DGI Phantom was introduced in a January 27, 2013, in a YouTube video, months before Jeff Bezos' *60 Minutes* interview that captured so much public attention over the prospect of small drones being used to deliver packages. Amazon sold 300,000 Phantoms and similar products within a little over a year. By the end of 2014 thousands of purchasers were clamoring for the FAA to speed up its process to provide a regulatory framework for commercial use.

The speed with which the new technology entered the marketplace and public consciousness caught almost everyone off guard. This included the FAA, which had been plodding along, thinking mostly about macrodrones, and the general public, who now were bombarded with news stories about the potential of the technology. At the same time they saw and heard often-inflated stories about risks to manned aircraft and worried about personal privacy intrusions.

Major interest groups had to scramble to figure out what their positions were on the subject. Elected officials similarly had to throw together some kind of position to respond to growing constituent concerns.

6.2.2 Public opposition

The general public is opposed to civilian drone use. This opinion has been shaped by an understanding of drones derived from military configurations used in the Iraq, Afghanistan, and Middle East conflicts, by concerns about governmental surveillance, and by a broadly-defined, amorphous, view of privacy.[1]

Opinion surveys data demonstrate the general public opposition. In a survey conducted in late January 2015 for Reuters, 45 percent of 2,000 online respondents disapproved of private ownership of drones altogether; 73 percent favored regulation; 64 percent said they would not want their next-door neighbor to have one; 71 percent said they should not be allowed to fly over private property; and 46 percent opposed use by news agencies.

68 percent, on the other hand, favored police use to solve crimes, and 62 percent said that police should be able to use them to deter crime.[2]

The primary concern seemed to be a broad concept of privacy—the use of civilian drones to pry into people's affairs—rather than fears of governmental surveillance. The privacy concern is amorphous. Some of it focuses on fears of drones being used to look through the windows of residences, but much of it seems to represent a generalized desire not to be annoyed.

Co-author Perritt had a brief confrontation in an early morning in August 2015, when he was flying a DJI Phantom at 150 ft over the public beach in Glencoe, Illinois. Hardly anyone was on the beach. After he had concluded his flights, he relaunched the drone and hovered it over a three-foot brick wall separating a public path from the bluff leading down to the beach. A man and a woman approached, both holding their cellphones out to take pictures of the drone. The man responded furiously to Perritt's greeting, saying, "You're violating our rights!"

"What rights?" Perritt asked.

"This is where we like to walk," the man said. It was unclear whether he was referring to the beach or the sidewalks that run through an adjacent park.

"It's a public space," Perritt responded.

"This will change," the man snarled, shaking his cell phone at Perritt.

The law of privacy, of course, has always been premised on protection of "a reasonable expectation of privacy." Judicial decisions deciding civil actions claiming the tort of invasion of privacy, and applying fourth amendment protections to governmental surveillance activities regularly reject privacy claims arising from allegedly private activities conducted in parks and other public spaces.

But what the law says does not exclude the political reality that widespread civilian drone flights will interrupt people's solitude, much as then future Supreme Court Justice Brandeis feared that the availability of small inexpensive still cameras in the final decade of the nineteenth century would have a similar effect: "Recent inventions and business methods call attention to the next step which must be taken for … securing to the, individual … the right 'to be let alone,'" he said.[3] He explicitly referred to "instantaneous photography" as an example of an "invention" that provoked his concern. Kodak had introduced its consumer-oriented Brownie camera, priced at $1.00, the year before.

Members of the public regularly approach the authors when they are flying microdrones. They mostly express curiosity, but many express vague concerns that drones are "creepy," even as they acknowledge that they can serve the public interest by supporting search and rescue, firefighting, agricultural productivity, event photography, infrastructure inspection, border security, and newsgathering.

An informal discussion over dinner by a dozen Chicago-area law professors attending an annual faculty party revealed mostly hostility toward civilian drones. Some participants reiterated the concern about drones peering through windows. Others worried about their potential use by terrorists, domestic and foreign. They wanted strict regulation, excluding them from airspace near airports and schools, and keeping them away from public gatherings like music concerts and sporting events.

The vigorous public debate over military and intelligence-agency use of armed drones to kill suspected terrorists in South Asia and the Middle East shapes public opinion about civilian drones. It is difficult for an ordinary person, who does not spend much time on the subject, to escape the idea that a drone is a killing machine—or at least an eye in the sky conducting general surveillance of all kinds of activities—most of which are benign.

6.2.3 Regulatory vacuum

While much of the literature on drone regulation focuses on what the FAA is doing with its section 333 exemptions and its NPRM, the reality is that most civilian drone use is unregulated. It is not clear how much regulation will emerge from state legislatures and local city councils, although many are testing the waters. No one knows for sure how the FAA is going to approach macrodrones, and the NPRM leaves big regulatory gaps, with respect to microdrones flown at night, beyond line of sight, or near crowds, where some of the most appealing legitimate applications exist.

Those gaps fuel public concerns about a threatening technology that is out of control. They also leave ample space for growing political conflict over the desirable content of regulation.

6.2.4 Lawlessness

Even if the public understood the broad consensus on limitations for microdrone flight—the ones basically set forth in the NPRM and in every section 333 exemption, they are regularly confronted with examples of microdrones being flown in complete disregard of these limitations.

The FAA is not enforcing its existing ban; nor does it seem to be enforcing the narrow scope of its section 333 exemptions. To be sure, section 333 exemption holders are likely to be conservative and comply, for the most part, with the limitations in their exemptions. They foresee commercial opportunity, have invested considerable effort and money in obtaining their exemptions, and do not want to lose them. Still, the conclusion is inescapable that many exemption holders sneak beyond the boundaries of what they officially are permitted to do. The FAA is not doing much about it.

The FAA's highly publicized "database" of "near miss" reports by pilots fuels the perception that the Wild West has returned—that pilots face a future in which they will be dodging drones, no matter where they fly. It does not matter that few of the near miss reports bear scrutiny; many of them turn out to be only distant sightings of drones; some of them are impossible because they report drones at 6,000 or 7,000 ft above the ground, where few of them are able to fly; and that many of the reported "drones" turn out to be birds or escaped birthday or weather balloons. Almost every sighting of an airborne object gets reported as a near collision between a drone and manned aircraft with passengers on board, especially once it gets filtered by popular journalism.

What matters in the political arena is public perception, and the public perceives that potential collisions between drones and aircraft carrying passengers are exploding.

6.3 Political dynamics

Political theory recognizes that policy decision-making in democratic societies results from the interplay of a number of different forces. The same thing is true of drone politics. To a considerable extent, how these forces interact with each other depends on whether the society is homogeneous or heterogeneous in ethnicity, ideology, and economic circumstances. Homogeneous societies tend to make decisions by consensus; heterogeneous societies make decisions by channeling competing interests through mediating institutions such as legislatures, political parties, political candidates, and interest groups.[4] Even in the smallest electoral districts, diverse and conflicting preferences exist about how particular issue should be resolved. But, in the United States at large, attitudes about drones and those who hold them are more heterogeneous than at the local level.

A baseline factor political behavior is *passion*. Political mobilization depends on how much segments of the society care about an issue. If someone does not care much about an issue, she will be largely indifferent to it and let other matters drive her involvement, choice of candidates, and interest group membership. She may have a weak preference about how an issue should be resolved, in which case she will not let that issue drive her political behavior.

Passion drives, not only preferences for candidates and adhesion to organizations; it also drives political contributions directly to candidates and to the political action funds of organizations. Political passion may arise instinctively from deeply held values obviously at play in a particular public policy issue; or it may be whipped up by political activists who persuade their targets that an issue is related to individual concerns. Such persuasion is the result of effective political entrepreneurship. ALPA, for example, is embarked on an aggressive campaign to limit drones by emphasizing the dangers of mid-air collisions with airliners; AUVSI is equally active in promoting public understanding of their potential to improve the public welfare, citing their use to fight fires. RC modelers are ferocious in their opposition to additional regulation of hobbyists, emphasizing personal liberty.

6.3.1 Political entrepreneurship

Meaningful political action, like business success, depends on effective entrepreneurship.[5] A political activist, whether a candidate for public office or advocate for an interest group, must assess accurately the needs and wants of constituents, craft a message that will resonate with them, pick channels for communicating that message to the constituents, and raise the money for the necessary communications. The political entrepreneurship that matters for drones is that necessary to achieve a hospitable legal and regulatory climate or, if one is opposed, to limit drone use as much as possible.

Accidents are inevitable, and some will be serious. The political entrepreneurship challenge for both sides is how to respond when such accidents occur. They will surely be newsworthy, and public opinion will be substantially shaped by them—and by the interpretations offered by opponents and proponents of different regulatory strategies.

Effective political entrepreneurs are opportunistic. A drone accident or complaint about drone activity provides an opportunity for drone opponents to spin the incident to magnify latent public concerns about safety and privacy. A story about successful drone use to support agriculture, public safety, search and rescue, or motion picture and television production provides an opportunity for drone proponents to spin the experience to make the point that drones stimulate economic growth and job development and make the lives better for ordinary Americans.

Good entrepreneurs create their own opportunities by building relationships with reporters and news and media executives and feeding them stories that they may find newsworthy or fill human interest story needs. A political candidate may find, in a drone incident with positive or negative ramifications, a rare opportunity to get out in front with a high profile public policy initiative that will resonate with her voters.

6.3.2 The art of the possible and the science of timing

Politics is often characterized as the art of the possible and the science of timing. The first part of this aphorism signifies that the substantive positioning of a campaign is an inexact process, but that timing of political action can be designed with greater precision based on a few rules of thumb. Interest analysis is uncertain because human reaction to events and stories about them proceeds from deeply ingrained, but often invisible, psychological forces operating within individuals. One of the determinants of public reaction to an event is how widely publicized it is and how competing stories about the circumstances are spun.

Public reaction to a drone accident or to a near miss with a manned aircraft accordingly depends on how widely it is known and on perceptions about the moral position of the participants. If a hobbyist recklessly flies a drone in the approach path of a major airport and buzzes an airliner on final approach, the reaction may be more adverse than if a sightseeing helicopter collides with a drone surveying a natural disaster. It is hard to know in advance.

How widely an event is known depends largely on how much the press and media cover it. That depends, in turn, on all the stories competing for time on the evening news broadcasts, on the front pages, and for blogs and Twitter posts. Once a story runs it often goes *viral* as social media algorithms promote a story. For example, "now trending" on Twitter and Facebook or a fan page spread stories rapidly. No one can predict in advance what combination of news stories might be breaking on a particular day.

The fortuity of news coverage and public awareness represents the art of the possible. So does the arousal of public passion. The science of timing reflects political entrepreneurship; it is opportunistic. If an aviation interest group with access to the public arena consistently has warned about the dangers of drones to manned aircraft, it can seize the moment when an accident or near miss is reported.

If a microdrone is being flown a few dozen feet over the heads of spectators at an outdoor concert and crashes into the crowd, the public reaction will depend on what else is competing for media and news attention on that day and on whether an anti-drone advocacy group believes it is a good story to whip up public support for its position. It also depends on whether the opposition party wants to take a swipe at the current government and its inability to maintain security against drones.

Thereafter, the political impact depends on timing as well. Certain legislation can pass or not pass under the same president depending on whether it is considered before or after midterm elections. An aphorism of presidential (and impliedly state-governor) politics is that the elected executive has the greatest power during the first hundred days of his incumbency—when the members of the legislature are frightened by his demonstrated political popularity.

6.3.3 *Crafting a message*

Effective political entrepreneurship requires crafting a message that will resonate with the public. The basic outlines of pro- and con- drone messages are reasonably well established. They will remain much the same over the months and years ahead, newly adorned with late breaking news about drone success or near disaster.

The basic negative message is:

- Drones are dangerous; it's only a matter of time before one brings down an airliner or kills someone on the ground.
- Drones represent a new kind of threat to personal privacy and to the security of private property.[6]
- They cannot be flown safely until they are equipped like manned aircraft, subjected to the same scrutiny of their hardware and software that make the national airspace safe, and until and unless they are flown by *real* pilots who know the regulations, understand weather and aeronautics, and have demonstrated their ability to fly safely in practice environments.[7]
- The technology is interesting, but its contribution to the economy and to ordinary life is marginal; there is no need to put lives at risk by being in a rush.
- Much of the excitement about drones in this and other countries is overblown and will turn out to be a passing fad as soon as people realize the limitations of what is being sold.

The positive message is:

- The US market economy and American democracy are beacons to the rest of the world because both have always been hospitable to new technologies.
- Drone technologies are just the latest in disruptive aviation technologies that have built the world's best system of air commerce. Stale ideas about regulation of older technologies should not stand in the way and put America behind in an international competition.
- Drones can make available the benefits of aerial observation and support in ways that have not been affordable before. A variety of industries already are benefitting, including agriculture, video entertainment, security of public utilities and transportation infrastructure, and news-gathering. Many others that are just being discovered as the free market is opened up to their use.[8]
- Drones do not present any significant safety threat; the small ones aren't capable of causing much damage because of their small size and weight;[9] big ones won't be flown in the National Airspace System until they are equipped with safety systems equivalent to those aboard manned aircraft. The claims of near misses are, for the most part, mere sightings of entirely benign activity. Every time a pilot sees a drone, he reports a "near miss."
- Existing safety features are breathtaking; drones can take off and land automatically, hover in one place when the operator releases the controls, refuse to go where they are prohibited by law, and automatically return to the launching point if something goes wrong (chapter 9 explores these features more fully.).
- Drones do not present any significant new threat to privacy. Ground-based photography is more intrusive; commercial operators want to get their job done, they do not want to waste time by spying on people.

Both messages can, of course, be refined and tailored to specific regulatory proposals. A drone proponent can explain that a drone is easier to see at night than in the daytime, and therefore that night operation should be allowed. A news organization that flies tethered drones can argue that those vehicles should not be subject to the limitations imposed on drones capable of free flight.

On the other side, opponents can emphasize the fact that highly advertised safety systems have poor reliability, and marshal evidence that flyaways of models heavily advertised as having such features are common.

6.3.4 Agendas, priorities, and intellectual capital

Every political candidate and every political advocacy organization has to decide what he, she, or it is going to talk about publicly. Candidates and advocates must decide on a platform. The list of possibilities is immense: global warming; gay marriage; the global counterterrorism effort; genetically modified agricultural products; the federal budget deficit; the minimum wage; and gun rights, to name a few. Drone regulation must compete for inclusion in the platforms.

Most advocacy organizations focus on particular subject matter, and this inherently limits the scope of their agendas. Political candidates are not inherently specialized-but an effective candidacy requires that they define themselves in terms of one or more signature issues. Even specialized groups have long lists of issues from which to choose for their agendas. The RTDNA and the National Press Photographers Association must decide whether to take a position on rules for hobbyist operators of large turbine powered fixed wing drones, or stick to rules for microdrones flown for commercial photography. They must decide whether to focus their lobbying and public education efforts on FAA requirements for DROP qualification and certification, or even more narrowly on medical certification for DROPs. Or, they can put more effort into shaping the rules for airworthiness certification of microdrone grid navigation systems and facilitating beyond line of sight (BLOS) and nighttime operations. Of course, interest groups can, and often do, collaborate, with one specializing in one aspect of an issue and a second concentrating on other aspects.

Most political candidates and political organizations are not single issue oriented; they typically have a half-dozen or so issues on their agenda. To some extent, it is feasible to pursue all of them at a baseline level of effort, but it is also necessary to set some relative priorities, because of resource limitations, and because it may turn out that an achievable result in the public-policy arena requires trading off concessions on one issue for gains on another. A regulatory decision-maker may be willing to go lightly on reliability requirements for microdrone navigation systems in exchange for tight rules to protect against DROP incapacitation.

And of course individuals and single entities comprising the public cannot pay equal attention to everything in the national public policy debate. Each individual, school board, church, or business entity cares more about some things than others, and their interests and priorities shift over time. Much of the political process is a struggle by political actors, such as candidates, to frame their platforms to resonate with what significant segments of the public care about.

Those are the ingredients of the art of the possible.

That leads to the question of *intellectual capital*. Elected officials, agency decision-makers, advocacy groups, and political candidates do not fashion public-policy solutions on the spur of the moment. Instead, public policy alternatives emerge gradually from a long process. The process begins with theoretical academic inquiries, discussion in public arenas, and eventual connection of new ideas with political and economic interests.

The term *intellectual capital* has multiple meanings. Often, it refers to proprietary business knowledge protected by an intellectual property law. In the context of this chapter, it also refers to the stock of policy ideas developed by universities, think tanks, and trade associations. Most legislators and executive branch policymakers are far too busy dealing with the tactical crises of the day to spend much time thinking up new ideas. When a policy issue, such a drone regulation, surfaces, the elected members of Congress, state legislators and city councilors do not go on a retreat and brainstorm about how to craft statutory or rule language to regulate drones. Decision-makers need off-the-shelf solutions for which the technical details are reasonably well worked out. They also need solutions for which the political reaction is predictable.

To get these off-the-shelf solutions, governmental policymakers, or more likely, their staffs (if they have them) reach out to lobbyists, academics, and law firms known to be interested in the subject and ask for their ideas. Eventually, these resources suggest statutory or regulatory language. To the extent that such resources do not exist, or if they exist but do not have very well-developed ideas, the policy development process is likely to go awry. This results in a poor match between regulatory strategy and reality, fueled by amorphous public ideas about what the problem is.

Thus public policy initiatives almost always are developed initially in universities and other research centers, often referred to as *think tanks*. Some of them are taken up by trade associations and advocacy groups, and eventually find their way onto someone's political agenda. The initial advocacy of an idea in the context of a political campaign often occurs when a candidate grabs one to define himself in the public mind or when a new high level agency head takes office and wants to redirect her agency's posture.

Germination of new policy ideas takes time, usually measured in decades. Examples are the idea of invading Iraq, first proposed by conservatives associated with the Republican Party ten years before the second President Bush embraced it as a response to the attacks of September 11. This is a perfect example of the science of timing.

The idea for a separate regulatory category for experimental and sport aircraft had been advocated by the Experimental Aircraft Association (EAA) for years, but only resulted in publication of a Notice of Proposed Rulemaking in 2002,[10] and promulgation of a final rule in 2004.[11]

The idea for eliminating the third class medical certificate for private pilots and allowing them to fly only on a driver's license, was first proposed by AOPA in 2002, and has not yet been adopted.

The rapid proliferation of microdrones useful for commercial applications has not yet allowed this kind of time for germinating ideas for their regulation.

Crystallization is a phase in the development of intellectual capital. A mere idea cannot define a political campaign or be promulgated as a regulation. It must be made concrete before opposing interests can work with regulators define its terms. Indeed, they need an idea in concrete form before they can be sure whether they are for it or against it. It is one thing to argue, as this book does, that microdrones should be regulated as consumer products and their sale prohibited unless they have certain autonomous safety features. But it is hard to know where anyone will stand on this proposal unless they know how it would work. For example, if microdrones are required to go through something resembling an airworthiness certification process for their avionics, advocates of faster deployment will be opposed. If FAA preapproval is not required, the same interests would be supportive—depending on the penalties if a vendor offers a microdrone, believing in good faith it meets the requirement, and turns out to be wrong.

Students of the political process long have understood that concentrated interests trump diffuse interests. Interests are concentrated when persons and entities sharing the interests cooperate with each other and pool their resources to affect public policy. Interests are diffuse when they are only one of many concerns affecting a large number of people who have no effective mechanism to cooperate with each other on the subject. Concentrated interests spend significant amounts of money on lobbying and contributions to political campaigns. Diffuse interests spend none, because they are not organized.

Concentration versus diffusion can be different at different levels of government. Pro-drone interests are concentrated at the national level and diffuse at the local level. Anti-drone interests are diffuse at the national level and concentrated at the local level.

Gradually drone proponents are establishing institutions to concentrate their interests in political arenas, as § 6.4.2 describes.

6.3.5 Civil society

Civil society comprises the press and media, political parties, political candidates, and private associations such as labor unions, trade groups, and self-styled public interest groups.

Civil society actors crystallize issues, aggregate them,[12] and invest in intellectual capital. They necessarily are entrepreneurial in their behavior.

They calibrate their positions in order to preserve existing involvement and attract new membership. They crystallize positions that may be only vaguely felt or articulated by members; in other words, they provide mechanisms to concentrate interests. They establish priorities, and they help elected officials and civil servants do factual and policy analysis. Especially in the United States, they channel campaign contributions to supportive officials and candidates for public office.

As electoral districts get larger, it become more difficult for an individual or small clusters of individuals to make their voices heard. Their interests become diffuse. They can vote, of course, but the prospect of their voting may have only modest influence on the initial formulation of issues and candidate selection. Accordingly, they tend to speak through organizations, frequently referred to as "civil society," which concentrate otherwise diffuse interests. Like political candidates, political action organizations such as trade associations must persuade their constituents to buy in to an agenda and that the organization will be effective in helping its constituents pursue it.

Grass roots initiatives are effective only if they organize into *movements*; widespread sentiment has little effect unless it can be translated into political action. Mediation of public sentiment occurs through political candidacies, more or less permanent trade associations and public interest organizations and through short term *political action committees* (PACs).

Political organizations such as parties and advocacy groups must deal with the difficulties explored by Theodore Olsen's *The Logic of Collective Action*. These complications include the free rider problem, the benefits of cheating, and agency costs.

The *free rider problem* arises because organized groups cannot limit the benefits of their activities to members who pay dues and otherwise support the group's activities by, for example, providing volunteer services. If a farm safety organization persuades OSHA to require devices on farm machinery that reduce the likelihood of an operator losing an arm or a leg because it gets entangled in the machinery, all farm equipment operators will benefit, whether or not they have supported the group's advocacy.

If a microdrone advocacy organization persuades the FAA to make the rules for microdrone moviemaking less onerous, every movie maker who wants to use microdrones benefits, whether or not she is a member of the organization.

The cheating problem arises because the participants in any agreement limiting competition may benefit collectively from the agreed-upon limits to competition, but each of them individually benefits more from violating the limits. Every economist knows that cartels break down over time for just this reason. The same difficulty confronts any private organization trying to set norms for conduct.

Suppose an association of aerial photographers forges an agreement among its members not to fly microdrones for imagery more than 400 ft above the ground or within 100 ft of crowds. The whole group benefits by reducing the prospect of accidents that will receive unfavorable publicity and by assuring that a particular photographer does not suffer competitive disadvantage by flying safely. But once the rules are in place and being adhered to widely, an individual photographer can gain a competitive edge by violating it. He may be able to get unique imagery through a close shot of the victim of an accident by flying lower, or a wider shot of the entire crowd by flying over it.

The *agency problem* arises when the professional leadership of a group uses the group's resources to pursue its own interests rather than the interests of the group membership. For example, the staff director of a drone advocacy group might use her participation in standard-setting discussions to bias the outcome in favor of a particular technology for which she holds a patent. Or, the spokesperson for an anti-drone privacy group might target a particular drone operator for criticism, not because of the actual conduct of that operator, but because of personal antagonism the spokesperson has for the operator.

And, of course, the officers of an organization can use its resources to increase their compensation, build fancy offices, and order limousine service, rather than for mission-related activities.

Fundamentally, the participants in collective action must agree on their agenda. Effective leaders of trade associations and other interest groups understand that, and they work hard to present issues and alternatives for resolving them so that a majority of their members can embrace the association's position. This is not always successful.

6.3.6 *Lawsuits pave the way*

Historically, public policy for new disruptive technologies was informed by litigation. Accidents would happen, victims would sue the users of the new technologies, and judges and juries would craft general common law rules that could eventually be codified into positive law in statutes or agency regulations. As statutes or rules began to emerge, users became subject to enforcement actions and then challenged the legitimacy of the rules on constitutional or other grounds. The debate in legislative arenas would focus significantly on whether the courts were getting it right or getting it wrong.

The litigation tool for developing drone regulation is mostly absent in the early stages of the debate. Some entrepreneurs and industries drawn by the potential of drones might have been expected to test the outer boundaries of the law, especially during 2014, when the FAA was constantly putting off issuance of its in NPRM and had not yet started the section 333 exemption process. That was a perfect time for innovators to provoke disputes by going ahead and doing what they wanted to for commercial purposes, notwithstanding the absence of a firm regulatory framework. But only a handful of disputes arose in 2013, most notably the *Pirker* case and the Texas equine case, discussed in chapter 8. Court dockets were remarkably free of drone cases.

Now that section 333 offers a pathway for legal commercial drone flight and the FAA has promised to finalize its proposed general rule by the end of 2016, few serious operators are eager to be test cases. Most of them think they can live with the limitations proposed in the NPRM. They are sitting back, waiting for generalized permission from the FAA.

6.4 Assessing interests

The political calculus of widespread civilian drone use proceeds from an accurate assessment of the interests involved. Each of the interests considered in the following subsections will shape drone policy and regulation in accordance with that group's passion on the issue and its degree of organization to present its views in each political arena.

6.4.1 *Aviation community*

The aviation community—the most influential interest in aviation rulemaking—is split. Some organizations, particularly those mostly comprising professional pilots, are instinctively opposed. They focus on midair collision risks arising from widespread use of drones, and also fear erosion of job opportunities because drones may replace some manned aircraft. Most pilots do not have career aspirations that include becoming DROPs. On the other hand, many aviation operators, growing in number, recognize that drones provide a new opportunity: a new line of business. Flight schools can offer DROP lessons as well as pilot lessons; aerial photography contractors can fill out their product lines by offering drone photography as well as helicopter photography; and ENG contractors can do likewise.

Other types of operators, however, such as EMS operators and rides-and-tour operators have no opportunity to offer drones in the near-term because they cannot carry passengers. Their concerns about collision risks place them in the opposition.

Aircraft manufacturers, both established ones who supply military drones, and startups, favor relaxation of restrictions on drone use, because the prospect of selling drones represents a new business opportunity. The Aerospace Industries Association, the major trade association of larger aircraft and space manufacturers, released a report in 2015 that advocated wider civilian use of drones.[13]

Airlines spent more than $1 billion on lobbying and political contributions from 1998 to 2015,[14] and airline pilot unions spent more than $2 million in 2014 alone.[15]

6.4.2 Drone communities

Although some overlap exists, advocates for an accommodating legal regime for drone operations are mostly outside the traditional aviation community. Only a handful of established helicopter and airplane operators have robust drone programs. Most of those offering drone services have sprung up independently or are professional photographers who have embraced a new tool. Few airplane and helicopter manufacturers also produce civilian drones; although large defense contractors have dominated the military drone business, and are working energetically to adapt their defense-funded drone technologies to develop products that can penetrate the civilian market for macrodrones. The most significant microdrone manufacturers are almost entirely independent of traditional aircraft manufacturers. The supply side of the market is almost as fragmented as the demand side; what consolidation is occurring mostly reflects acquisition of microdrone manufacturers by firms interested in their use as a tool for established non-aviation activities, such as consumer and commercial photography (GoPro is developing a microdrone product, and Panasonic is teaming with DJI).

The organization of pro-drone voices reflects the novelty of civilian drones and the early passivity or outright opposition by most elements of the aviation community. The most significant drone interest groups with a national footprint are the Association for Unmanned Vehicle Systems International (*AUVSI*), the UAS America Fund, the UAV Coalition, and the Academy of Model Aeronautics (*AMA*). The AMA has the longest history. Formed in 1936, it has a membership of 175,000 people, with 2,500 chartered, local clubs, and has safety rules for its members that largely have been internalized by the FAA and the Congress. It is capable of turning out its members by the thousands to put pressure on legislators and rulemakers, as was evident in its dominance of the rulemaking making record on the FAA's microdrone NPRM. It can mobilize its members relatively easily because, like many of those involved in a other complex hobbies centered on technology, they are passionate about their hobby.

The AMA opposes new drone regulations; it is perfectly satisfied with the status quo. Currently, the FAA follows a forbearance enforcement policy despite the literal application of the FARs to RC model aircraft hobbyists, and the Congress has temporarily withheld FAA authority to regulate model aircraft, including drones. The AMA is far less concerned about expanding commercial uses of drones and, like everyone else, does not really know what to do about the hundreds of thousands of casual users, unaffiliated with an RC model organization.

Logically, the AMA would prefer policies that drive the casual users into its membership ranks. Accordingly, it embraces measures like the Chicago ordinance which would essentially ban recreational drone use except by operators who join the AMA, obtain liability insurance through it, and mostly operate in conjunction with AMA-chartered clubs. The Chicago ordinance is considered more fully in chapter 8.

AUVSI is a well-funded international advocacy group centered in the United States. Presently dominated by drone and drone systems manufacturers, it nevertheless has many members—mostly individuals or small businesses—who are or want to be drone operators, along with major potential customers, such as CNN. Nearly 10,000 people and more than 600 exhibitors attended its 2015 expo in Atlanta The exhibit hall rivaled that of the main helicopter trade association, Helicopter Association International (HAI), in its scope. The organization has every hallmark of long-established aviation groups like HAI. It offers safety, technology, and business opportunity workshops; it has regional chapters, and it is expanding its educational activities with a variety of web-based tutorial videos, text articles, blogs, and webinars.

AUVSI is actively working at reshaping general public opinion to appreciate the benefits that civilian drones can provide and to gain a more nuanced understanding of their risks. The participation of senior FAA policymakers and senior members of Congress in AUVSI activities reinforces the conclusion that it is well-connected where it counts.

UAS America Fund has narrower scope than AUVSI, concentrating more on supporting particular activities that prove concepts, and by generating intellectual capital for policymaking. Its detailed proposal for a micro-UAS carve out from the general regulatory regime, analyzed in chapter 8, is a case in point. It submitted a formal petition for rulemaking that proposed such an approach just before the FAA released its NPRM. Its intervention with the

Office of Management and Budget (OMB) and the White House was instrumental in ensuring the FAA's careful attention to the proposal and in the NPRM. It regularly works with the FAA leadership, other executive branch policymakers, and with key congressional staff members.

The Small UAV Coalition characterizes itself as a coalition of "leading technology companies to pave the way for commercial, philanthropic, and civil use of small UAVs in the US and abroad." Its fifteen full members include DJI, 3DR, Parrot, Amazon, Google, GoPro, and the agricultural and survey sensor and analytical software vendor PrecisionHawk. Unlike AUVSI, it is not a mass-membership organization. Increasingly, conflicting priorities of its members make it difficult for it to speak in a clear voice or to offer concrete proposals.

6.4.3 Agriculture

Evidence supports the proposition that the agricultural community favors a regulatory environment that is hospitable to drones—both microdrones and macrodrones—to inspect crops and to apply chemicals to them. Because drone operations almost certainly will occur over land owned by the operator or by someone the landowner contracts with, as crop-dusting occurs now, farmers have few privacy or trespass concerns.

Some farmers shrug when the subject of drones comes up. "What can they do for me?" they ask skeptically. Until they hear a convincing answer, some parts of the agricultural community will stay on the sidelines of the public policy debate and will not change whom they vote for based on their position on drones.

But attitudes are shifting in favor of drones. Workshops and presentations by the National Farm Bureau characterize drones as the next big thing in agriculture and criticize FAA delay in approving their use to support agriculture. The Farm Bureau testified in favor of drone deployment in 2015 before the Senate Commerce Committee, and agriculture support is among the top three uses allowed in the FAA's grant of section 333 exemptions.

Agricultural interests spend substantial amounts of money on lobbying and political contributions.[16]

6.4.4 Public safety community

The law enforcement community favors flexibility for public safety agencies to be able to use drones, but tends to be opposed to wide public use of drones. They have no difficulty understanding how a microdrone can be helpful in enforcing a perimeter for a live shooter incident, conducting surveillance over a suspect, or enhancing accident reconstruction.

Surveillance of the police is another matter. The experience in Ferguson, Missouri is illustrative, where local law enforcement induced the FAA to impose a Temporary Flight Restriction to keep news helicopters out. Similarly, in Illinois, law enforcement interests ferociously opposed a lifting of a state ban on cell-phone photography of arrests at police stops. Law enforcement interests can be expected to oppose any use of drones to monitor law enforcement activities.

The police also may have to enforce regulations of drone use. They are likely to favor clarification of the rules for drone use under federal, state, and local law—as long as it preserves a considerable range of discretion for those who enforce it.

6.4.5 *Real estate brokers*

Real estate brokers are on record as favoring wider use of drones to capture aerial imagery of properties that are for sale. No obvious reason exists for them to dilute their advocacy. The National Association of Realtors endorsed the 2015 NPRM, and has encouraged its members to seek section 333 exemptions before using drone imagery to enhance their marketing efforts. Real estate photography is among the most common uses approved by the FAA in section 333 exemptions, and many users in this industry fly drones without regard to FAA restrictions.

The real estate industry spent more than $1 billion on lobbying and political contributions from 1998 to 2015.[17]

6.4.6 *News organizations*

Various associations of broadcasters, newspapers, and journalists have been active in the public policy debate advocating elimination of restrictions on press and media use of drones to collect newsworthy imagery. Indeed, they argue that they have a First Amendment privilege to engage in such use.

The National Association of Broadcasters, the Radio and Television Digital News Association, and other press organizations, filed petitions and comments with the FAA encouraging broader privileges to fly ENG drones. CNN has taken the lead on deploying drones and is participating in an FAA sponsored R&D effort to explore mitigating measures that would permit BLOS operations and operations over people.

Commercial TV and radio stations spent $31.7 million on lobbying in 2014 alone.[18]

6.4.7 *Package delivery services*

Amazon and Google dominate the e-commerce marketplace. Their size makes them concentrated interests unto themselves. Both of them have been aggressive in exploring the possibility for delivering small packages by drones. Amazon and Google also are significant pioneers in technologies and not only have the resources to influence the public discourse about drones—think of the reach of Jeff Bezos' interview with *60 Minutes* in November 2013—but also have the resources and the talent to do significant demonstration programs of technologies on their own. For example, both of them could deploy and demonstrate the feasibility of neighborhood navigation discussed in chapter 3's NAMID concept.

Both companies have been clear that they favor wider deployment of drones, and no reason is apparent why either of them would temper that position in the future. As the dialogue matures, they may be joined by others who have small packages to deliver, such as Domino's Pizza, or a startup, "Tacocopter," which plans a drone taco delivery service.

Amazon has been particularly active in the political area, pointing out that the FAA's belated approval of its application was overtaken by technology, emphasizing that it had to go overseas to test its ideas, and joining the FAA in an experimental and demonstration program for package delivery. The image of drones delivering packages from Amazon fascinates the public, and thus is likely to influence the political debate beyond its objective merits.

6.4.8 Utilities

Electric power distribution enterprises, railroads, and pipeline operators already conduct aerial surveillance that can be conducted in part by drones. The remoteness of much of their infrastructure makes them prime candidates for early relaxation of rules restricting drone flight.

On the other hand, they may be skeptical about the ability of drones to do the job, especially as they consult their established aviation departments and aerial surveillance contractors.

Powerline and pipeline inspection is among the most common uses approved by the FAA in its growing inventory of section 333 exemptions.

All of these industries are concentrated and sophisticated in making their views heard in political arenas. Electricity and gas utilities,[19] and railroads[20] spent more than $2 billion between them on lobbying and political contributions from 1989 to 2015.

6.4.9 RC hobbyists

Radio-controlled model aircraft hobbyists comprise one of the most active and influential interest groups in the aviation field. Their numbers are large, they are passionate about their hobby, and they strongly resist governmental restrictions. Nearly half of the 4,000 or so comments filed on the FAA's sUAS NPRM were from RC hobbyists opposing any FAA regulation of their recreational flying activities.

6.4.10 Video entertainment

The first industry to obtain section 333 exemptions for commercial microdrone flight was the motion picture and television production industry. Hollywood is influential because it has regular access to the eyes of the entire citizenry. Most consumers of video entertainment are far more interested in the content than how it was made. Nevertheless, as more movies and television programs are made with drones instead of traditional tools of cinematography, the phenomenon is likely to capture some favorable public attention and shift the debate in favor of some forms of commercial drone use.

6.4.11 Anti-war and anti-government groups

Pacifist groups and those with a conspiratorial view of government initiatives of any kind, while smaller in numbers, are quite passionate in their advocacy. Their staunch opposition to military use of drones makes them largely immune to objective arguments to distinguish military and civilian uses.

It may be, however, that some of these groups on the right could be convinced to favor drones for border enforcement against entry of illegal aliens.

6.4.12 National security groups

Groups concerned about national security are be split. They favor drone use to monitor infrastructure and other critical facilities for possible terrorist attacks, but they also are concerned about drones falling into the hands of actual and potential terrorists. They fuel growing concerns about hackers and open-source coders, defeating safety features, seizing control of drones, or blocking control links, thus posing threats to security, privacy, and rule compliance.

They emphasize the possibilities for weaponizing even small drones, and press for drone interdiction technologies.

6.5 Arenas

6.5.1 Legislative

The United States Congress has consistently prodded the FAA to act more quickly to integrate drones into the national airspace system. The FAA Revitalization and Reform Act of 2012 imposed deadlines for such integration and authorized the FAA to grant exemptions and issue special rules while a final general rule is being developed. The FAA must be reauthorized by October 1, 2015, and development of legislation to do that provides an opportunity for drone advocates. The chairman of the House Aviation Subcommittee, Frank Lobiondo, and Cory Booker, a member of the Senate Aviation Subcommittee, have been prominent advocates of wider deployment of drones.

6.5.2 Government agencies

The FAA already has drones on this agenda. The primary arena for formulating drone policy is the FAA's rulemaking process, begun in earnest with the 2015 NPRM for microdrones. Some 4,000 comments were filed on the proposal, most of them favoring the agency's general approach. The section 333 exemption process represents another arena. More than 5,000 exemptions have been granted, and thousands of petitions for exemption are pending.

Of course it has other things on its agenda as well, but the FAA is paying close attention to voices of the interests that it traditionally has cared about. All administrative agencies suffer some degree of regulatory capture, because the groups that they regulate have such a strong incentive to forge close relationships with agency personnel. This means that the FAA will instinctively pay more attention to what aviation interests say about the drone issue than others with which it has not dealt with much in the past, such as agricultural interests and real estate brokers. Indeed the FAA usually operates in reliance on some kind of a consensus forged by different aviation interests working out their differences through public or private advisory committees.

6.5.3 State and local government

In local legislative arenas—state legislatures and municipal city councils—the concentrated voices will be the drone opponents, and the drone adherents will have to scramble even to know when an anti-drone proposal is on the agenda.

On September 9, 2015, California Governor Jerry Brown vetoed Senate Bill 142, which would have made it a trespass to fly less than 350 ft over private property without permission from the landowner. The drone industry had lobbied fiercely, but unsuccessfully, to prevent passage. It was more persuasive with Governor Brown, whose veto message said that the statute "could expose the occasional hobbyist and the FAA-approved commercial user alike to burdensome litigation and new causes of action."[21]

State and municipal government regulation of aviation is, for the most part, preempted by federal law. Nevertheless, as chapter 8 explains, the Supremacy Clause of the United States Constitution allows limited state and municipal measures on matters of traditional state or local concern such as control of private property, redress for personal injuries, and protection of intangible personal interests such as privacy.

Regulation by state legislatures, by municipal legislative bodies, or by administrative agencies at the state and local level alters the political dynamics. Pro-drone interests are more likely to be concentrated at the federal level and diffuse at the local level. Drone opponents are more likely to be concentrated at the local level and diffuse at the national level. This means that, if a city council, a zoning board, or a park district, has jurisdiction to adopt some kind of drone regulation, the politics is likely to lead to more restrictive measures than if the United States Congress or the FAA were to consider the same proposal.

Accordingly, pro-drone interests are likely to argue federal preemption to deflect state or local initiatives, while anti-drone interests are likely to navigate their way around preemption by carefully targeting their proposals.

Chapter 8 provides a list of state and local legislation already adopted or in the advanced proposal stage.

6.6 Political action

The challenge for opponents or advocates is to construct coalitions of interests that oppose wider civilian use of drones, or that favor wider civilian use, depending on their objective. Which existing proponents could be persuaded to raise the priority they give to drones and put more resources into their advocacy efforts? Which groups with significant influence are sitting on the sidelines, and what arguments can be made to get their support? Are there any concessions that can be made to groups in opposition to weaken their advocacy against? What promising political candidates can be persuaded to put drones on their agendas? What kinds of advertising might help bring significant segments of the general public around to positions of support? What should their content be? What are the most effective channels? Radio? Television? Explicit advertising or earned news or feature coverage? What markets are most important in terms of the political calculus, recalling that an upsurge of public support in particular districts can alter the position of elected representatives from that district.

What can drone operators be encouraged to do that would attract public support? For example, an operator with a suitably equipped microdrone might be encouraged to assist in a search and rescue effort for a lost person, or respond to a natural disaster—including drone-related publicity stunts.

Is it feasible to organize a quick response capability so that adverse press or media stories can be countered quickly and favorable stories spread further?

How can the commercial marketplace and the political arena be interlinked most effectively? Can visible segments of commerce be persuaded to introduce drones and report favorable experiences in the public arena? What suggestions for political activism might commercial users or aspirants embrace?

What kinds of outreach to skeptical groups whose positions have not gelled yet might be effective in bringing them around into positions of support? For example, it is reasonable to believe that many aviation groups, certainly including commercial operators, and maybe some groups of pilots, may be open to persuasion. The operators could be encouraged to recognize that drones represent a new line of business instead of merely threatening existing lines of business. Some pilots might be encouraged to view drones as opening up a career option as they reach retirement age or have difficulty maintaining their medical certification. Early-career pilots in training can be linked up with drone operators who need rated pilots to service as DROPs.

6.7 Determining the content of positions

It is not enough to ask whether someone is for or against civilian drones. Civilian drone use is not an ideology. Anyone interested in the subject of drones will favor them in some circumstances and oppose them in other circumstances. A pilots' group may have its concerns about the risk of midair collisions sufficiently mitigated by a height restriction. A drone manufacturer may be persuaded to support a regulatory approach that imposes little aircraft certification cost. Open architecture for drone telemetry and a tail number registration system may alleviate some of the concerns of privacy groups. Line of sight restrictions may undermine support by utilities. RC model hobbyists may support restrictions if they offer an exemption for organized RC model club activities.

6.8 Impact of accidents

Aviation accidents happen. When they do, public response often reshapes regulation. The collision of two airliners over the Grand Canyon in 1956 and a later collision of an airliner and a military fighter incited a public uproar and caused the Congress to enact the Federal Aviation Act of 1958, replacing the Civil Aeronautics Authority with the FAA, and merging air traffic control of civilian and military aircraft. The Colgan Air crash near Buffalo in 2009 caused the FAA to tighten training and aeronautical experience requirements for airline pilots.

As drones proliferate, they will be involved in accidents, regardless of the regulatory regime they operate under. When this occurs, the public will react, and there will be calls for additional drone regulation. What result is produced will depend as much on the pre-existing agendas of interest groups as on technical analysis of the accident.

6.9 Law should lag technology

Law is often criticized for being behind technology. That is not a weakness; it is a strength. For law to be ahead of technology stifles innovation. What is legal depends on guesses lawmakers have made about the most promising directions of technological development. Those guesses are rarely correct. Law should follow technology, because only if it does so will it be able to play its most appropriate role of filling in gaps and correcting the directions of other societal forces that shape behavior: economics, social pressure embedded in the culture, and private lawsuits.

Here's how law should work: a new technology is developed. A few bold entrepreneurs take it up and build it into their business plans. In some cases it will be successful and spread; in most cases it will not. The technologies that spread will impact other economic players. It will threaten to erode their market shares; it will confront them with the choice of new technology if they wish to remain viable businesses; it will goad them into seeking innovations in their legacy technologies.

The new technology will probably cause accidents, injuring and killing some of its users, and injuring the property and persons of bystanders. Widespread use of the technology also will have adverse effects on other, intangible interests, such as privacy and intellectual property. Those suffering injury will seek compensation from those using the technology and try to get them to stop using it.

Most of these disputes will be resolved privately without recourse to governmental institutions of any kind. Some of them will find their way to court. Lawyers will have little difficulty framing the disputes in terms of well-established rights, duties, privileges, powers, and liabilities. The courts will hear the cases, with lawyers on opposing sides presenting creative arguments as to how the law should be understood in light of the new technology. Judicial decisions will result, carefully explaining where the new technology fits within long-accepted legal principles.

Law professors, journalists, and interest groups will write about the judicial opinions, and gradually, conflicting views will crystallize as to whether the judge-interpreted law is correct for channeling the technology's benefits and costs. Eventually, if the matter has sufficient political traction, someone will propose a bill in a city council, state legislature, or the United States Congress. Alternately, an administrative agency will issue a notice of proposed rulemaking, and a debate over codification of legal principles will begin.

This is a protracted, complex, unpredictable process, and that may make it seem undesirable. But it is beneficial, because the kind of interplay that results from a process like this produces good law. It is the only way to test legal ideas thoroughly and assess their fit with the actual costs and benefits of technology as it is actually deployed in a market economy.

Waiting to use civilian drones until the FAA is able to crystallize guesses about what problems they may cause is exactly the wrong approach. Far better would be to carve out a space in which people can do whatever they want with drones and develop a body of experience and data about what their real benefits and costs are. Then and only then is it time to write regulations.

Of course the space within which this freewheeling market-driven experimentation can take place should be selected so that, on the one hand, it is big enough for a genuine body of experience to develop, but on the other hand so that catastrophes with great cost of human life and property are unlikely to occur.

Notes

1 See, e.g., ACLU (American Civil Liberties Union), *Domestic Drones*, (2015), www.aclu.org/issues/privacy-technology/surveillance-technologies/domestic-drones (expressing concerns about domestic drone use impact on privacy).

2 Ipsos Poll Conducted for Reuters (Jan. 29, 2015), www.reuters.com/article/us-usa-drones-poll-idUSKBN0L91EE20150205.
3 Samuel D. Warren & Louis D. Brandeis, *The Right to Privacy*, 4 *Harv. L. Rev.* 193, 195 (1890).
4 V. O. Key, *Politics, Parties, and Pressure Groups*, (1964) (analyzing empirical data on role of interest groups in electoral and legislative politics).
5 James Q. Wilson, *Political Organizations* (1995) (explaining the essential role of political entrepreneurship in activities of trade associations and other advocacy groups); Jeff Fishel, *Ambition and the Political Vocation: Congressional Challengers in American Politics*, 33 *J. Politics* 25, 34 (1971) (analyzing the motivation of political challengers).
6 Ted Poe, *The Spying Drone Over a Virginia Neighborhood*, 159 *Cong. Rec.* H3294, 13th Cong., 1st Sess. (Jun. 12, 2013), www.congress.gov/congressional-record/2013/6/12/house-section/article/H3294-1 (civilian drones used inappropriately).
7 ALPA, *When You Fly…We Can't*, (Oct. 1, 2015), www.alpa.org/news-and-events/air-line-pilot-magazine/2015/102015/when-you-fly (statement by airline pilots' union emphasizing dangers posed by drones).
8 Rep. Frank LoBiondo, *Unmanned Aerial Systems: Let American Innovation Take-Off*, www.lobiondo.house.gov/unmanned-aerial-systems-let-american-innovation-take.
9 Henry H. Perritt, Jr. & Eliot O. Sprague, *Drones*, 17 *Vanderbilt J. Ent. & Tech. L.* 673, (2014) (explaining importance of kinetic energy in crash damage, proportional to weight).
10 FAA, Certification of Aircraft and Airmen for the Operation of Light-Sport Aircraft, 67 Fed.Reg. 5368 (Feb. 5, 2002).
11 FAA, Certification of Aircraft and Airmen for the Operation of Light-Sport Aircraft, 69 FR 44772 (Jul. 27, 2004).
12 Richard H. Pildes, *Romanticizing Democracy, Political Fragmentation, and the Decline of American Government*, 124 *Yale L.J.* 804, 828 (2014) (explaining the role of political parties in aggregating interests); Nathaniel Persily, *Toward a Functional Defense of Political Party Autonomy*, 76 *N.Y.U. L. Rev.* 750, 793 (2001) (identifying aggregation of interests as one of four critical functions of political parties); John H. Aldrich, *Why Parties?: The Origin and Transformation of Political Parties in America* at 10 (1995) (explaining political party challenge in aggregating diverse interests).
13 AIA, *Unmanned Aircraft Systems: Myth vs. Reality*, (Jun. 25, 2015), www.aia-aerospace.org/research_reports/unmanned_aircraft_systems_myth_vs_reality/
14 www.opensecrets.org/lobby/top.php?indexType=i.
15 www.opensecrets.org/lobby/induscode.php?id=LT100&year=2014.
16 Taxpayers for Commonsense, *Political Power of the Agribusiness & Crop Insurance Lobbies*, www.taxpayer.net/images/uploads/downloads/Political_Power_of_Farm_And_Crop_Insurance_Lobbies_Fact_Sheet.pdf.
17 The real estate industry spent $1.2 billion on lobbying from 1998 to 2015. www.opensecrets.org/lobby/top.php?indexType=i.
18 www.opensecrets.org/lobby/induscode.php?id=C2100&year=2014.
19 Electric utilities spent $2 billion on lobbying from 1998–2015. www.opensecrets.org/lobby/top.php?indexType=i.
20 Railroads spent $34.3 million on lobbying in 2014. www.opensecrets.org/lobby/indusclient.php?id=M04&year=2014.
21 Office of the Governor, *To the Members of the California State Senate* (Sep. 9, 2015) (veto message), www.gov.ca.gov/docs/SB_142_Veto_Message.pdf.

7 Regulation around the world

7.1 Introduction

Regulation of drones involves specific application of a *rule of law*. The institutional arrangements to deliver a rule of law vary widely among states committed to a rule of law. But all of them proceed from the premise that governmental decision makers must be held accountable and that any governmental action against citizens must be subject to challenge and scrutiny through some kind of independent, formal, review.

Rule of law does not mean immunity from political influence. Indeed, international promoters of political reform usually talk about rule of law and democratization in the same breath. Democratic values insist upon mechanisms to ensure that every governmental entity is subject to political accountability, ultimately to elected representatives or some other representative of the will of the people. Different systems and different political theories ensure political accountability in distinctive ways. In the West, the ultimate authority must be subject to popular election. This includes both the executive and the legislature, in presidential systems; and the legislature, in parliamentary systems, which selects the executive. In party-dominated systems, typified now mainly by China, political accountability ultimately is to the party, which enjoys legitimacy, not through an electoral process, but from its revolutionary success mobilized by the people (according to Marxist theory).

In Europe, British Commonwealth countries, the United States, Japan, and India, regulators are appointed directly or indirectly by executive branch officials. Judicial review of their decisions makes agency activities visible and therefore subject to legislative revision.

In the United States, political control is sought by allowing political appointment of a relatively deeper layer of senior agency officials outside the civil service than is typical in other democratic systems. In party systems, political accountability is assured by a party structure that parallels the administration, geographically, and topically.

This chapter begins by explaining why regulatory agencies exist, identifies the hallmarks of such agencies, and explains how they fit in the traditional tripartite grouping of governmental institutions: legislative, executive, and judicial. It explains general concepts of administrative law that provide an analytical framework for understanding the institutional structure for drone regulation in particular states. Then it provides a brief overview of drone regulation in major countries, identifying the agency responsible for drone regulation, the mechanisms for political accountability and judicial review, and the philosophy and content of existing drone regulation.

7.2 Administrative law

Drone rules are made mostly by administrative agencies acting under legislative authority delegated to them by elected legislatures. Administrative law determines whether the content of the rules is permissible and whether the procedures used to develop the rules were adequate. In most systems of administrative law, courts—either the regular courts or specialized ones—determine whether an aviation agency has satisfied the requirements of administrative law.

7.2.1 Theories of administrative law

Administrative law systems around the world follow three models for judicial review of agency decisions: the English model, the French model, or the German model. Variations in socialist legal systems are considered in § 7.4. In the English model, agency decisions are reviewed by the regular courts under the same rights, powers, and privileges doctrines applicable to purely private disputes. In the French model, agency decisions are reviewed by a special tribunal that is part of the administration under public law doctrines that align rights, powers, and privileges differently from private law. In the German model, agency decisions are reviewed by the separate judicial branch, but not by the regular courts. Review is the responsibility of specialized administrative courts within the judicial branch. The German model focuses on constitutionally enshrined individual rights.

Understanding these models requires examining the history of the legal systems that produced them—long before anyone was thinking about airplanes, helicopters, or drones.

The differences between the English and French models result from the different dynamics of political struggles between the crown and regionally based elites in France and England. In England, the genius of the Norman conquest of 1066 was that the crown left a decentralized system of lawmaking and law enforcement intact while replacing Saxon barons with Norman ones. While much political conflict over the following nine centuries grew out of struggles between the crown and the barony (think of the Magna Carta in 1215), disputes over the exercise of power usually sought to rein in the exercise of local authority, where most control resided. Accordingly, the common law courts in England were royal institutions, intended to provide central-government recourse to abuses by local barons. The phrase *common law* referred, not to the sources of legal doctrine, but to the premise that it was common to all of England. Central–local political tension resulted in sharply circumscribing the jurisdiction of these royal courts. It also was the case, even as common law courts became more independent of the crown, that legal actions to review governmental acts were shoehorned into the existing categories for private disputes: if a bailiff beat you up, your recourse was a common law action for battery. If he took your property illegally, your recourse was a common law action for conversion. If he burst into your house, you could sue him for trespass to land.

Prerogative, specialized, courts such the Star Chamber established by the Stuart kings were short-lived. In the run-up to the English Revolution, the 1610 Case of Proclamations established the superiority of the common law courts over the crown.

Judicial review in France evolved differently. There, rather than a centuries-long dynamic involving a flow of power back-and-forth between the crown and the barony; the course of French political development, until the French Revolution in 1789, was a unidimensional progression of efforts by the crown to centralize authority. The *parliaments*, the regular courts, were controlled not by the crown, but by the barons. They regularly nullified royal action, and so French monarchs sought to set up a central machinery that would have exclusive jurisdiction to review governmental decisions. The result was the *Conseil de Roi*, transformed by Napoleon into the *Conseil d'Etat*. It was a part of the administration—the executive branch—rather than the judicial branch. Its role was to superintend administration by state officials to make sure they implemented policy decisions made by emperor (during Napoleon's time) or by the popularly elected parliament (after his time) and, later, to make sure they did not infringe individual rights. The two concerns resulted in the crystallization of a separate body of law separate from that used to resolve private disputes, known as *public law*.

Important differences also exist with respect to staffing of the three systems. In England, lawyers become members of the bar through apprenticeship-like professional training working under existing members of the bar. University training was a late arrival. Judges are recruited almost entirely from the practicing bar. English lawyers and judges pride themselves on being generalists.

French training of members of the Conseil d'Etat is dramatically different. Almost all members graduate from the Ecole d'Administration, the only school for training officers of public administration. Even after they take up judicial duties on the Conseil d'Etat, they typically also have other duties as public officers. They are elite members of the bureaucracy and adherents to a legal culture separate from that of judges in the regular courts.

In Germany, judges on the specialized administrative courts are judges who have been through the process for training all German judges. In that respect, Germany resembles England. But German law students specialize early, choosing to become private practitioners, judicial officers, or public prosecutors.

One should not exaggerate the effect of different theoretical traditions, however, the different systems are more similar than different.

Which theoretical model prevails depends on how globalization and development occurred in the 19th and 20th centuries. States that were part of the British Empire follow the English model. States that were subjected to the Napoleonic conquest or French colonialization follow the French model. The German model represents a hybrid alternative. Post World War II American dominance implanted aspects of the English model in Japan and Germany.

As with so many other areas of law, globalization and the human rights movement have resulted in considerable convergence. As §§ 7.5–7.7 explain, drone regulation is converging as well. Over time, the content of drone regulation will be similar from nation to nation, except where explicit policy reasons justify differences. The systems for producing them already are similar.

7.2.2 The role of administrative agencies

An important reality in complex, modern, states and economies is that the quantity and sophistication of governmental decisions preclude relying on direct democracy and the electrical process to make all of the necessary governmental decisions. Some degree of specialization and delegation of authority is necessary. It is hard to imagine the British Parliament, the United States Congress, or the German Bundestag having the time or the expertise to write the hundreds of pages of rules or to make the hundreds of enforcement decisions that the British, US, and German aviation agencies do. Everywhere in the developed world, administrative agencies distinct from legislatures, executives, and courts make rules and enforce them.

As governmental power has gravitated to such agencies, legal systems have developed bodies of *administrative law* to hold agency decision-makers accountable to the political process. Consistently, administrative law insists that bodies with rulemaking and enforcement power act only within powers explicitly delegated to them by top-level governmental authority. Second, top agency officials must be appointed and removed in a manner that preserves their direct or indirect accountability to a chief executive who herself is either directly elected (in a presidential system), accountable to an elected legislature (in a parliamentary system), or answers to top party bodies (in a party-dominated system). Then, decisions by the administrative agency must be subject to review through some kind of judicial or quasi-judicial mechanism.

It is popular to complain about "bureaucracy," by which the complainers usually mean a complex set of rules that often prevent them from doing what they want. But any modern state and economy must bureaucratize in order to have a rule of law. Bureaucracy means the channeling of decision-making through formal organizational structures and circumscribing it by rules. Bureaucracy exists in both public and private sectors. Anyone who has ever worked for a large for-profit or nonprofit corporation knows that the number, complexity, and details of rules that come from private authority are as great as those that come from the Federal Aviation Administration.

Bureaucracy provides several benefits. Its rules make decision-making at lower levels more efficient, because decision-makers do not have to solve every problem and decide the appropriate approach to it *de novo* (with nothing to start with). It allows central authority to propagate its policy decisions down through multiple levels of subordinates. It prevents lower level personnel from expending too much or most of their energy on jurisdictional battles and sorting out conflicting decisions, because the rules define areas of exclusive jurisdiction for each.

7.2.3 Elements of administrative procedure

Despite differences and details, several realities and principles are universal in aviation regulation. In the United States basic principles are codified in the federal Administrative Procedure Act.[1] Similar statutory frameworks exist in other countries at the national and state or provincial level. The frameworks require some form of opportunity for interested parties and the general public to participate in the formulation of rules that have the force of law—resembling, in that regard, statutes adopted by the legislature. In the United States, the norm is for the agency to publish a draft rule in a *notice of proposed rulemaking* (NPRM) and allow at 30 to 60 days for interested parties and the public to submit comments criticizing, supporting, or suggesting changes in the draft. The FAA's February 2015 NPRM for microdrones is an example. When it publishes its final rule, the agency must demonstrate that it has taken into account the comments it received. If it fails to show that, or fails to offer a rational explanation for deciding contrary to the weight of the comments, its final rule may be overturned by judicial or other reviewing authorities.

When agencies apply rules to particular circumstances, a process known as *adjudication* in the United States, they must afford the basics of adjudicative procedural due process. At minimum, that requires notice and an opportunity to be heard before a neutral decision-maker, and typically also involves an opportunity to present evidence, to be represented by a lawyer, to have a decision based exclusively on the formal record developed in the hearing, and to have a reasoned explanation for the ultimate decision including fact-finding based on evidence in the record and conclusions of how the law applies to the facts thus found. The highly publicized *Pirker* case in the US is an example. Raphael Pirker was assessed civil penalties by the FAA in an enforcement action for flying his drone recklessly on the University of Virginia campus. He demanded and received a hearing. The administrative law judge dismissed the action, finding that drones were not aircraft under US aviation law. The FAA appealed to the National Transportation Safety Board, which overturned the lower decision, and reinstated the civil penalties. The Pirker decision is analyzed in greater depth in chapter 8.

Most systems also provide a mechanism for reviewing agency decisions, either through the regular courts, through specialized courts, or by high-level administrative bodies. In common law countries like the British Commonwealth and the United States, the regular courts conduct judicial review. In France, a chain of judicial review ending up with the Conseil d'Etat performs similar functions. Judicial review typically is supplemented by hierarchical review within larger agencies.

This basic approach applies to governmental regulation of drones, just like it applies to governmental regulation of anything. A government official may not legally punish someone for operating a drone or tell him to stop unless the punishment or order is backed up by administrative due process. Neither rules nor enforcement actions are valid unless they are within an agency's delegated authority, unless they are rationally related to a legitimate objective within the agency's authority, and unless the required process has been followed.

For judicial review to be meaningful, the reviewing bodies must have standards against which to measure administrative agency decision-making. The standards focus on procedure as much as substance—as long as the substance stays within the boundaries of the rulemaking authority delegated to the administrative body.

The details of aviation regulatory agencies vary considerably from state to state. Some agencies are more specialized than others. In some countries the drone regulatory agency has authority only over aviation; in others its authority extends to transportation generally; in a few it regulates all commercial activities.

Some drone regulatory agencies are more independent than others. The agency may be a separate department or ministry within the executive branch. It may be an agency separate from the executive. It may be a distinct department within an executive department of ministry. It may simply comprise specialized personnel within an organizational component with broader responsibilities.

Then, there is the matter of procedure. Most administrative law systems distinguish between adjudication and rulemaking. More formal procedures, including live hearings, are typically available in the adjudication process. Less formal practices, such as paper processes for public comment on proposed rules, typify rulemaking processes.

Some agencies have mechanisms for persons outside the agency formally to propose a new rule or a modification of an existing rule. If the agency declines to act on a proposal, the proponent may be able to go to court review the agency's failure to act. In other systems, a private rulemaking initiative must be pressed informally, through the political process.

When the agency decides that a new rule or modification of existing rule is appropriate, it usually must give notice to the public. This occurs through a centralized register of some kind in many systems; in others, the agency gives public notice itself.

Not surprisingly, details for how interested persons submit comments on an agency proposal vary. In some places they send the comments directly to the agency; in others, such as the United States, they post them electronically to a government-wide Internet server.

Sometimes the agency holds a live hearing on a proposed rule. Usually, the agency has discretion whether to hold a hearing or to proceed on written documents only. If a hearing occurs, pre-established rules of procedure and rules of evidence apply.

Once a rule is finalized and promulgated, agencies must justify its content, usually by discussing the consideration it gave to comments that were filed. How much explicit analysis of filed comments is required varies considerably.

Once the rule is promulgated, some mechanism always exists for obtaining review by an independent body—a court or a higher level administrative body. That review encompasses the rule's validity, the justification for it, and the procedures used to develop it, under varying levels of deference to the agency as the primary decision-maker.

When a rule is enforced by the imposition of penalties, its validity often can be challenged in the enforcement proceedings. That is what Pirker did in the US drone enforcement proceeding.

Agency rules may be enforced by the regular police or only by agency inspectors and enforcement agents.

7.3 Models for drone regulation

The following subsections provide an overview of drone regulation—both in structure and content—in major developed countries with active aviation industries. Regulation in the United States is considered separately in chapter 8.

7.3.1 Australia

Australia, although a member of the British Commonwealth, is an independent sovereign. Australia's constitution originated in the Commonwealth of Australia Constitution Act 1900, enacted by the British Parliament. The Statute of Westminster,[2] the Statute of Westminster Adoption Act 1942, and the Australia Act 1986 established Australia as sovereign, independent of the British Crown.

The Australian constitution derives from the Commonwealth of Australia Constitution Act, enacted by the Parliament of the United Kingdom on July 9, 1900.[3] Under it, the Australian Parliament has plenary legislative power.[4] Judicial review of all administrative-agency decisions is presumptively available under the ADJR Act, section 39B of the Judiciary Act of 1903, or under section 75(v) of the Constitution.[5] Some agency decisions are reviewable by the Administrative Appeals Tribunal, but only if the agency's substantive legislation so provides.[6] The Australian legal tradition distinguishes between judicial review and merits review. Merits review, the province of the AAT, is broader, as a number of aviation cases demonstrate.[7]

The federal government has the power to regulate interstate aviation under the "trade and commerce" clause of the constitution.[8] It also has the incidental power to regulate intrastate aviation to the extent necessary to protect the safety of, or to prevent interference with, interstate or foreign air navigation.[9] Aviation in Australia is regulated by the Civil Aviation Safety Authority (CASA).[10]

CASA regulates drones under subpart 101F of the Civil Aviation Safety Regulations (CASR).[11] The regulations apply only to large UAVs and to small UAVs operated for "purposes other than sport or recreation."[12] The regulations classify[13] drones into three categories: *large*, *small* and *micro*. A large UAV is defined as a fixed-wing configuration weighing more than 150 kg (330 pounds) or a rotary wing configuration weighing more than 100 kg (220 pounds).[14] A micro UAV weighs less than 100 g (0.22 pounds).[15] A small UAV weighs between 100 g and 100 or 150 kg.

Micro UAVs are exempt from regulation.[16] Large UAVs must have special airworthiness certificates,[17] but small UAVs are not subject to airworthiness certification.[18]

The rules provide for *UAV controller*[19] and *UAV operator*[20] certificates. A DROP is a UAV controller; the organization that is responsible for operating the drone is a UAV operator. Controller certificates are required only for operation of large UAVs. UAV controller applicants must qualify for radio operator certificates, pass aviation and instrument rating knowledge tests, complete a training course conducted by the UAV manufacturer, and have at least five hours experience in operating UAVs.[21] Small UAVs may be operated by anyone.

Operation of either large or small UAVs requires an operator's certificate.[22] Certificated operators must meet requirements for organization, facilities, personnel, and procedures.[23] The requirements are not trivial. An operator that operates more than one UAV may not be certified, for example, unless the operator has a full-time chief controller.[24]

UAVs may not be operated above 400 ft AGL except in approved areas.[25] Anyone may apply to CASA to designate approved areas for UAV operations. Approved areas are subject to limitations to protect air navigation and may be granted for specific flights or indefinitely.[26]

Operating limitations also prohibit operations closer than 30 meters (approximately 100 ft) to a person not involved in the operation.[27] Small UAVs may be operated only within approved areas or above 400 ft AGL outside populous areas with CASA approval.[28] Large UAVs may be operated only with CASA approval, which must be granted if the operator is certificated and the operation does not contravene the terms of the certification.[29] CASA approval for large UAV operations may be conditioned on requirements for daytime-only operations area limitations, and radio advisory broadcasts.[30]

UAVs may not be operated "over a populous area at a height less than the height from which, if any of its components fails, it would be able to clear the area."[31] The CASA may waive this restriction as to certificated UAVs, considering the redundancy of the drone's critical systems, the fail-safe design characteristics of the vehicle, and the security of its communication and navigation systems.[32] Operations in controlled airspace require maintaining a listening watch and periodic broadcasts on ATC frequencies.[33]

7.3.2 Brazil

The Brazilian regulatory authority is the National Agency for Civilian Aviation—Agência Nacional de Aviação Civil (ANAC), a component of the Brazilian Secretariat of Civil Aviation.[34] Judicial review of ANAC decisions is available in the regular courts.[35]

The Brazilian approach to drone regulations resembles the early United States: drone flight—by any size drones—is permitted only in geographic areas approved in advance by the ANAC. The agency is struggling to define appropriate regulation for drone flight out of the sight of the DROP. DROPs in Brazil must have "pilot's licenses."

A 2012 presentation by the agency projected the availability of special drone flight authorization in segregated areas by 2013, more general regulations for licensing, operating rules, and aircraft certification by 2014, and type certification by 2017.[36]

7.3.3 Canada

The Canadian Aeronautics Act[37] delegates to the Minister of Transport (Transport Canada) authority to regulate aviation.[38] Nominally, it gives the "Governor in Council" authority to establish requirements for flight crew members and other aviation personnel; design, manufacture, distribution, and maintenance of aircraft; and operating rules.[39] The Ministry has promulgated general regulations for aviation.[40]

Constitutionally, Canada is a federal state within the British Commonwealth. Canada's constitution comprises a collection of acts of the British Parliament.[41] Under these acts the Queen of England, "acting by and with the advice and consent of the [Canadian] Senate and House of Commons" has general legislative authority over Canada. The Canadian parliament, however, has exclusive legislative authority with respect to most matters, including the regulation of trade and commerce.[42] The ten provinces of Canada have sole authority to exercise general legislative, executive, and judicial powers with respect to matters within each province.[43] The federal government, however, exercises exclusive power over certain enumerated matters, including aviation.[44]

The Federal Court of Canada has jurisdiction to review decisions by the Transportation Appeal Tribunal.[45] The standard of review is "reasonableness."[46]

A Special Flight Operations Certificate (SFOC) from Transport Canada is required for drones weighing more than 25 kg (55 pounds).Under amended regulations issued at the end of November 2014, no such certificate is required for commercial flights of drones weighing 2 kg (4.5 pounds) or less, and for certain commercial operations of drones weighing between 4.5 pounds and less than 55 pounds.[47]

By the end of 2015, Transport Canada had granted roughly 1,600 permits under which some 110 firms were conducting regular drone operations. The government has committed itself to shepherd development of a robust drone industry in Canada and is actively courting US companies frustrated with the slow legal pace in the United States.[48]

On May 28, 2015, Transport Canada published notice of proposed amendments to the drone regulations.[49] The proposal would maintain the SFOC requirements for drones weighing more than 25 kilograms (55 pounds) and others operated BLOS. The notice expressed intent to align Canadian regulation of microdrones with FAA regulations.[50] The proposed regulations would become effective at the end of 2016, when the existing interim regulations expire.

It would eliminate the distinction between recreational and commercial flight;[51] although it would provide an exemption for model aircraft operated within the activities and under the rules of model aircraft organizations.[52] It would impose more stringent requirements on small UAV operations in and around urban and built-up areas and those close to aerodromes, and less stringent requirements on small UAV operations in remote areas.[53] For urban operations, the proposal would require pilot permits, as distinguished from pilot licenses, conditioned on self-certification of medical fitness; training by a flight school, a UAV training provider, a third party; passage of a Transport Canada written or test specific to UAV operations; practical delivering by the manufacturer, operator or by a third party, providing the person providing such training held a UAV pilot permit; and passage of a skill tests or proficiency check conducted by qualified UAV operators, manufacturers or third parties.[54] Pilot permits would not be required for operations in remote areas.[55]

It would require that manufacturers of a small UAV (complex operations) system be required to declare that the UAV system meets a design standard for UAV systems developed by a Transport Canada industry working group.[56] It would not require type certificates or certificates of airworthiness.[57]

Operations over urban or congested areas, but not operations in remote areas, would have to be conducted within line of sight in visual meteorological conditions (VMC), outside certain controlled airspace, and to take off and land only from property where the operator has permission of the property owner to conduct drone operations.[58] Remote-area operations could be conducted only in the daytime, within line of sight of the operator and outside controlled airspace.

7.3.4 China

The Chinese regulatory authority is the Civil Aviation Administration of China (CAAC), an organizational component within the Chinese Ministry of Transport. It shares responsibility for aviation regulation with the Central Military Commission. Review of administrative agency decisions is available under the Administrative Supervision Regulations promulgated by the State Council; which invests the Ministry of Supervision/CPC Central Discipline Commission with authority to review decisions of agencies outside the Party. Administrative reconsideration bodies provide another route for review. In theory, the Administrative Procedure Law allows parties to bring suit in the regular courts when their rights are violated by administrative agencies, but the uneven quality of the judiciary makes this an unattractive route for seeking review.

The CAAC follows an approach to drone regulation similar to that of Canada.

Licenses are required only for operators of drones weighing more than 7 kg. For drones weighing more than 116 kg, both a pilot's license for the operator and certification of the drone are required. The intermediate category may fly without advance approval, subject to regular air traffic control. Flight in controlled airspace requires advance approval. A considerable backlog of operator applications exists, but a senior Chinese official was quoted in late 2014 as promising no enforcement actions against those flying without a license during the transition period.[59]

The relative maturity of Chinese drone regulation has permitted Amazon and local competitors to experiment with drone package delivery more extensively in China than elsewhere.[60]

Chinese law is especially sensitive to perceived threats to national security, equally from domestic sources as from foreign. Reaction to such a threat is typically aggressive, usually including military involvement. In one highly publicized incident a drone conducing aerial survey operations near the Beijing airport induced a panic. The operator was prosecuted for endangering public security, not for violating aviation regulations.[61]

Penalties for violations of the law in China typically are more severe than in other states, although enforcement is uneven and often politically motivated.

7.3.5 European Union

The European Union has evolved from a free-trade zone established after World War II into a weak political federation. Its guiding legal principle is that harmonization of law across its 27 members will facilitate free movement of goods, capital, and labor across European borders. Its central institutions—the European Council, the European Commission, the European Parliament, and the European Court of Justice—promote that goal by enacting and enforcing law and by adjudicating disputes over its application. The European Council and European Commission constitute the executive branch of European Union government. The Commission is supervised by a multimember body with representatives of state members of the Union, and is subdivided into an array of directorates handling specialized subject matter. The Commission has the power of legislative initiative—most new statutory law originates in the Commission—but its proposals do not have the force of law until they are approved by the European Parliament and the European Council. The European Parliament is the legislative body, comprising representatives directly elected by the populations of member states. Its role has strengthened from being merely advisory to having the power of legislative veto and now to a certain amount of legislative initiative. The European Court of Justice (ECJ) is the judicial arm of the Union. It has power to review decisions by the other Union bodies for conformity with law, to decide whether member state legislation or judicial decisions contravene European law, and to resolve disputes between the other Union institutions.

Incorporation of the separate European Covenant of Civil and Political rights into European Union law has significantly broadened the scope of the ECJ subject matter jurisdiction. Now, a person claiming that a decision by the Commission is ultra vires (outside the body's authority), invalid because a failure to follow necessary procedures, or violative of human rights can obtain ECJ review. The ECJ is bifurcated into a subordinate body, the Court of First Instance, and a review body, the ECJ itself.

EU administrative law is relatively undeveloped as it applies to judicial review of EU administrative agencies. Most application and enforcement of EU law is the responsibility of national agencies and courts.[62] Therefore, most EU administrative law focuses on national administrative law.

The European Parliament and Council established the European Aviation Safety Agency (EASA) in 2002,[63] an independent body with responsibility for assisting the Commission in preparing legislation, assisting member states in implementing EU legislation, and issuing certification specifications and certificates.[64] Its decisions are subject to review by a specialized Board of Appeal, and thereafter to review before the ECJ.[65] The directive makes clear its goal: "the establishment and uniform application of common rules in the field of civil aviation safety and environmental protection."[66] To that end, it envisions transfer of functions from national aviation authorities to the agency.[67]

Thus responsibility for aviation regulation in Europe is divided between the European Aviation Safety Agency (EASA) and national aviation authorities in the member states of the EU.

In general, the EASA has responsibility for regulating aircraft weighing more than 150 kg, while the national authorities are responsible for lighter aircraft. In 2014, the EASA regulated aircraft airworthiness and certification, training and certification of mechanics and repair stations, while national authorities regulated pilot training and certification. They shared responsibility for airspace regulation and operating rules.

In 2013, the EASA published a roadmap for integration of drones into European airspace.[68] It extolled the economic benefits of wider drone use in Europe to boost industrial competitiveness, to promote entrepreneurship, and to create new businesses, stimulating growth and job creation. It concluded that "RPAS rules must also be as light as necessary, in order to avoid an unnecessary burden on the emerging industry."[69] Seven member states—Czech Republic, France, Ireland, Italy, Sweden, Switzerland

and UK—had regulations for drone flight in place while other members—Belgium, Denmark, The Netherlands, Norway, and Spain—were still formulating their plans.[70] The roadmap projected a multiyear process, extending to 2035, for finalizing all the necessary regulatory elements to eliminate the fragmentation inherent in national efforts and to harmonize drone regulation across Europe.[71]

It projected that the first part of the regulatory framework would be in place by 2016. The report noted the unworkability of the 150 kg line of demarcation between EASA and national authorities and expressed expectation that legislation, then in the draft stages, would be enacted to give the EASA full responsibility for drone regulation across Europe.[72] The roadmap was accompanied by an appendix detailing the necessary research and development and the schedule for implementing a progressively more complete regulatory framework. It suggested that the evolution of European standards for DROPs, drone aircraft certification, and would be strongly influenced by the recommendations coming out of JARUS (see § 7.6).

It identified five scenarios that warrant different regulatory treatment:

- Very low level (VLL) operations below 400 ft and within 1,500 ft from the DROP, who keeps the drone within his line of sight
- Extended visual line of sight (E-VLOS), in which the DROP is supported by one or more observers who keep the drone within their line of sight
- Beyond visual line of sight (B-VLOS), in which flight beyond visual line of sign is supported technologically, e.g. by FPV imagery
- VFR and IFR operations above 400 ft within radio range
- IFR and VFR operations above 400 ft beyond radio range.[73]

It projected initial VLOS operations in 2013, with integration complete in 2014–2015.[74]

In March 2015, EASA published a *Concept of Operations* for drones. Long on philosophy and short on specifics, the report indicated that there are 2,495 operators and 114 manufacturers of very small-to-small drones with a maximum take-off mass up to 150 kg. The European regulatory approach proposes three categories of operations:

1. The first category of "open" requires an authorization for the flight but requires users to remain within defined boundaries similar to the Canadian exemptions currently in place (for example, within visual line of sight, at safe distance from aerodromes and from people). The "open" operation category of drones would not require an authorization by an Aviation Authority for the flight. It would, however, stay within defined boundaries for the operation, e.g. distance from aerodromes or from people.
2. The second operation category of "specific" requires a risk assessment that will lead to an authorization with specific limits. The EASA specific operations category for UAVs is similar to the current Canadian SFOC

process. The "specific" operation category requires a risk assessment that will lead to an Operations Authorization with specific limitations adapted to the operation.

3. The third operations category is "certified" and is required for operations with a higher associated risk or which might be requested on a voluntary basis by organizations providing services such as remote piloting or equipment such as "detect and avoid." A December, 2015 Technical Opinion by EASA reiterates the categories and provides additional detail. For the foreseeable future drone regulation in Europe will remain fragmented, despite continued EASA pressure for harmonization.

7.3.6 France

Aviation in France is regulated by the Direction Generale de l'Aviation Civile, within the Ministry of Environment, Development, and Energy.[75]

French drone regulation distinguishes between model aircraft flown for recreational purposes and drones flown for other purposes.[76] Model aircraft are largely unregulated.

Basic requirements for non-recreational drone operations were established by a decree published on May 10, 2012, in the official Journal of the French Republic.[77] The requirements were in place for a trial period of 18 months, and were expected to be followed by revised regulations. Under them, manufacturers must obtain a DGCA type design certificate, the requirements for which vary according to seven weight classes. Operators must be approved according to mission type and the manufacturer and model of aircraft. DROPS must receive training and get a DNC (statement of skill level).[78] The decree established four categories of drone flight. The categories permit drones weighing no more than 25 kg to fly up to 150 m above the ground, no more than 100 meters from the DROP as long as he keeps it within line of sight in an unpopulated area (S1); drones up to 25 kg to be flown up to 50 m above the ground within 1 km via line of sight or FPV when no person is on the ground within the area of flight (S2); drones weighing less than 4 kg to operate up to 150 ft above the ground, within 100 m of the DROP in urban areas or close to people and animals (S3); and establish a miscellaneous category not meeting the criteria for the first three (S4).[79]

According to a list published by DAGD in July 2014,[80] some 600 enterprises have received permission to operate microdrones in agriculture, banner towing and other advertising, surveying, and photography—categories for which standardized approval is available.

7.3.7 Germany

Germany's Luftfahrt-Bundesamt (LBA, "Federal Aviation Office") is responsible for aviation regulations.[81] It is a component of the Federal Ministry of Transport and Digital Infrastructure.

Judicial review of agency decisions is the responsibility of specialized administrative courts within the judicial branch of government. Those courts sharply limit their intrusion into the discretion of administrative agency decision-makers. They are more concerned with protecting individual rights than with confining agencies to their delegated authority.[82] Agencies have broad discretion to choose their own procedures for rulemaking, for example. German judicial review of agency decisions generally is deferential—when the authorizing statute gives the agency discretion—and review is limited to ensuring that the agency followed specified procedures, correctly investigated the facts, did not violate the principle of equality (similar to US equal protection), applied general standards of evaluation, and did not consider irrelevant matters.[83]

Germany regulates commercial drone flight but not hobbyist flight. The commercial operation of drones weighing more than 25 kg or out of the line of sight of the operator general is prohibited.

Regulation of drone flight is decentralized, with German states and municipalities having authority to determine if particular operations do not pose a risk to air traffic, public safety, or public order.[84] Drones weighing less than 5 kg with electric propulsion systems can be operated in most states under a general permission lasting two years. Applications must include information about the applicant and the drone, proof of insurance, description of the operator's experience, a map of the intended operating area, and consent of the property owner. Operations must be limited to 100 m above ground level.[85]

7.3.8 India

India is a federal constitutional democracy with a mainly ceremonial president elected by a directly elected bicameral (two-house) parliament and legislative assemblies of the states, acting jointly.[86] Executive power is exercised by a prime minister and cabinet accountable to the legislature. The judiciary is nominally independent.[87] There is no administrative procedure act, but Indian courts have reasoned that governmental decisions are inherently reviewable by the courts.

The Directorate General of Civil Aviation (DGCA) within the Ministry of Civil Aviation regulates civil aviation in India. The Indian government is planning to replace the organization with a Civil Aviation Authority (CAA), modeled on the lines of the US Federal Aviation Administration (FAA).

In October 2014, the DGCA issued a public notice banning civil use of drones "for any purpose whatsoever," pending development of regulations for certification and operations.[88]

7.3.9 Italy

Italy is governed through a typical parliamentary system[89] in which the majority party in the Parliament, or a coalition of parties constituting a majority, selects the prime minister and cabinet who head the executive branch.[90] A President of the Republic is elected by a joint session of the parliament.[91] He has mainly ceremonial powers.[92]

Under the Italian Constitution, the judicial system is independent of the legislative and executive branches.[93] Following the French model,[94] the Council of State has judicial review powers over public administration.[95]

The Italian Civil Aviation Authority (ENAC) is responsible for regulating aviation safety.[96]

ENAC's Regulation for Remotely Piloted Aerial Vehicles was issued under the authority of article 743 of the Italian Navigation Code on December 16, 2013.[97] It allows for operation of drones weighing less than 150 kg by operators possessing ENAC authorization. It provides separate rules for those weighing 25 kg or less and those weighing more than 25 kg and less than 150 kg.[98] Simplified procedures are authorized for drones weighing less than 2 kg.[99]

Non-critical operations are those that:

- Do not involve overflight of congested areas, gatherings of people, urban areas and infrastructures; in restricted areas; or over railway lines and stations, highways, and industrial plants
- Occur in airspace designated as V70
- Occur at least 150 meters over and 50 meters horizontally from persons and property not under the direct control of the operator
- Occur in daylight, in uncontrolled airspace, and
- Remain at 8 kilometers from airports and from approach and departure paths.

Other operations are designated as critical.[100]

Restricted-type certificates are available for drones manufactured as series, obviating the need for a separate airworthiness certificated for each aircraft.[101] Depending on the airspace in which the drone will fly, transponders and the capability of communicating with ATC may be required.[102] Autonomous landing capability may be required to cope with emergencies.[103] A plate identifying the drone system and providing operator data must be affixed to the drone and to the DROPCON.[104]

Operators may obtain operating authorization only after establishing an adequate organization with flight manuals and after conducting a test program to demonstrate safety.[105] For non-critical operations the operator must file a declaration describing the technical and performance features of its aircraft, results of experimental tests, its operations, and including copies of flight, maintenance, and operating manuals.[106] Declarations for non-critical operations are simply verified, while applications for critical operations receive more thorough evaluation, including further analysis and testing at the option of the ENAC.[107]

Operators must demonstrate that their DROPs "have the qualifications required to conduct the planned activity"[108] and must have a technical manager to oversee operations, airworthiness, and training.[109] ENAC may authorize operations for short phases of flight outside the line of sight of the DROP.[110] DROPs must hold either pilot licenses or have attended manufacturer-provided training programs for the specific drones they are to fly.[111]

Liability insurance meeting the requirements of EC Regulation No 785/2004 must be provided.[112]

7.3.10 Japan[113]

The constitution of Japan[114] establishes a parliamentary system under the symbolic authority of the Emperor of Japan.[115] The parliament, known as "The Diet," has two houses, the House of Representatives and the House of Councillors.[116] Executive power is vested in a Cabinet headed by the Prime Minister.[117] The Prime Minister is selected by the Diet[118] and has authority to appoint and remove other members of the cabinet.[119] The judiciary is independent of the political bodies.[120]

Administrative appeals mechanisms are rare. Judicial review of agency focuses on formal legality: whether the action was statutorily authorized, whether the delegation is clear about its boundaries and its purpose. One commentator summarizes the standard of review as follows:

> [A]gency decision-making is subject to reversal for abuse of discretion if the decision lacks an adequate factual basis due to fact-finding errors, if the agency's assessment of the facts is "obviously unreasonable," if the agency has ignored relevant factors or considered irrelevant factors ...[121]

Civil aviation in Japan is regulated by the Civil Aviation Bureau, a component of the Ministry of Land, Infrastructure and Transport (MLIT).[122] The Ministry is authorized by statute[123] to register aircraft,[124] to certify airmen,[125] and to regulate aircraft operations.[126]

The Civil Aeronautics Act allows drone flight under restrictions set by the MLIT.[127] MLIT regulations require advance approval for flight of "model aircraft" and "other items" into controlled airspace or higher than 150 meters above the ground within airways.[128] After a drone carrying trace amounts of radioactive material landed on the roof of the Prime Minister's office on July 2, 2015, pressure ensued for tightening drone rules and resulted in the MLIT's releasing proposed rules that would prohibit drone flight "except during daytime."

7.3.11 UK

The United Kingdom of Great Britain and Northern Ireland (UK) is a sovereign state in which significant governmental authority is exercised separately for its three components, England, Scotland, and Northern Ireland. The UK Parliament is the supreme legislative body, operating under a parliamentary system in which executive authority is exercised by a prime minister and cabinet selected by the majority party, or a coalition of parties in the House of Commons. A second house of parliament, the House of Lords, exercises only limited legislative authority. The judiciary for England, Scotland, and Northern Ireland is separate; but in all three cases, it is independent of both parliament and executive. Under the Constitutional Reform Act of 2005, the House of Lords was divested of its appellate judicial function, which now reposes in a Supreme Court of the United Kingdom.

Judicial review of governmental decisions is a long-standing tradition, as 7.2.1 explains. Contemporary judicial review is deferential to both substantive decisions by and procedures used by "expert regulatory bodies."[129]

The UK Civil Aviation Authority (CAA) derives its authority from the Civil Aviation Act and the Air Navigation Order.[130]

CAA regulation of drones distinguishes between microdrones, defined as small unmanned aircraft (SUA), with empty-fuel weights less than 20 kg; and macrodrones, defined as unmanned aircraft with empty-fuel weights more than 20 kg but less than 150 kg. As § 7.3.5 explains drones weighing more than 150 kg are regulated by the EU's CASA. Additional rules are applicable to small unmanned surveillance aircraft (SUSA), defined as SUAs carrying any form of surveillance or data-collection equipment. SUAs do not require airworthiness approval or registration. Their operation does, however, require operating permits and evidence of pilot qualification.[131] The general rules prohibit flying in congested areas. The aircraft must be kept within the visual line of sight (normally within 400 meters horizontally and 400 ft vertically) of its remote pilot (the 'person in charge' of it). Operations beyond these distances must be approved by the CAA, according to a submission by the operator to prove that it can do this safely. An earlier, 2012, CAA guidance summarizes basic principles applicable to drone operations,[132] some of which are superseded by the 2014 regulations.

Approval of specific operating plans requires operators to apply on standard forms.[133] The CAA makes a standard form available to apply for exemptions to fly an Unmanned Aircraft under 20 kg; aerial work permissions to fly a Small Unmanned Aircraft; or Small Unmanned Surveillance Aircraft Permissions. Categories of operation for which application may be made include forestry, agriculture, construction, infrastructure survey, photography, filming and media, research and development, and security and emergency services.[134]

Applicants must submit information about the vehicles to be flown,[135] the areas of operation,[136] qualifications of DROPs,[137] information about the applicant's operations manual,[138] and proof of liability insurance.[139] Either a pilot's license or equivalent pilot experience to be evaluated on a case-by-case basis, is required.[140] Applications must be submitted at least 28 days before operations are to commence.[141]

Macrodrone operations require submission and approval of more elaborate data including vehicle airworthiness information.[142]

In a notice issued in November 2014,[143] the CAA announced a specialized regulatory regime for drone operations over congested areas. It permits operators of small unmanned aircraft (SUA), defined as microdrones with empty-fuel weights less than 20 kg, and small unmanned surveillance aircraft (SUSA), defined as SUAs carrying any form of surveillance or data-collection equipment, to obtain annual authority to operate their drones with fewer limitations if they submit a satisfactory Congested Areas Operations Safety Case Template (CAOSC) to the CAA. Drones weighing less than 7 kg need not use the CAOSC to obtain standard permission. CAOSCs must include hazard identification and risk assessment for all aspects of the intended application.

7.4 Socialist legal systems

Socialist law is a hybrid of western civil law traditions overlaid by Marxist political theory. Some commentators deny that it is a distinct legal system, but its peculiarities deserve exploration. The Russian Revolution of 1918 and the Chinese revolution 30 years later devalued formal judicial adjudication because it was associated with the regimes that were being swept away. Rather than having courts adjudicate private disputes, revolutionary principles said that disputes should be resolved directly by the people, acting in a decentralized fashion through peoples' tribunals. The people, thus constituted for adjudication purposes, would discover on their own the correct revolutionary principles to be applied to resolve disputes.

As the Communist Party strengthened in both states to manage centrifugal forces resulting from too much decentralization of political power, party supervision of courts substituted for a hierarchical system of appeals within formal state institutions.

Then reforms took root in Russia after Joseph Stalin died in 1953, and in China after Mao Zedong died in 1969. Stalin had consolidated ultimate political power after Vladimir Lenin died in 1924 and suppressed even suspicions of dissent by removal from office, incarceration or execution. There was no such thing as administrative procedure distinct from political maneuvering within the Communist Party. Mao similarly exercised exclusive personal power for the rest of his lifetime after leading the military displacement of the corrupt Nationalist government in 1949.

Both Stalin and Mao undertook complete transformation of their societies, emphasizing industrialization. Stalin's reign of terror was largely successful in leapfrogging Russia into a major technological power. Mao succeeded in consolidating a fragmented Chinese political system, but his 1957 Great Leap Forward and 1966 Cultural Revolution devastated a generation of Chinese leadership and left the Chinese economy gasping. Both of them killed tens of millions through famine and widespread political purges.

After Stalin and Mao died, the party in both states recognized that many private disputes without significant political consequences could be handled safely in state-run courts; only political cases needed active party involvement.

As a result, socialist legal systems in Russia, China, and in many postcolonial states that embraced socialism, took on the coloration of western legal systems, with hierarchical judicial branches within formal state governmental apparatus. The party remained on the sidelines, ready to jump in as necessary.

Marxist ideology plays a role, apart from revolutionary impulses transmitted by revolutionary elites, but is now being crowded to the fringes by globalization. Socialism devalues the ownership of private property at the center of legal and economic doctrine and emphasizes, instead, the role of the state, not only in planning and supervising economic production, but also in vigilantly resisting counterrevolutionary impulses. "Counterrevolutionary impulse" often is a euphemism for any criticism of the political leadership.

Socialist law has a complexity far beyond a caricature of Marxism, defined by ownership of the means of the production by the working class. Nevertheless, socialism places more control of economic activity in the hands of the government and central economic planners. It allows only limited scope for market-based economic discourse. Even when markets are allowed, market participants are primarily community-organized and controlled entities, much preferred over purely private entrepreneurship.

That means a starting point for dispute resolution that is more comfortable with the state than and market-oriented systems. Nevertheless, a socialist state, like any other complex state, recognizes that decision-makers may run amok, and that the effective functioning of the system requires that they be held accountable through some kind of dispute-resolution mechanisms.

In party-based and party-dominated states like the pre-1989 Soviet Union, or post-revolution China, accountability to independent institutional reviewers is largely accomplished through a party apparatus that that parallels the governmental institutional structure. If a citizen feels wronged by the action of a governmental decision-maker, he appeals to the party official with responsibility for that agency. The process is informal and political rather than procedurally formal and legal.

Even in party-dominated socialist legal systems, however, tribunals following formal procedures exist with authority to review decisions by other governmental actors.

Legal systems working out of this tradition tend to place more emphasis on criminal law, and relatively less emphasis on adjudicating purely private disputes. Within the private dispute resolution mechanisms, property and contract play less of a role than tort law.

Nevertheless, integration into the global economic order—a condition that almost all states aspire to—requires some mechanism for enforcing contracts, and the human rights movement has forced at least the appearance of judicial mechanisms to vindicate individual rights and blunt state excesses.

But in both Russia and China, the dominant surviving socialist regimes, judicial decision-making remains unpredictable, with the possibility always remaining open of nontransparent intervention by the party and by local political elites. Enforcement of judicial decisions generally is weak, depending even more on party and local political will. Pathways for traditional, formal, appeal are murky. Administrative law, in the sense of recourse to the courts to correct administrative agency abuses, excesses, and mistakes, is in its infancy.

The most important thing to appreciate about socialist legal systems is not so much the ideological background of Marxism as it is the role the party plays in active oversight.

In all of these resumes, political adroitness is far more important than good lawyering.

7.5 International law

International aviation law is more directly enforceable by national courts than other international legal regimes. The international legal system is premised on the idea that sovereign states have exclusive power to regulate activities taking place on their territory. In the natural state, the sovereign powers are unlimited, but states may subject exercise of their sovereignty to international law. This may occur through consistent and widespread state behavior, in which case a norm of *customary international law* arises, which states must follow. The concept of customary international law resembles the common law in some respects but also differs from it importantly. It resembles the common law in the way that its norms arise and are discovered. The key question for both legal regimes is, "what is customary?" It differs from common law in that customary international law pits relatively vague international norms against sovereignty. Common law may have equally vague standards, but it is applied by a sovereign within its own territory, through its own judicial institutions. Customary international law as a concept is much criticized by commentators who question whether it should be incorporated into domestic US law at all.[144]

Universally accepted, however, is another branch of international law: *conventional law*, based on treaties—"conventions." Treaties or conventions are written documents ratified by member states, in which they agree to exercise their sovereignty according to the treaty. The treaty represents a yielding or ceding of sovereignty, sometimes to international institutions that are established in the treaty, as in the case of the UN Charter and in other cases without specialized international machinery, as under the Geneva Convention on the law of armed conflict.

Two competing theories of international law operate: the *monist* and the *dualist* theories. In the monist theory, international law is integrated with national law and applied directly by national courts. In the dualist theory, international legal norms are not directly applied by national courts. They are applied only if the sovereign's legislature adopts them explicitly and makes them a part of national law. The United States adheres to dualism.[145] Many European states embrace monism.[146]

The International Civil Aviation Organization (ICAO) is an intergovernmental body established in 1944 by the Convention on International Civil

Aviation (the *Chicago Convention*).[147] It comprises representatives from some 191 member states and global aviation organizations. Its goal is to develop international standards and recommended practices (SARPs) which guide, but do not bind, states in developing national aviation regulations. An Assembly of all the member states meets every three years and elects members of a Council, which exercises ongoing authority under the convention. The Council has power to adopt SARPs and incorporate them as annexes to the convention.

Some 10,000 SARPs are now in effect, ranging in subject matter from personnel licensing to airport design. Each is associated with one of 19 annexes to the convention.

Drafting or amending a SARP culminates with a recommendation by ICAO's Air Navigation Commission[148] to the Council. The commission has 22 members; each elected by the member states. Current representatives come from United States, United Kingdom, Spain, the Russian Federation, France, Germany, a consortium of Scandinavian states, China, Canada, Brazil, and Australia, among others. Only after the Council adopts a commission recommendation does it become an official SARP.

Some ICAO standards are directly enforceable in the United States.[149]

ICAO published a report on drones in 2011.[150] The report envisions a multi-year effort by national authorities and by ICAO as it develops SARPs and non-binding guidance material in advance of SARPs.[151] It notes that drones are well suited for civilian aviation operations that are "dull, dirty, or dangerous."[152] It further observes:

> The demand for small civil RPA flying visual line-of-sight (VLOS) … for law enforcement, survey work, and aerial photography and video will continue to grow. Larger and more complex RPA—able to undertake more challenging tasks—will most likely begin to operate in controlled airspace where all traffic is known and where ATC is able to provide separation from other traffic. This could conceivably lead to routine unmanned commercial cargo flights.[153]

In 2015, ICAO issued a comprehensive manual on drone regulation, including detailed recommendations on aircraft airworthiness and DROP certification standards and on operating rules for both integrated and segregated drone operations.[154] It contains extensive discussion of system design, DROP certification requirements, and mitigation of lost control link risks and of failure of ATC voice communications.

ICAO sponsored a symposium on remotely piloted aircraft systems in March 2015 at ICAO headquarters in Montreal. A senior official of the Air Navigation Bureau made a presentation in which she projected ICAO guidelines by 2018 for operating drones in non-segregated airspace, and refined operational procedures by 2014 for lost control-link and detect-and-avoid technologies.[155]

7.6 International coordination and consultation

An international body, Joint Authorities for Rulemaking on Unmanned Systems (JARUS) provides a forum within which national aviation regulators regularly consult on the content of their drone regulations. It is open to all national aviation authorities.[156] Present members include the United States, Great Britain, France, Germany, Brazil, Israel, Italy, and the Scandinavian countries, among others. The body has no rulemaking or treaty negotiation authority, but only provides a forum for consultation. It has working groups on Operational and Flight Crew Licensing Requirements, Airworthiness, Detect and Avoid, Control and Communication, and UAS System Safety.

Among other things, JARUS published a 2012 draft Certification Specification for Light Unmanned Rotorcraft Systems, defined as those weighing less than 750 kg, and excluding manned aircraft.[157] and a 2013 guidance on control links.[158] The control-link guidance usefully distinguishes between "telecommand and telemetry" functions.[159] It articulates performance specifications, which, among other things, require national regulators to establish certain link parameters concerning:

> 3.1.3.1 Communication transaction time
> The maximum time for the completion of the operational communication transaction after which the initiator should revert to an alternative procedure.
>
> 3.1.3.2 Continuity
> The probability that an operational communication transaction can be completed within the communication transaction time.
>
> 3.1.3.3 Availability
> The probability that an operational communication transaction can be initiated when needed.
>
> 3.1.3.4 Integrity
> The probability of one or more undetected errors in a completed communication transaction.[160]

7.7 Convergence and divergence

In certain respects, regulatory approaches around the world are converging, as regulators borrow ideas from one another. Aviation agencies in the US, Europe, Canada, and Australia, are explicit in their goal of achieving greater harmony in drone regulation. Movement in that direction is apparent with each new proposed rule.

In other respects, however, they will remain different, reflecting different political concerns and different legal traditions for holding government accountable. Regulation will continue to be a formal process, but the pathways to shape regulatory ideas will continue to differ.

Notes

1 5 U.S.C. §§ 553–704.
2 1931, 22 & 23 Geo. 5 c. 4.
3 www.aph.gov.au/About_Parliament/Senate/Powers_practice_n_procedures/~/link.aspx?_id=956BE242B820434A995B1C05A812D5E1&_z=z.
4 Australian Constitution § 5; ch. I, pt I, § 1.
5 Australian Administrative Law Policy Guide sec. 3.6.1, www.ag.gov.au/LegalSystem/AdministrativeLaw/Documents/Australian%20Administrative%20Law%20Policy%20Guide.pdf.
6 § 25 Administrative Appeals Tribunal Act 1975, No. 91, 1975, www.comlaw.gov.au/Details/C2014C00742/Html/Text#_Toc402786254; www.aat.gov.au/LawAndPractice/JurisdictionAndTimeLimits/JurisdictionList.htm (AAT list of statutes with respect to which it has jurisdiction, including Civil Aviation Act of 1988).
7 Hon. Justice Duncan Kerr Chev LH, *Excellence in Government Decision-Making* (Jun. 21, 2013), www.aat.gov.au/Publications/SpeechesAndPapers/Kerr/FreedomToBeFair21Jun2013.htm. e.g. Hazelton and Civil Aviation Safety Authority [2010] AATA 693 (Sep. 10, 2010) (requiring reconsideration of denial of medical certificate to pilot); Repacholi Aviation Pty Ltd and Civil Aviation Authority [2006] AATA 578 (Jun. 30, 2006) (overturning decision to cancel pilot's license); McIver Aviation Pty Limited and Civil Aviation Safety Authority [2005] AATA 391 (May 3, 2005) (requiring reconsideration of conditions imposed on banner-towing operation).
8 Australian Constitution § 51(i).
9 Airlines of New South Wales Pty Ltd v. New South Wales (No 2), [1965] HCA 3, (1965) 113 CLR 54 (High Court of Australia) (validating some regulations as within federal power and invalidating others).
10 www.casa.gov.au/scripts/nc.dll?WCMS:HOMEPAGE::pc=HOME; Civil Aviation Act 1988, www.comlaw.gov.au/Details/C2014C00640.
11 www.comlaw.gov.au/Details/F2012C00363/Html/Volume_3.
12 CASR § 101.235(1).
13 Civil Aviation Safety Regulation part 101.
14 101.240 Definitions for Subpart [check].
15 CASR § 101.240 (providing definitions).
16 CASR § 101.235(3).
17 CASR § 101.255.
18 CASR § 101.255 (requiring airworthiness certification to large UAVs, implying that small UAVs need no such certification); Ibid. § 101.280(2) (limiting operation of "a UAV that is not a certificated UAV" over populous areas).
19 CASR division 101.F.3 (certification of UAV controllers).
20 CASR civision 101.F.4 (certification of UAV operators).
21 CASR § 101.295.
22 CASR § 101.270 (requirement for UAV operator's certificate).
23 CASR § 101.355(1) (setting forth requirements for certification).
24 CASR § 101.340(2).
25 CASR § 101.085.
26 CASR § 101.030.
27 CASR § 101.245.
28 CASR § 101.250.
29 CASR § 101.275(1B).
30 CASR § 101.275.
31 CASR § 101.280(2) (non-certificated UAVs, i.e. small UAVs); Ibid. § 101.280(3) (certificated UAVs).
32 CASR § 101.280(4).

33 CASR § 101.285.
34 Law No. 11.182 (Sep. 27, 2005); Brazilian Aeronautical Certification Regulations, www2.anac.gov.br/ingles/legalRegulatory.asp.
35 Art. 37 of Federal Constitution.
36 www.icao.int/Meetings/UAS/Documents/11_Honorato-Roberto_ANAC_Brazil_Presentation%20v1.0.pdf.
37 R.S., 1985, c.A-2.
38 Aeronautics Act § 4.2.
39 Aeronautics Act sec. 4.9.
40 www.tc.gc.ca/eng/acts-regulations/regulations-sor96-433.htm.
41 Constitution Acts, 1867–1982, www.laws-lois.justice.gc.ca/eng/Const/index.html.
42 Constitution Acts § 91.
43 Constitution Acts § 92.
44 Quebec (Attorney General) v. Canadian Owners and Pilots Association, 2010 SCC 39, File No. 32604 (Canada 2010) (invaliading provincial statute regulating airport siting).
45 Marina District Development Co. v. Canada, 2013 FC 800, docket no. T-324-12 (Federal Court of Canada 2013); Transportation Appeal Tribunal of Canada Act (S.C. 2001, c. 29).
46 Dunsmuir v. New Brunswick, [2008] 1 S.C.R. 190, 2008 SCC 9, docket no. 31459 (Supreme Court of Canada) (deferential standard of judicial review of agency decisions for reasonableness involves "existence of justification, transparency, and intelligibility" and whether decisions falls within range of outcomes defensible on facts and law delegated to agency).
47 www.tc.gc.ca/eng/civilaviation/standards/general-recavi-uav-2265.htm?WT.mc_id=8r1cy#safety; www.news.gc.ca/web/article-en.do?nid=900449 (announcing Nov. 2014 changes); Transport Canada, *Review and Processing of an Application for a Special Flight Operations Certificate for the Operation of an Unmanned Air Vehicle (UAV) System*, Staff Instruction (SI) No. 623-001 (Nov. 19, 2014), www.tc.gc.ca/eng/civilaviation/standards/general-recavi-uav-4161.html; Transport Canada, *"Flying an unmanned air vehicle?,"* www.tc.gc.ca/media/documents/ca-standards/Info_graphic_-_Flying_an_umanned_aircraft_-_Find_out_if_you_need_permission_from_TC.pdf. Operators must give Transport Canada basic information like contact information, model, and boundaries for the intended flight. wwwapps.tc.gc.ca/Saf-Sec-Sur/2/NPA-APM/doc.aspx?id=10294 ("Agriculture surveys, cinematography and film, and police investigations are the leading and most mature market applications of UAVs in Canada").
48 *In Canada, Wooing Drones for Work*, N. Y. Times, Apr. 6, 2015 at p. B6 ("Bits").
49 Canadian Aviation Regulations Advisory Council (CARAC), Notice of Proposed Amendment (NPA): Unmanned Air Vehicles, wwwapps.tc.gc.ca/Saf-Sec-Sur/2/NPA-APM/doc.aspx?id=10293 (May 28, 2015) [hereinafter "NPA"].
50 NPA at p. 5.
51 NPA at p. 9.
52 NPA at p. 11.
53 NPA pp. 12–13.
54 NPA at p. 16.
55 NPA at pp. 23–24.
56 NPA at p. 18.
57 NPA at p.19.
58 NPA at pp. 19–20.
59 europe.chinadaily.com.cn/business/2014-08/18/content_18436980.htm.
60 www.ibtimes.com/china-beat-amazon-prime-air-commercial-drone-delivery-market-1491636.
61 europe.chinadaily.com.cn/business/2014-08/18/content_18436980.htm.

62 Peterbroeck c.s.vs. Belgian State, case C-312/93 at para. 12 (ECJ 1995) (holding that member states must provide adquate procedures to protect persons affected by their application of EU law); Rewe-Zentralfinanz vs. Landwirtschaftskammer für das Saarland, Case 33/76 at para. 5 (ECJ 1976) (member states must designate courts with jurisdiction and assure adequate procedures when their institutions apply EU law).
63 Regulation (EC) No 1592/2002 of the European Parliament and of the Council of 15 July 2002 on common rules in the field of civil aviation and establishing a European Aviation Safety Agency.
64 Ibid. at para. 12.
65 Ibid. at para. 15.
66 Ibid. at para. 18.
67 Ibid. at para. 21.
68 European RPAS Steering Group, Roadmap for the integration of civil Remotely-Piloted Aircraft Systems into the European Aviation System (Jun. 2013), ec.europa.eu/DocsRoom/documents/10484/attachments/1/translations/en/renditions/native [hereinafter "European Roadmap"].
69 European Roadmap at 5.
70 European Roadmap at 7.
71 European Roadmap at 5.
72 European Roadmap at 7.
73 European Roadmap at 13.
74 European Roadmap at 14.
75 www.developpement-durable.gouv.fr/Drones-civils-loisir-aeromodelisme.
76 Ibid.
77 Decree 11.04.2012 (May 10, 2012).
78 www.federation-drone.org/les-drones-dans-le-secteur-civil/la-reglementation-francaise/.
79 http://www.activedrone.com/reglementationdronedgac/reglementation francaisepilotedrone-en.html.
80 www.developpement-durable.gouv.fr/IMG/pdf/EXPLOITANTS_DRONES_AU_02072014_cle093db9-1.pdf.
81 www.lba.de/EN/Home/home_node.html.
82 Nuno Garoupa & Jud Mathews, *Strategic Delegation, Discretion, and Deference: Explaining the Comparative Law of Administrative Review*, 62 *Am. J. Comp. L.* 1, 19–20 (2014) (describing institutional structure and applying theoretical model to the culture of German administrative law).
83 Jan S. Oster, *The Scope of Judicial Review in the German and US Administrative Legal System*, 9 *German L. J.* 1267, 1271 (2008).
84 § 16 paragraph 1 point 7 Air Traffic Regulations (LuftVO).
85 Bundesministeriuym für Verkehr, Bau, und Stadtentwicklung, Kurzinformation über die Nutzung von unbemannten Luftfahrtsystemen, www.riot.ch/wp-content/uploads/2014/01/BMVBS-Kurzinformation-unbemannte-Luftfahrtsysteme-10.2013.pdf.
86 Constitution of India, art. 54, www.lawmin.nic.in/olwing/coi/coi-english/Const.Pock%202Pg.Rom8Fsss(9).pdf.
87 Const. of India, chap. IV.
88 Directorate General of Civil Aviation, Public Notice, File No. 05-13/2014-AED (7 Oct. 2014), www.dgca.nic.in/public_notice/PN_UAS.pdf.
89 Part II, Title I, Const. of Italy, www.senato.it/documenti/repository/istituzione/costituzione_inglese.pdf (establishing bicameral parliament).
90 Italian Const. Part II, Title III, art. 94.
91 Italian Const. at Part II, tit. II., art. 83.
92 Ibid. art. 87.
93 Italian Const. Title IV, art. 104.

94 Vera Parisio, The Italian Administrative Procedure Act and Public Authorities' Silence, 36 Hamline L. Rev. 3 (2014) (describing Italian system for judicial review of agency action, and criticizing lack of reviewability of agency inaction).
95 Italian Const. Title IV, art. 103.
96 www.enac.gov.it/servizio/info_in_english/.
97 www.enac.gov.it/repository/ContentManagement/information/N1220929004/Reg%20SAPR%20english_022014.pdf.
98 Ibid. art. 6(1).
99 Ibid. art. 8(18).
100 Ibid. art 8(5).
101 Ibid. art. 8(4).
102 Ibid. art. 18.
103 Ibid. art. 18(2).
104 Ibid. art. 8(2).
105 Ibid. art 8(7)–8(8).
106 Ibid. art. 8(9).
107 Ibid. art 8(10)(non-critical); Ibid. art. 8(11) (critical).
108 Ibid. art. 13(1)(a).
109 Ibid. art. 13(1)(b).
110 Ibid. art. 16(5).
111 Ibid. art. 17(2)–(3).
112 Ibid. art. 20.
113 The authors appreciate research assistance for this section from Chicago-Kent research librarian Scott Vanderlin.
114 www.japan.kantei.go.jp/constitution_and_government_of_japan/constitution_e.html
115 Const. of Japan, ch.1 (role of Emperor).
116 Const. of Japan ch. IV.
117 Const. of Japan, ch. 5, arts 65–66.
118 Const. of Japan, ch. V, art. 67.
119 Ibid. art. 68.
120 Const. of Japan ch. VI, art. 76.
121 Cheng-Yi Huang & David S. Law, Proportionality Review of Administrative Action in Japan, Korea, Taiwan, and China at p. 6 (Oct. 28, 2014), Washington University School of Law Legal Studies Research Paper Series Paper No. 14-08-07, papers.ssrn.com/sol3/papers.cfm?abstract_id=2496220.
122 www.mlit.go.jp/koku/15_hf_000020.html.
123 Civil Aeronautics Act (Act No. 231 of 1952), www.cas.go.jp/jp/seisaku/hourei/data/caa.pdf.
124 Ibid. arts. 3–9.
125 Ibid. arts. 22–36.
126 Ibid. arts 57–99–2.
127 "Notwithstanding the provisions of Articles 65 and 66, any aircraft equipped with apparatus which enables it to fly without being boarded by a pilot may, when permitted by the Minister of Land, Infrastructure, Transport and Tourism, engage in flight without being boarded by any pilot under the provisions of the said articles. (2) The Minister of Land, Infrastructure, Transport and Tourism may, in granting permission set forth in the preceding paragraph, impose flying restrictions on the said aircraft, when he/she deems it necessary to prevent any dangerous effects on other aircraft." § 87, Civil Aeronautics Act, Act No. 231 of 1952, as amended by Act No. 118 of 2006.
128 §§ 209-3, 209-4, Ordinance for Enforcement of the Civil Aeronautics Act, Ordinance of the Ministry of Transport No. 56 of July 31, 1952, www.japaneselawtranslation.go.jp/law/detail/?printID=&ft=1&re=02&dn=1&x=63&y=20&co=02&ia=03&ky=airport&page=6&vm=02.

129 easyJet Airline Co. v. CAA, Case No: CO/4884/2008, [2009] EWHC 1422 (Admin) (26 Jun. 2009) para. 58.
130 CAP 393 Air Navigation Order (Apr. 2015), publicapps.caa.co.uk/docs/33/CAP%20393%20Fourth%20edition%20Amendment%201%20April%202015.pdf, UK Statutory Instrument 3015/2009.
131 CAA Basic Principles of Unmanned Aircraft, www.caa.co.uk/Commercial-Industry/Aircraft/Unmanned-aircraft/Small-unmanned-aircraft/ (summarizing requirements of Article 166 of ANO 2009) [hereinafter "Article 166 Summary"].
132 CAA, CAP 722: Unmanned Aircraft System Operations in UK Airspace (Mar. 2015), publicapps.caa.co.uk/docs/33/CAP%20722%20Sixth%20Edition%20March%202015.pdf.
133 SRG1320 Application for Operation of a Small Unmanned Aircraft (SUA) in UK Airspace (Aug. 2015), publicapps.caa.co.uk/docs/33/SRG1320Issue05.pdf.
134 SRG1320, at item 7.
135 Ibid., item 5.
136 Ibid., item 7.
137 Ibid., item 4.
138 Ibid., item 6.
139 Ibid., note 8.
140 Article 166 Summary at note 2.
141 Ibid., item 11.
142 SRG1321 Application for Operation of an Unmanned Aircraft (UA) over 20kg in UK Airpsace (Sep. 2014), publicapps.caa.co.uk/docs/33/SRG1321Issue03.1.pdf.
143 Civil Aviation Authority, Information Notice No. IN-2014/184 (Nov. 12, 2014), www.caa.co.uk/docs/33/InformationNotice2014184.pdf (superseding IN-2014/179).
144 Curtis A. Bradley & Jack L. Goldsmith, *Customary International Law as Federal Common Law: A Critique of the Modern Position*, 110 *Harv. L. Rev.* 815 (1997).
145 Curtis A. Bradley, *Breard, Our Dualist Constitutional, and the Internationalist Conception*, 51 *Stanford L. Rev.* 529 (1999) (arguing against monist theory for the United States).
146 Melissa A. Waters, *Creeping Monism: The Judicial Trend Toward Interpretive Incorporation of Human Rights Treaties*, 107 *Colum. L. Rev.* 628, 640–641 (2007) (noting adherence of European states in the civil law tradition to embrace monism).
147 www.icao.int/about-icao/Pages/default.aspx.
148 www.icao.int/about-icao/AirNavigationCommission/Pages/default.aspx.
149 Paul Stephen Dempsey, *Compliance & Enforcement In International Law: Achieving Global Uniformity In Aviation Safety*, 30 *N.C. J. Int'l L. & Com.* Reg. 1, 22–24 (2004) (providing detail on ICAO; analyzing US cases on direct enforcement of ICAO standards); British Caledonian Airways v. Bond, 665 F.2d 1153, 1161 (D.C. Cir. 1981) (finding some, but not all, ICAO standards to be directly enforceable).
150 ICAO, Unmanned Aircraft Systems (UAS) (2011), ICAO Circular 328-AN/190, www.icao.int/Meetings/UAS/Documents/Circular%20328_en.pdf.
151 2011 ICAO Report para. 2.10.
152 ICAO 2011 Report at para. 3.12.
153 Ibid. para. 3.15.
154 ICAO, Manual on Remotely Piloted Aircraft Systems, Doc. 10019, AN/507 (2015), www.wyvernltd.com/wp-content/uploads/2015/05/ICAO-10019-RPAS.pdf.
155 Catalin Radu, *ICAO Vision* (Mar. 23–25, 2015), www.icao.int/Meetings/RPAS/RPASSymposiumPresentation/Day%201%20Session%201%20Catalin%20Radu%20-%20ICAO%20Vision.pdf.

156 www.icao.int/Meetings/UAS/Documents/09_Honorato-Roberto_JARUS_Brazil_Presentation%20v1.0.pdf.
157 rpas-regulations.com/index.php/community-info/jarus/file/969-jar-doc-01.
158 jarus-rpas.org/sites/jarus-rpas.org/files/jar_02_doc_-_jarus_rpas_c2_link_rcp_-_10_oct_2014_1.pdf.
159 Ibid. sec. 2.2.1.
160 Ibid.

8 Regulatory philosophies and constraints in the US

8.1 Introduction

The Federal Aviation Administration (FAA) oversees aviation in the United States. The FAA has two distinct responsibilities. One is to regulate aviation; the other is to operate the air traffic control system.

US aviation regulation stands on three conceptual pillars: aircraft certification,[1] pilot certification, and operating rules. In order to fly legally, (1) a pilot with an appropriate certificate for the operation must fly (2) a certificated aircraft (3) in conformity with the operating rules. If the flight is noncompliant with any of the three requirements, the operation is illegal, subjecting the operator to civil monetary penalties, license suspension or revocation, and possible injunction. Criminal penalties under federal law are limited to security violations.

To satisfy its mandate to integrate drones into the national airspace system, the FAA must develop regulations in each of these three areas. The statute recognizes that different regulatory requirements are appropriate for microdrones and macrodrones, as does the FAA. As of the end of 2015, no standards existed for certification of drone aircraft; no standards or procedures existed for certification for drone operators (DROPs), and no operating rules existed for drone flight. In February 2015, however, the FAA issued a proposed regulation for microdrones and while it considered several comments on the proposal, it issued more than 5,000 section 333 exemptions allowing commercial drone operation on a case-by-case basis.

8.2 Basic regulatory concepts for aviation

Drones are classified as *aircraft* under the extremely broad statutory definition, and therefore they could be flown by persons holding pilots' certificates in conformity with the operating rules, except that the vehicles themselves cannot have certificates because no mechanism exists for obtaining one.

Moreover, pilots are privileged to fly under their airman certificates only for those categories, classes and types of aircraft for which they hold a rating. Most light airplanes and helicopters do not require type ratings, but a pilot who is certificated only for the rotorcraft category or only for the airplane category, or for both, does not have the privilege of flying the newer multicopters that predominate, because these vehicles are neither rotorcraft nor airplanes. Therefore under a strict reading of the rules, a pilot holding even the highest rating for any existing category and class does not have privileges extending to drones.

So the development of a general regulatory regime for drones is underway. A rational regulatory regime for any human activity should contain only those requirements and restrictions associated with reduction of identifiable risks. This is so for aircraft and airman certification and for operating rules. For example, aircraft certification should ensure that an aircraft will not experience a structural failure when it encounters turbulence of a magnitude that can be expected where and when it flies. When it is flown within its normal operating envelope, it should not be prone to an upset, such as a spin, from which an ordinary pilot will be unable to recover. Its performance, such as takeoff and landing distances, at various weights, should be known by pilots.

Airman certification is aimed at assuring that an individual possesses the requisite knowledge and skills to fly an aircraft safely under all of the flight conditions he is likely to experience, including emergencies. At minimum, this includes being able to take off and land the aircraft without creating hazards to himself or to other persons or property. It means that he should be able to continue to fly safely even if certain systems, such as navigational equipment, electrical power, or some flight instruments, fail. He should be able to keep the aircraft under control and land it safely even if one or more engines fail. On the other hand, certain emergencies are beyond recovery by any pilot. If a wing or a main rotor blade separates from an airplane or helicopter, no pilot can recover. The risk of these extreme conditions occurring is addressed in aircraft certification and not pilot certification.

Pilots must not only have the requisite skills, as discussed immediately above; they also must be knowledgeable about operating rules, the general characteristics of the aircraft they are certificated to fly, and certain basic theoretical subjects, such as aerodynamics and meteorology. Pilots must operate aircraft only within their published speed, weight, and turbulence limitations, according to clearances from air traffic control. To do this, they must know the aircraft limitations and ATC procedures.

Operating rules are aimed primarily at reducing the likelihood of collisions between aircraft and between aircraft and objects on the ground and of reducing the risk of operating an aircraft outside its certificated capabilities. Part of the strategy for collision avoidance is to require communication with air traffic control personnel and compliance with their instructions. For example, the operating rules prohibit pilots from flying in controlled airspace without communicating with ATC facilities and complying with clearances.

8.3 Organizational structure of regulation

Drone regulation in the United States is largely a product of rulemaking and adjudicatory powers delegated to the FAA in the Federal Aviation Act, subject to policy oversight by the Office of Information and Regulatory Affairs in the White House Office of Management and Budget. Like all regulation, it must satisfy the requirements of the federal Administrative Procedure Act.[2] The FAA has begun the process of establishing regulations for drones through a proposed general rule and several thousand specific exemptions for commercial activities while it collects more data through six test centers. The following subsections address each of these structural aspects of drone regulation.

8.3.1 Administrative procedure

The federal Administrative Procedure Act applies broadly to administrative action at the federal level, regardless of whether it is taken by an independent agency such as the FCC or by an agency contained within an executive department such as the FAA. Under the APA, rules may not become final until they have been subjected to a notice-and-comment process, originated by publishing a *notice of proposed rulemaking* (NPRM) in the Federal Register. Agencies sometimes like to precede an NPRM with a less formal *Advance Notice of Proposed Rulemaking*, *Request for Comments*, or *Notice of Inquiry*. Any person is entitled to submit a comment. The collection of all the comments submitted on a particular NPRM comprises the rulemaking docket. When the agency decides to finalize a proposed rule, it must explain its evaluation of comments and justify the final rule in light of the comments. This is known as *informal rulemaking*. Article III courts review agency compliance with these requirements and regularly send rules back to the agency for reworking their content or the agency's justification. This is so even though courts owe considerable deference to agency policymaking and factfinding.

A small subset of rules must be developed through a more formal adjudication-like process, when the governing statute—the one giving agency substantive authority—so prescribes. This is known as *formal rulemaking*.

Enforcement actions, known as *adjudication*, are handled separately, through a process modeled on civil litigation.

8.3.2 Federal Aviation Act

Aviation regulation in the United States is the responsibility of the Federal Aviation Administration (FAA). The FAA is an administration within the executive Department of Transportation.[3] It is headed by an Administrator, appointed by the President, subject to Senate approval.[4] Although the Secretary of Transportation has general authority over the personnel and activities of the FAA,[5] the Administrator is the final authority for FAA rules.[6] In promulgating or amending rules, the Administrator must comply with the Administrative Procedure Act, must act on petitions for rulemaking within six months, and must issue final rules no later than 16 months after the last day for public comment on proposed regulations.[7] Major rules, including those that raise significant legal or policy issues, must be approved by the Secretary in advance of promulgation.[8]

The Administrator is obligated to "encourage the development of civil aeronautics and safety of air commerce ..."[9]

The statute gives the FAA authority to prescribe air traffic regulations[10] and to set minimum standards for aircraft and their propulsion and other systems;[11] for inspections of aircraft and their systems;[12] issuance of airman certificates, type certificates, production certificates, and airworthiness certificates.[13]

The statute prohibits any "person" from operating a civil aircraft without an airworthiness certificate, from serving as an airman without an airman certificate, from employing a person as an airman who does not have an airman certificate, from operating an aircraft in violation of a regulation or the terms of a certificate,[14] and from operating an aircraft unless it the aircraft is registered with the FAA.[15]

The Federal Advisory Committee Act does not apply to aviation rulemaking committees designated by the Administrator.[16]

8.3.3 FAA

The FAA is a component of the Department of Transportation. The Administrator of the FAA reports directly the Secretary of Transportation, who has delegated to the Administrator most aviation functions statutorily given to the Secretary.

Aviation regulation in the United States stands on three building blocks: certification of the airworthiness of aircraft, certification of pilots and other airman, and issuance and enforcement of operating rules.

Airworthiness certification is a multi-year, complex process that reviews every detail of a new aircraft design and all its subsystems, culminating in a *type certificate*. A type certificate is specific to a particular model or cluster of similar models of an aircraft. Aircraft designers must establish that the basic design of an *airframe* (the aircraft without its propulsion system) meets requirements for stability and control and for structural strength in various flight conditions. Materials and the way they are connected to form fuselage, airframe, control services, and landing gear must meet strength and operability requirements. The designer must demonstrate how the aircraft recovers from unusual attitudes and publish appropriate flight limitations in the airworthiness certificate and in the manuals that are mandatory for pilots, other aircrew members, and mechanics. Instruments for determining aircraft attitude such as compasses, altimeters, airspeed indicators, rate-of-climb indicators, turn indicators, and artificial horizons must meet technical requirements established by international organizations and by the Defense Department. Automation systems, such as autopilots, flight planning and navigation systems and fully automatic digital engine control systems (FADECs) must meet similar technical requirements and demonstrate their reliability, with redundancy and backup systems provided as necessary to meet reliability targets. Aircrew interfaces are part of this certification.

Propulsion systems must meet similarly detailed, but separate, certification requirements, prescribing the design and capability of components such as compressor and turbine blades, shafts, burner cans, pistons, cylinders, fuel delivery pumps, and ignition subsystems. The designer must calculate the theoretical safe time between overhaul (TBO) and disclose it in conjunction with sale.

Most of the actual testing is accomplished by manufacturers rather than by the FAA, but the airworthiness certification rules prescribe the test protocols in detail. Almost all design is an iterative process. A design is tested, sometimes leading to the discovery that it does not meet goals, necessitating redesign and retesting. By some estimates, airworthiness certification of a new complex airplane design costs $20 million or more. The FAA is under considerable pressure from the aviation community to reduce the costs for airworthiness certification, which is seen as a barrier to innovation.

Despite decades long efforts to simplify the process for experimental and homebuilt aircraft, the airworthiness and type certification process for them is burdensome as well. The FAA evaluates the plan for amateur-built aircraft[17] Builders must submit applications for registration 60–120 days because contemplated completion of assembly.[18] Before granting an airworthiness certificate, the FAA inspects amateur-build aircraft[19] including "an onsite, visual, general airworthiness certification inspection of the aircraft,"[20] and recommends involvement of designated airworthiness representatives (DARs) before that occurs.[21] The inspection may require some disassembly.[22] The FAA inspection includes review of inspections by certificated mechanics and other participants covering the aircraft, engine, and propeller or rotor blade(s). Builders typically must provide photographs documenting construction details. The inspection and records review substantiates sound workmanship methods, techniques, and practices.[23]

The inspection is followed by a flight test "appropriate for the applicant to show the aircraft is controllable throughout its normal range of speeds and maneuvers and that the aircraft has no hazardous operating characteristics or design features."[24] Flight tests must accumulate 25–40 hours[25] and follow recommended flight test procedures.[26] The FAA may require additional flight test hours.[27]

The airman certification process, as explained in chapter 4, comprises a progressive system of ground and flight instruction under curricula prescribed and approved by the FAA, culminating in a written knowledge test and a practical flight test administered by an FAA inspector or by a private *designated pilot examiner*. Distinct requirements exist for varying levels of certificate, ranging from sport/recreational pilot at the bottom, through private pilot, commercial pilot, and airline transport pilot at the top. It all starts with a student pilot certificate, which requires no examination, but does require registration with the FAA, a medical certificate, and close supervision by a CFI.

The necessary flight instruction is conducted by hundreds of mostly private flight schools scattered around the country. The vast majority of them are small private operators using young commercial pilots with CFI ratings as instructors while they build time to move up the career ladder.

Each category of certificate authorizes the holder to operate only certain classes and categories of aircraft and, for more complex categories, only certain types or models of aircraft. An instrument rating, qualifying the pilot to fly under instrument flight rules, is a separate certificate that can be attached to any of the basic levels.

The FAA traditionally has adopted regulations or modifications of existing regulations through a network of industry advisory committees. Indeed, it is rare for the FAA to take rulemaking action without it representing a consensus of industry groups. Industry consensus is relatively stable regarding most aspects of aviation regulation, but when a new technology and new capabilities such as that associated with drones presents itself to the FAA, its advisory committee process is slow to react. Many different elements of interest-based positions take time to crystallize, as explained in chapter 6, and the advisory committees provide a framework for interest-group representation.

Certificated airmen fly certificated aircraft within an elaborate body of operating rules prescribing procedures for taking off and landing, flying from point-to-point so as to minimize the risk of collision with other aircraft, receiving and following instructions from FAA air traffic controllers in visual and instrument meteorological conditions, and avoiding terrain and other obstacles depicted on a comprehensive set of aeronautical charts issued by the FAA and its contractors in paper and electronic form.

Public aircraft, defined in the Federal Aviation Act,[28] are aircraft operated by federal, state, or municipal governmental entities and their contractors.[29] Public aircraft are exempt from the airworthiness certification and licensed-pilot requirements of the FARs, and from many, but not all, of the operating rules. Some of the operating rules in Part 91 apply only to *civil aircraft*, thereby excluding public aircraft. For example, 14 C.F.R. § 91.7, requiring an airworthiness certificate, applies only to civil aircraft, while other operating rules apply to all aircraft.[30] 49 U.S.C. § 44711 prohibits operation of a *civil* aircraft without an FAA airman certificate, thus exempting public aircraft from the requirement to have licensed pilots. The practical effect is to allow public aircraft to be operated by pilots holding either military pilot certification or civilian pilots' licenses.

Compliance with FAA rules is assured by the leverage it has over pilot, mechanic, and aircraft certificates and by a system of civil penalties that can be imposed by the agency, subject to higher-level administrative and judicial review. Penalties against operators and mechanics are reviewed by higher level bodies within the FAA before they get to court; actions against pilots are subject to administrative review by the National Transportation Safety Board.

8.3.4 OIRA

The Office of Information and Regulatory Affairs (OIRA) is an organizational unit within the Office of Management and Budget in the Executive Office of the President.[31] Under a series of executive orders, all regulatory agencies must submit proposed rules and amendments to existing rules to OIRA for clearance before they can be published in the Federal Register—the first step in the public rulemaking process. The purpose of OIRA review is to assure greater political accountability by regulatory decision-makers. It gives the elected president of the United States leverage to influence agency decisions.

Continuing controversy exists over the role of OIRA. Even now, independent regulatory agencies such as the Federal Communications Commission and the Consumer Product Safety Commission need not submit draft rules to OIRA. Little doubt exists, however over the president's power to require executive branch agencies like the FAA, which are part of executive departments directly accountable to the president through his power to appoint cabinet secretaries, to submit to OIRA review.

OIRA review may result in a return letter, asking the agency to perform additional analysis before issuing a rule, a review letter, urging additional analysis, or a prompt letter, asking an agency to give priority to an issue.[32]

OIRA review is less transparent then agency rulemaking activities.[33] The Administrative Procedure Act does not apply to OIRA because it is a part of OMB, which is part of the immediate presidential governing apparatus. OMB emerged from the Bureau of the Budget, established in the 1921 Budget and Accounting Act to give the president control over the federal budget process. The Bureau of the Budget was renamed Office of Management and Budget and received additional authority to oversee the management of federal agencies under President Nixon in 1970.

8.3.5 Structure of drone regulation

The FAA Modernization and Reform act of 2012 contains specific provisions requiring the FAA to promulgate regulations to integrate drones into the National Airspace System. Under these provisions, the FAA was required to, and did, issue a Comprehensive Plan and a Roadmap setting forth a schedule of drone integration activities extending into the 2020s. The Congress envisioned classifying drones by weight and operational capability. Section 333 obligates the FAA to consider whether the risks and characteristics of microdrones could make them eligible for earlier operation under a greatly simplified regulatory regime. The statute also requires the establishment of test centers at various places around the country, where research can be connected and data collected to inform the FAA's regulatory decision-making. The FAA established six test centers in 2013.

The FAA's position, ratified by before National Transportation Safety Board in November 2014, is that drones, at least when they are flown for commercial purposes, are aircraft, and therefore, no one may operate them unless they hold airworthiness certificates, unless they are operated by someone holding a pilot certificate, and unless they comply with the requisite flight rules. No specific category of unmanned aircraft exists within the airworthiness certification rules, no category of drone operator exists within the airmen certification rules, and no specific part of the operating rules addresses drone flight, regardless of size.

The 2012 statute envisioned that the FAA would have an initial set of rules in place by 2015, permitting microdrones meeting certain requirements to be operated legally for commercial purposes. By implication, the Congress expected that macrodrones would be subject to something resembling the rules for manned aircraft, developed over a longer period of time, allowing more detailed attention by the FAA and by interest groups to their content.

The Notice of Proposed Rulemaking for microdrones, issued in February 2015, represents the FAA's first tangible regulatory step toward finding the appropriate place for microdrones in the NAS. A few months earlier, the FAA began issuing exemptions under the authority of section 333 of the 2012 Act, waiving airworthiness certification for microdrones, allowing commercial operations on an applicant-by-applicant basis, and imposing operating limitations.

The final rule makes only a few changes to the proposal, generally increasing flexibility for drone operation beyond what was proposed. The FAA invited waiver applications to develop a record to lifting some of the limitations in the final rule.

8.4 Risk assessment

Crafting a regulatory regime for drones must begin with risk assessment. The risks associated with unmanned aircraft flight are similar in some ways and different in other ways from those associated with manned aircraft flight. Risks also are different for different weight categories of drones.

The risks to life and property when a drone collides with another aircraft or crashes into persons or property on the ground are proportional to the drone's mass (weight) and the square of its speed as chapter 5 explains.

Accordingly, a risk-based approach to drone regulation begins with classifying drones into different weight and speed categories. The 2012 statute and the FAA's planning documents do this in a crude way by distinguishing between sUAS and UAS with a dividing line 55 pounds. England, France, and recommendations by the UAS America Fund impose greatly simplified requirements for very light microdrones—those below 3 pounds in the UAS

America Fund proposal, more stringent requirements for those in the heaviest category, and moderate requirements for one or more intermediate classifications. The FAA's NPRM makes it clear that the agency is considering a finer breakdown of weight categories in its final rule, specifically including the UAS America Fund proposal.

The regulators must decide what kind of regulation is appropriate for each category: what kind of aircraft standards, operator qualifications are necessary and how they should be assured, and what operating rules are necessary.

Macrodrones have weights approaching, and in many cases, exceeding those of manned aircraft associated with the same types of mission. They, therefore, pose the same risks of death, injury, and damage if collisions occur as many aircraft, with the important exception that aircrews and passengers are not endangered by a collision. Accordingly the logic is compelling that macrodrone regulation should resemble airplane and helicopter regulation more than microdrone regulations do.

8.5 Penalties and enforcement

For violating the Federal Aviation Act itself, FAA regulations, or any term of a certificate, the FAA may impose civil penalties,[34] or revoke, suspend, or modify a certificate.[35] Certain violations—mostly of security provisions—also are crimes. The statute imposes civil penalties up to $25,000 per violation on any person who violates registration or safety statutory provisions, FARs or the terms of a certificate.[36]

Enforcement procedures differ depending on whether the FAA proposes a civil monetary penalty or a suspension or revocation of an airman certificate.[37]

It also depends on whether the respondent is an airman—a pilot, a mechanic, or a flight attendant[38]—or someone else. For actions against airmen, the administrative review lies in the National Transportation Safety Board[39] rather than in higher levels of the Department of Transportation. For others, the administrative review process remains within the Department of Transportation. In all cases, the respondent ultimately is entitled to judicial review of the final agency action, whether that action is taken by the NTSB[40] or by the Department of Transportation.[41]

The enforcement process begins with a complaint, which can be filed by anyone.[42] A resulting citation issued to a respondent must specify the violation alleged and propose specific penalties. The person to whom the citation is issued is entitled to respond.[43] At that point, as at any point in the process, the matter might be settled, or it may go to the next level.[44]

Section 13.15 of the FAA penalty rules governs civil penalties imposed without administrative assessment. Section 13.16 governs administrative assessment of penalties against a non-airman, while section 13.18 governs administrative assessment against airmen. Section 13.17 covers seizure of aircraft, while section 13.19 covers certificate actions.

The procedure under section 13.15 applies only if the amount in controversy involves more than $50,000 for individuals and small businesses and $400,000 otherwise. The Administrator first sends a civil penalty letter to the respondent, which must contain a statement of the charges and the statutory or regulatory provision on which they are based, and an offer to settle for a specific amount. The respondent then may present defenses to the Administrator. If there is no settlement and the Administrator rejects the defenses, he presents the matter to the Attorney General for collection of the civil penalty.

In smaller cases, administrative assessment is the next step if the matter is not settled. The process begins with a notice to the respondent specifying the violation and proposing an amount of civil penalty. The respondent may file an answer and, optionally request a conference with the FAA, submitting additional information or documents. If the conference is unsuccessful in resolving the matter, or if none is requested, the procedural paths diverge for airmen and non-airmen.

Under section 13.16, non-airmen may appeal a final notice of penalty by docketing it for a hearing before an administrative law judge (ALJ) within the Department of Transportation. Administrative law judges under the Administrative Procedure Act are quasi-judicial personnel, specially selected and entitled to independence and job security vis-à-vis the agency that uses them to conduct its hearings.[45]

The ALJ makes an initial decision, which becomes the final decision of the FAA if it is not appealed further. If the respondent appeals to the FAA decisionmaker—typically someone in the FAA Chief Counsel's office to whom the Administrator has delegated authority—the FAA decisionmaker must review the record and issue a final order affirming, modifying, or reversing the ALJ's initial decision. The FAA may not, however, increase the amount of the civil penalty.[46]

Airmen, under section 13.18, follow a similar path up to the point after an informal conference, if the respondent requests one. Thereafter, an airman, rather than requesting a hearing before an ALJ, requests a final order, which enables him to request a hearing before the NTSB.[47] The respondent is entitled to a live hearing before an ALJ at the NTSB.

Other civil penalty assessments affirmed by an ALJ[48] are subject to limited administrative review before the Administrator.[49]

Once an ALJ is assigned, she must receive pleadings and, if requested by either party, must conduct a live hearing, generally following the Federal Rules of Civil Procedure and the Federal Rules of Evidence, although ALJs have discretion to deviate from the stringent application of those rules in order to improve efficiency and promote justice.[50]

After hearing the evidence presented by both parties, the ALJ issues an initial decision, which becomes final agency action eligible for judicial review, unless one of the parties seeks higher level administrative review.

Review in these agencies is deferential; the reviewing body accepts the facts found by the ALJ unless they are unsupported by substantial evidence in the record taken as a whole. On matters of law and policy, however, the reviewing body engages in de novo (fresh) decision-making. Things are more complicated when the NTSB is the reviewing body, because it owes deference to the FAA on matters of policy.[51]

Once administrative review is final, either party may petition for review in the United States Court of Appeals for the D.C. Circuit, or in the judicial circuit where the respondent lives or does business.[52] NTSB penalty decisions are subject to judicial review under section 46110.[53] Judicial review of Administrator decisions is available under section 46110.[54] The reviewing court must accept FAA findings of fact as conclusive, but only if they are supported by substantial evidence.[55]

Judicial review occurs under the usual APA standards: the court must uphold the agency decision unless the challenger can show that it was unsupported by statutory authority, was arbitrary and capricious or otherwise not in accordance with law, violated constitutional rights, or that it was unsupported by substantial evidence in the record taken as a whole.[56] Under the United States Supreme Court decision in *Chevron*,[57] agencies are entitled to deference on their interpretations of the law that governs them, but only when there is an indication that Congress gives them that statutory interpretation power. Once the court of appeals has decided the petition for review, the Supreme Court has the power to review the case further if it issues a writ of certiorari. Whether to grant a writ is entirely discretionary.

An alternative pathway of review exists in the United States district courts. Those courts have jurisdiction to enforce penalties imposed by the FAA,[58] and to issue injunctions against violations.[59] The Government must commence a district court action to collect a civil money penalty because only courts can take property to satisfy citizen obligations to the government; agencies lack that power. When the FAA wants to enforce a civil penalty in district court, it must go through the Justice Department, which typically means that the US attorney in the judicial district where the respondent is found will file a civil action against the respondent. The Justice Department has discretion whether to accept the FAA's recommendation to file an action.[60]

District courts also have authority to entertain actions for injunctions against certain illegal agency action,[61] and to compel agency action unlawfully withheld. When a district court exercises these powers, it is obligated to evaluate the lawfulness of agency action, which necessarily draws it into scrutiny of specific issues and conclusions supporting the action. The district court has the power—indeed the obligation—to decide a defense presented by the respondent claiming that the penalty is invalid because the agency acted illegally, arbitrarily, in contravention of factual evidence, or otherwise in violation of law.

When certificate suspension or revocation is involved, the FAA itself has the power to execute the penalty when all agency and judicial review is complete: it just suspends or revokes the certificate, and that makes it so. An airman or operator conducting activities under a revoked or suspended certificate commits a new violation. In contrast, when money penalties are involved, the administrative agencies like the FAA lack the power to take property from an individual or enterprise to satisfy the penalty; only courts can do that. That is why the FAA must go to district courts to enforce penalties. When a civil penalty award has been reduced to a judicial judgment, however, the usual mechanisms for executing judgments by attaching property and selling it are available.

Incarceration is not available as a penalty in civil actions, although a participant might commit contempt of court, for example, by refusing to turn over assets attached by the court in conjunction with judgment execution, in which case, he can be incarcerated for civil or criminal contempt.

8.6 Integration into National Airspace System

8.6.1 The FAA's ban on commercial drone operations

Outside the final drone rules, the FAA takes the position that it is illegal to fly a drone commercially without special authorization from the FAA under a section 333 exemption or otherwise. One affected by this ban would look in vain for any express statutory prohibition on commercial drone operation. This section explains how the ban arises. 49 U.S.C. § 44711 prohibits "a person" from: operating a "civil aircraft in air commerce" without an airworthiness certificate or in violation of such a certificate; operating such an aircraft as a pilot without an appropriate airman certification or in violation of regulations employing an airman such as a pilot without an appropriate certificate. 49 U.S.C. § 40102(a) defines "air commerce" as "foreign air commerce, interstate air commerce, the transportation of mail by aircraft, the operation of aircraft within the limits of a Federal airway, or the operation of aircraft that directly affects, or may endanger safety in, foreign or interstate air commerce." "The same section defines aircraft" as "any contrivance invented, used, or designed to navigate, or fly in, the air." A "civil aircraft" is "an aircraft except a public aircraft." A "public aircraft" is an aircraft owned or operated for a federal, state, or local governmental entity. 49 U.S.C. § 44701 authorizes the FAA Administrator to prescribe regulations and standards for "civil aircraft in air commerce," and other matters to promote safety. Section 44702 authorizes the Administrator to issue airman (pilot), aircraft airworthiness and type, and operator certificates. Drones are "contrivance[s] used and designed to "navigate, or fly in, the air." Thus to operate one that does not have an airworthiness certificate, to operate one without a pilot certificate, or to operate one in violation of air traffic regulations is illegal.

Under its interpretation of these statutory provisions the FAA prohibits commercial flight of drones unless civilian operators comply with the final microdrone rule or obtain a section 333 exemption or an experimental airworthiness certificate and a certificate of waiver and authority (COA), both of which limit operations to specific operators.

The final rule adds a new Part 107, which covers microdrone airworthiness, DROP certification, and operating rules and exempts microdrones from other parts of the FARs that would impose conflicting requirements. Everything outside the scope of Part 107 remains subject to manned-aircraft rules unless a COA is obtained.

8.6.2 The 2012 Act

The FAA Modernization and Reform Act of 2012,[62] requires the FAA to "develop a comprehensive plan to safely accelerate the integration of civil unmanned aircraft systems into the national airspace system," by November 10, 2012,[63] and to publish a five-year "roadmap," by February 14, 2013.[64] Specific provisions[65] require that the plan must provide for the integration of drones into the national airspace system no later than September 30, 2015,[66] promulgate a final rule to permit operation of microdrones by 18 months after release of the comprehensive plan, and to publish a notice of proposed rulemaking (NPRM) for implementation of the plan by the same date, with a final rule to be effective no later than sixteen months after release of the NPRM.[67]

The 2012 Act requires the FAA to decide if microdrones represent a separable category that can be regulated in a simpler regime than is necessary for macrodrones.[68]

8.6.3 FAA Comprehensive Plan and Roadmap

The 2012 legislation[69] requires the Secretary of Transportation, in consultation with interested parties, to "develop a comprehensive plan to safely accelerate the integration of civil unmanned aircraft systems into the national airspace system."[70]

The FAA released its comprehensive plan on November 6, 2013,[71] one year late, and its roadmap on November 7, 2013,[72] nine months late. The agency's failure to meet the statutory deadlines for these milestones and for release of the NPRM was due, initially, to internal conflicts between regional inspectors and central office policymakers over how accommodating the regulatory environment should be for early operations of microdrones. Later, after the NPRM was released, the delay was attributable to the magnitude of the task of putting together detailed regulatory language and justifications for it, and processing thousands of comments and section 333 exemption petitions.

The FAA's Comprehensive Plan explicitly assumes that routine UAS operation should not require exceptions or unique authorizations.[73] The Roadmap concluded that a regulatory framework for microdrones could precede a framework for macrodrones.[74]

8.6.4 Part 107

The FAA released a rule on June 21, 2016 that allows microdrones to be flown in the national airspace system without any special waiver or exemption.[75] The rule largely adopts the proposed rule published in February, 2015 and embraces a risk-based approach, resulting in relatively light regulation. Under it, microdrones must be flown below 400 ft, within the line of sight of the DROP, and only in the daytime. No conventional pilot certificate is required, but DROPs must pass an FAA-specified knowledge test.

The aircraft themselves need not have airworthiness or type certificates. They may not be flown in class A, B, C, D, or E airspace without prior ATC approval for specific flights. A DROP must inspect the drone before each flight to ensure that it can be flown safely.

8.6.4.1 Weight limits

Part 107 defines a broad category of microdrones weighing up to 55 pounds, using the weight limit specified in the statute.[76] The agency initially intended to subdivide microdrones into different weight categories, with different requirements for each, but ultimately concluded that such an approach would be unduly burdensome and complex and that sufficient data is lacking to adopt principled requirements for the different categories.[77]

Although it rejected it for the final rule, the NPRM provided significant detail on a possible *micro UAS category,* based on the Canadian rule, a proposal by the UAS America Fund and a recommendation from the FAA's Aviation Rulemaking Committee. If the subcategory is adopted, it would limit heights to 400 ft AGL, speeds to 30 knots, and horizontal distance to 1,500 ft from the DROP. Instead of being required to take a knowledge test, DROPs would self-certify. Flights over people would be permissible. The vehicles must be made of frangible materials.[78]

8.6.4.2 Airworthiness and type certification

Part 107 makes it clear that the FAA bent over backwards to avoid subjecting microdrones to traditional airworthiness and type certification. The NPRM explained that it considered a *permit to operate* (PTO) requirement based on detailed description of airframe, control station, and communications link; design following consensus standards; manufacture in compliance with design, reinforced by quality assurance procedures; and complete flight testing.[79] It concluded that such an approach would not be proportionate to the risks posed by microdrones and would not result in significant safety benefits.[80]

Moreover, the traditional airworthiness and type certification process consumes 3–5 years. Subjecting microdrones to such a process would not be practically feasible and would likely result in a particular model becoming technologically obsolete by the time it was certificated.[81] Instead, the FAA concluded that its other risk-mitigating requirements such as DROP certification, limitations on areas of operation, and pre-flight inspection would adequately mitigate risk, without airworthiness or type certification.[82]

The final rule subjects microdrones to the same registration and tail-number display requirements as manned aircraft, allowing microdrones that are too small to accommodate the display size requirements to display their tail numbers at the largest practicable size.[83]

8.6.4.3 Remote pilot certification

In all of the section 333 exemptions granted before release of the NPRM, the FAA required that microdrone DROPs have at least private-level pilots' certificates. Part 107 backs away from that requirement and imposes instead a requirement for a new *remote pilot certificate*, based on microdrone specific knowledge requirements. The NPRM concludes that requiring a conventional pilot's license would impose an unnecessary burden on microdrone operations.[84] Learning how to fly an airplane or helicopter involves acquisition of knowledge and skills that are not applicable to microdrone flight. Moreover, the knowledge and skills required for manned aircraft does not include tools necessary for safe microdrone operation, such as application of the see-and-avoid rule to confined operations and dealing with loss of the control link.[85]

Instead of requiring a pilot's license, the FAA has created a new category of airman[86] known as a *remote pilot*. To be certificated a DROP candidate must pass an initial aeronautical test, and be retested every two years.[87] Unlike pilots, they need not demonstrate any particular level of aeronautical experience or flight proficiency.[88] Demonstration of flight proficiency through a checkride is required of pilots so that they can fly without jeopardizing the safety of people on board, as well as avoiding harm to persons or property on the ground. In order to do this, a pilot must go through a "complex process that includes adjusting aircraft attitude with flight controls, reducing engine power, and scanning for other traffic in order to land the aircraft on the ground after takeoff."[89] In contrast, a DROP can land a multicopter microdrone "simply by pressing the altitude joystick down until the rotorcraft descends to the ground."[90] In other words, microdrones are much easier to fly than airplanes or helicopters.

Moreover, a DROP has the option of sacrificing the microdrone in response to an emergency. A pilot does not have that option unless she wants to risk killing herself and her passengers.[91]

The justification for skills testing of pilots has "at best, a limited applicability" to microdrone operations. Accordingly, Part 107 does not include such certification requirements. Instead, DROPs must pass a knowledge test. Whether a candidate acquires the necessary knowledge through self-study, a commercial training course, or by pilot training is up to the candidate.[92] The areas that would be tested are:

(1) Applicable regulations relating to small unmanned aircraft system rating privileges, limitations, and flight operation
(2) Airspace classification and operating requirements, obstacle clearance requirements, and flight restrictions affecting small unmanned aircraft operation
(3) Official weather sources and effects of weather on small unmanned aircraft performance
(4) Small unmanned aircraft loading
(5) Emergency procedures
(6) Crew resource management
(7) Radio communication procedures
(8) Determining the performance of small unmanned aircraft
(9) Physiological effects of drugs and alcohol
(10) Aeronautical decision-making and judgment,
(11) Airport operations, and
(12) Maintenance and preflight inspection procedures.[93]

The NPRM explains the relationship between each of these areas of knowledge and safe operation of microdrones.[94]

Recurrent knowledge testing would be required every two years, based on the requirement that pilots undergo periodic flight reviews.[95] This would help ensure that a DROP who does not fly regularly, who flies only certain types of missions, or who has simply forgotten what he learned for the initial knowledge test remains current.[96]

One of the justifications for insisting on pilots' licenses in the section 333 exemptions was the need for security screening of DROPs. The final rule satisfies this requirement by subjecting DROP candidates to a TSA *security threat assessment* (STA).[97] This requirement would be backstopped by requiring identity verification through in-person contact with an FAA office, a flight school, or a certified flight instructor.[98]

DROPs would not be required to hold medical certificates, but the rule would prohibit a DROP from flying a microdrone whenever he knows or has reason to know of any physical or mental condition that would interfere with safe operation,[99] such as drug or alcohol impairment, cardiovascular crises, vision problems, or unusual stress that might interfere with the exercise of sound judgment.

8.6.4.4 Operating rules

The most basic operating rules limit microdrone flight to heights less than 400 ft AGL within the line of sight of the operator.[100] The FAA considered but rejected the idea of imposing a quantitative horizontal distance limit, because of a lack of sufficient data to set a principled limit.[101] It makes it clear that a moving limit associated with operation from a moving vehicle is not permissible.[102]

In addition the rules require that, before each flight, the DROP familiarize himself with the operating area and assess risks to persons and property on the ground and in the air, explicitly considering local weather conditions, local airspace and flight restrictions, and other ground hazards.[103]

The rules exclude microdrones from Class B, C, and D airspace and near airports in Class E airspace unless the DROP obtains prior authorization from ATC or the airport manager.[104] It also excludes them from prohibited and restricted airspace.

Before flight, the DROP must inspect the microdrone, following manufacturer-recommended procedures and conducting on-the-ground tests necessary to assure proper operation of safety-critical systems and components. This implicitly requires ensuring that manufacturer-recommended maintenance is complete.[105] The DROP must ensure that the microdrone has enough power to fly its intended mission.[106]

The DROP need not ensure design conformity or reliability probabilities associated with airworthiness certification of manned aircraft, but should look for dents, corrosion, misalignment, loose wires, binding controls, lose fasteners, and excessive wear.[107] In particular he must check the control link by using the DROPCON to verify proper flight control deflection,[108] or, in the case of a multi-copter, to confirm proper asymmetric thrust inputs necessary for controlled flight.

The NPRM discusses risks associated with loss of the control link—what the FAA calls *loss of positive control.*[109] It explains that the FAA considered imposing requirements for technology that would automatically terminate flight if the control link is lost.[110] Ultimately, it concluded that operational alternatives exist to mitigate the risk without imposing technological equipment and airworthiness certification requirements.[111] Imposing such requirements would require considering the different flight characteristics of different configurations of microdrones—longer glides by fixed-wing aircraft as opposed to straight-down or at steeper angles for rotorcraft losing power.[112] Limiting microdrone flight to relatively small confined areas that can be policed by the DROP before flight adequately mitigates the risk.[113]

As for manned aircraft, Part 107 prohibits careless or reckless operation of microdrones so as to endanger the life or property of another.[114]

8.6.4.5 Comments solicited

As significant as the rules proposed are the subjects on which the FAA explicitly invited comment. It discussed the UAS America Fund proposal for a segmented approach based on weight, and concluded that data is insufficient to define different weight categories and to prescribe differences in the way they may operate. It invited comment, however, on the idea that there should be a special category at the low-end, designated as *Micro UAS*, in which microdrones weighing less than 4.4 pounds could be flown without the DROP certificate required for the larger ones.[115]

For the full category, up to the 55 pound limit, the FAA solicited comment on whether pilots' licenses should be required, giving ALPA and the agriculture pilots an additional opportunity to insist on this barrier to entry. It also invited comment on whether skills qualification and aeronautical experience should be required, as they are for pilots' licenses, going beyond the knowledge test requirement.

The NPRM invited comment on whether specific autonomous safety technology should be mandated, such as return to home, which the NPRM called loss of positive control. It also asked for input on whether the line of sight requirement might be relaxed for systems equipped with appropriate first-person-view systems. It invited comment on whether the separate observer requirement included in all of the section 333 exemptions should be a part of the general rule.

It asked for comments and proposals for operation of microdrones from moving vehicles.[116]

8.6.5 Interim frameworks

8.6.5.1 Section 333 exemptions

The FAA has the power to establish special rules for drones and to exempt them from regulations generally applicable to air commerce. Section 332 of the 2012 Act of 2012, commands the Secretary of Transportation to issue a final rule for operation of sUAS in the national airspace system. Until such a rule becomes final, section 333 of the 2012 Act authorizes the Secretary of Transportation to allow operation of certain unmanned aircraft systems "notwithstanding any other requirement of this subtitle." "This subtitle" refers to Subtitle VII of title 49 of the United States Code, which includes sections 40101 to 50105—all of the provisions relating to aviation. That is the FAA's statutory authority for the so-called "section 333 exemptions." The FAA also has other authority for issuing exemptions: 49 U.S.C. § 40109(b) authorizes the FAA Administrator to grant exemptions from any regulation promulgated under sections 40103 (b)(1) and (2) (air traffic regulations and management of US airspace), 40119 (security R&D), 44901 (passenger and freight screening), 44903 (air transportation security), 44906 (foreign air carrier security), and 44935–44937 (security training of personnel, including air carrier pilots) when he decides that the "exemption is in the public interest."

The FAA used the section 333 authority to determine that an airworthiness certificate is not required for a UAS to operate safely in the National Airspace System (NAS). Having made that determination, it proceeding to authorize, by through an application process, certain unmanned aircraft operators to perform commercial operations prior to the finalization of the Small UAS Rule, which "will be the primary method for authorizing small UAS operations once it is complete."[117]

In September 2014, the FAA granted the first six section 333 exemptions to motion picture and television producers and their contractors, a move that the New York Times heralded as a "a leap forward."[118]

By the end of 2014, however, the agency had approved only about a dozen section 333 exemptions. In April 2015, the FAA announced a "summary grant" process to speed up the processing petitions for exemptions.

Now, the agency issues a summary grant when it finds it has already granted a previous exemption similar to the new request.

The summary grants relax the requirement in the initial grants that the DROP hold at least a private pilot's license, now allowing operations under section 333 exemptions by recreational or sport pilot certificate holders and pilots with a valid driver's license to satisfy the medical requirement, rather than an FAA medical certificate.[119]

By mid-2016, the FAA had granted more than 5,000 summary section 333 exemptions to operators in various industries. Chapter 1 shows the pattern of grants to operators in different industries.

All section 333 exemptions are accompanied by a "blanket" 200-ft nationwide COA limiting flight to less than 400 ft AGL and further restricting flight near airports, in restricted airspace, and above densely populated areas. An operator who wants to operate outside the parameters of the blanket COA must apply for a separate COA specific to the airspace required for its operation.[120]

The combination of an exemption and its associated COA impose limitations similar to those in Part 107, except for the pilot-certificate requirement rather than the remote pilot certificate proposed in Part 107.

8.6.5.2 Registration requirement

In December 2005, the FAA issued an interim final rule requiring registration of all microdrones weighing more than 0.55 pounds.[121] The rule, subject to amendment as comments are received, required both commercial and hobbyist and recreational users to register their vehicles, with somewhat different requirements. Hobbyists may not fly their drones unless they register by February 18, 2016. Commercial operators may choose between traditional paper-based registration and the new web-based registration after March 31, 2016. Hobbyist user registration covers all drones flown by that particular hobbyist, who is issued an FAA registration number that must be affixed to each grown. Commercial operators must register each drone separately and obtain a registration number for each drone. Hobbyists need provide only basic name, address, and email information. Commercial operators must also provide information on manufacturer, model, and serial number.

8.6.5.3 Special Airworthiness Certificates

Before the FAA began its section 333 exemption process, microdrones, even tiny ones like the Hubsan X4—a 35-gram, $250 toy—could not be flown for commercial purposes unless they had a special airworthiness certificate. The application for special airworthiness certificates[122] provides, among other things:

- Only manufacturers can obtain airworthiness certificates for production flight testing[123]
- Flight is limited to geographic areas specified in the SAC[124]
- Applicants must submit "flight manuals" and checklists, evidence of a training program for crewmembers, who must be licensed pilots or have completed an FAA-approved training program[125]
- FAA inspection of aircraft, control stations, and support equipment[126]
- The microdrone must be equipped with a transponder[127]
- Two-hour advance coordination with ATC[128]
- Annual inspection by a certified mechanic.[129]

Civilian operators are only permitted experimentation, demonstration, and training operations under experimental airworthiness certificates.

Existing microdrones, even the larger ones, cannot satisfy these requirements. One might try to defend the FAA's position by arguing that the requirements are necessary to ensure safety of other aircraft and persons or property on the ground. But this is not so.

The contents of the application for a special airworthiness certificate, especially Appendix A of the order, are manifestly unsuited for microdrones, given their size, payload, and flight profiles. The transponder requirement, the geographic limitations, and the reference to chase planes are entirely inconsistent with the features of microdrones. The section 333 exemption process, and prompt adoption of the NPRM is far better suited to the realities of microdrones.

8.6.5.4 Certificates of Waiver and Authority

Except for Part 107 and the section 333 process, federal, state, and local governmental entities may not fly drones in the national airspace unless they obtain a Certificate of Waiver of Authorization (COA) from the FAA.[130] The 2012 Act requires the Secretary to expedite issuance of COAs for public UAS.[131]

Section 334 of the FAA Modernization and Reform Act of 2012[132] requires the FAA, within 90 days of enactment, to enter into agreements with government agencies to "simplify the process for issuing certificates of waiver or authorization" for small drones operated:

(i) within the line of sight of the operator;
(ii) less than 400 ft above the ground;

(iii) during daylight conditions;
(iv) within Class G airspace; and
(v) outside of 5 statute miles from any airport, heliport, seaplane base, spaceport, or other location with aviation activities.[133]

The FAA reported that it "and the Department of Justice's National Institute of Justice have established an agreement that meets the congressional mandate. Initially, law enforcement organizations will receive a COA for training and performance evaluation. When the organization has shown proficiency in flying its UAS, it will receive an operational COA. The agreement also expands the allowable UAS weight up to 25 pounds."[134]

Pursuant to the Congressional mandate, the FAA simplified its process for considering governmental requests for COAs.[135] It expanded the default period of authorization to 24 months from 12, increased allowable weight to 25 pounds, and announced that it would issue COAs to law enforcement agencies for training and performance evaluation, to be followed by operational authority once the applicant establishes proficiency.[136] It established a web-based application procedure, and provided for expedited procedures for one-time approvals of time-sensitive disaster relief missions.[137]

Law enforcement and other public-safety agencies also are eligible for section 333 exemptions.

8.6.5.5 Remote-controlled model aircraft

Section 336 of the 2012 Act prohibits the FAA from promulgating rules for model aircraft, which it defines as "unmanned aircraft ... (2) flown with visual line of sight of the person operating the aircraft; and (3) flown for hobby or recreational purposes.[138] The prohibition applies only if:

(1) the aircraft is flown strictly for hobby or recreational use;
(2) the aircraft is operated in accordance with a community-based set of safety guidelines and within the programming of a nationwide community-based organization;
(3) the aircraft is limited to not more than 55 pounds unless otherwise certified through a design, construction, inspection, flight test, and operational safety program administered by a community-based organization;
(4) the aircraft is operated in a manner that does not interfere with and gives way to any manned aircraft; and
(5) when flown within 5 miles of an airport, the operator of the aircraft provides the airport operator and the airport air traffic control tower (when an air traffic facility is located at the airport) with prior notice of the operation (model aircraft operators flying from a permanent location within 5 miles of an airport should establish a mutually-agreed upon operating procedure with the airport operator and the airport air traffic control tower (when an air traffic facility is located at the airport)).[139]

Section 336 essentially embraces advisory restrictions for remotely controlled model airplanes under a 1981 FAA Advisory Circular.[140] On September 2, 2015, the FAA replaced the circular with a new one FAA, Model Aircraft Operating Standards, AC 91-57A.[141] Subpart E of Part 107 excludes model aircraft from restricted airspace, and declares that model aircraft operations that endanger the safety of the National Airspace System, "particularly careless or reckless operations" or those that "interfere with or fail to give way to any manned aircraft" may be subject to FAA enforcement action.

Hazards from recreational flight of microdrones or small fixed-wing model aircraft are as great when they are flown for recreational purposes as when they are flown for commercial purposes. YouTube videos featuring FPV flying[142] make it clear that many model aircraft operators regularly fly heavy aircraft far beyond the line of sight and into clouds, where they pose collision hazards.

Model aircraft meeting all the statutory definitional requirements are not covered by Part 107 requirements for commercial operations, but Subpart E reiterates the FAA's statutory authority to bring enforcement actions against model aircraft operators who endanger safety.[143]

The safe harbor for model aircraft is written around traditional practices of well-organized and long-established model aircraft hobbyist organizations such as the Academy of Model Aeronautics. In a traditional model aircraft club, RC hobbyists get together at designated fields as a group and cooperatively fly their aircraft, usually with one person serving as the pilot, and the second serving as an observer. Adherence to safe practices depends on the culture of a particular group and the dynamics of interaction on a particular day, but the clubs have rules, both general and specific for operations for any particular field, and the club members generally follow them, exerting social pressure on anyone who deviates. A visit to an RC hobbyist field, encounters hobbyists with their RC airplanes flying them in pairs, talking and joking about their plans and past exploits. They all know each other. It unlikely that one of them would stray too far from the norm and risk getting kicked out of the club.

If a club member gets interested in drones and buys a DJI Phantom, he is likely to fly it in this fashion—unless he decides to try to make money with it. Then the pathway of the section 333 exemption process and the eventual final rule for sUAS are open to him. His habit of compliance with RC club rules and his general awareness of the FAA probably will cause him to comply rather than just to ignore the restrictions on commercial microdrone flight.

None of this poses any significant new threat to other aircraft or to the citizenry in general. Hobbyists have a good safety record, and commercial microdrone operators are unlikely to put their exemptions and certificates at risk by flouting the FAA's detail rules for commercial operations—whatever their eventual content.

The threat comes from a new quarter: from the thousands of people who got microdrones as Christmas or birthday presents, but have no prior connection with an RC model club or any prior interest in tinkering with model aircraft. Some of them have entrepreneurial instincts; many already are entrepreneurs, especially photographers, freelance journalists, civil engineers, surveyors that mostly account for the 5,000 section 333 exemptions that have been granted, and the much larger number of pending petitions. But the vast majority of these casual purchasers do not plan on starting a business or making arrangements to fly their drone as a part of an RC club activity. They are going to take it out into their backyards, local parks, and nearby schoolgrounds and fly for fun. When they go to a sporting event, a music festival, or some other recreational gathering, they will think about taking their drone for the same reason they take their cameras. They will take their drones on their vacations for the same reasons they take their cameras on vacation: it will be a good way to get some good imagery for their Facebook pages and to record videos to put on YouTube.

This is the source of the greatest threat, not RC hobbyists or commercial microdrone operators. The drone that landed on the White House lawn[144] was not being flown for commercial purposes; it was flown for fun in connection with an alcohol-fueled party. Likewise the incident in King County Washington[145] involved recreational, rather than commercial, drone flight. Arguably, these consumer operations fall outside the statutory safe harbor for RC hobbyists because they are not:

> operated in accordance with a community based set of safety guidelines and within the programming of a nationwide community-based organization.[146]

The Conference Report on the 2012 Act explains:

> In this section the term 'nationwide community-based organization' is intended to mean a membership based association that represents the aeromodeling community within the United States; provides its members a comprehensive set of safety guidelines that underscores safe aeromodeling operations within the National Airspace System and the protection and safety of the general public on the ground; develops and maintains mutually supportive programming with educational institutions, government entities and other aviation associations; and acts as a liaison with government agencies as an advocate for its members.[147]

There is enough ambiguity in the language of section 336, however, to support broad claims that the Congress has placed all forms of consumer drone activity beyond the FAA's reach. The statute may be amended, of course. But any proposed amendment is like to the face ferocious opposition from the RC hobbyist community and therefore is uncertain of passage.

8.6.6 2015 reauthorization

The statutory authority for the FAA under the 2012 Act expires at the end of September 2015. As 2015 drew to a close, the Congress was considering new reauthorization legislation.

Two bills were introduced in the Senate that will shape consideration of the new legislation. S.1314, introduced by Senator Booker,[148] would authorize commercial microdrone operations until the FAA promulgates its final rule. S.1618, introduced by Senator Feinstein,[149] would extend the FAA's authority to "consumer drones"—those flown for hobbyist and recreational purposes—and mandate that the FAA prohibit the sale of drones unless they meet certain safety standards. New York Senator Chuck Schumer announced his intention to seek a provision in the new legislation that would require geo-fencing as a pre-condition of sale.

8.6.7 Beyond the final microdrone rule

Part 107 embraces certain philosophical precepts likely to remain intact, even as the details of the proposed rules change in the course of the comment-review process or possible litigation after promulgation:

Incrementalism. The content of the rule reflects regulatory accommodation of reality: it is much easier to craft reasonable requirements for safe operation of microdrones than to craft requirements for macrodrones. The NPRM explicitly describes the FAA's approach as incremental and the NPRM itself as only the first step in a more comprehensive plan for integration. This obviously refers to drones beyond the 55-pound weight limit statutorily defining the sUAS, but incrementalism also is suggested within the scope of sUAS covered by the rule. The agency invites comment on its original idea of dividing the sUAS category into as many as eight size-based groups. It also provides considerable detail on how a separate set of regulation requirements would regulate the smallest group: *micro-sUAS*.

A fundamental shortcoming in Part 107 is its failure to address specifically the differences between state-of-the art multicopters, which predominate at the low-end of the 55-pound weight class, and fixed-wing sUAS, which predominate at the high end. This omission provides a further opportunity for incrementalism.

Smaller microdrones merit faster regulatory approval than larger ones for three reasons: they present lower risk; there are more of them already in the market, which provides more data on their operations; and simplified regulatory approaches such as DROP self-certification can give the FAA more time to develop new infrastructures for matters such as DROP testing. In particular, their unique risks must be differentiated from those of larger aircraft and those with less automation—as the FAA did to a considerable extent in Part 107. The explosion in the availability of sUAS mainly relates to multicopters such as the DJI Phantom, the DJI Inspire, and the Cinestar 8HL. These aircraft, electrically powered and equipped with sophisticated navigation and flight control systems, have flight characteristics and operator control issues quite different from those of fixed-wing sUAS, many of which are gasoline- or diesel-engine powered and require more distance to take off and land. The final rule should take into account these profound differences.

Much remains to be done; finalizing the NPRM is only one step to integrate drones into the National Airspace System. Pressure is building for regulatory certainty for larger drones. Pressure is already intense to allow newsgathering operations close to people on the ground and at night, for BLOS operations. The FAA will eventually act on them separately from promulgating the final rule. Then, of course, there is drone delivery of packages. Even as the FAA addresses those matters, the general microdrone regulation itself is likely to be modified to treat different weight classes differently.

The good analysis in the NPRM about the effect of weight on risk should be refined to deal with the different categories of risk within the entire class of less-than-55-pound sUAS. The analysis also should recognize that larger vehicles in this overall class will have different flight characteristics and different levels of performance, and that their cost will increase as size increases. Concurrently, the ability of vendors to absorb greater requirements will be greater as price increases.

The requirements for safe operation logically should be ratcheted up as size increases, generally following the FAA's concept of multiple "groups."[150] At least three subcategories or groups are plausible: one for the largest vehicles, say, those above 20 pounds, an intermediate category, with weights between 8 pounds and 20 pounds, and a micro UAS category as suggested in the NPRM, but with an upper weight limits of 8, instead of 4, pounds. The upper weight limit of 8 pounds for the micro group matches the weight of the bird that must be fired into an engine for an airline transport aircraft as part of it certification testing.[151]

The boundaries separating the three weight groups within the sUAS below-55-pound group is essentially arbitrary, however, until more data is available on risk. The NPRM says that the FAA originally considered as many as five subcategories, defined by weight and operating characteristics. Five may be the right number, as opposed to three, but a proliferation of subcategories is problematic because data is lacking on how each category poses different types of risk. Risk assessment at this point is inherently theoretical and speculative. Unlike the situation for manned aircraft, no drone accident statistics exist as the basis for analysis.

While it is premature to specify exactly what should be required for largest vehicles, the following requirements are reasonable:

- ADS-B out for vehicles that will fly more than 500 ft AGL
- Practical test and aeronautical experience requirements for BLOS operations
- More onboard automation, with greater functionality

The macrodrone group, comprising vehicles above the 55-pound cutoff for the NPRM is more likely to involve fixed-wing designs because of their longer endurance and larger payloads. Vehicles in this category are more likely to carry flammable fuels. The higher useful load of macrodrones also makes it more likely they will carry potentially hazardous payloads, such as chemical applications for agriculture.

For this category, different operating rules and operator requirements may be appropriate for rotary wing, as opposed to fixed-wing configurations, just as they are for manned aircraft.

Other specific categories such tethered drones also appear to warrant separate regulatory attention. A tethered drone resembles a kite or a tethered balloon, and the FAA's existing rules for those categories of aircraft represent a more logical starting point than the rules for free-flying manned aircraft.

Self-certification regarding the requisite knowledge should be allowed for DROPs flying the smaller categories. Self-certification of operator qualification will permit flights to begin sooner, while the FAA is constructing the knowledge testing infrastructure for operators of larger vehicles. The rules should require that a candidate certify, not only that he possesses the requisite knowledge, but also that he successfully completed some kind of formal course in the requisite areas of knowledge. A certificate from a flight school or from an organization offering training and testing would suffice to meet this requirement. The FAA has the power to audit satisfaction of this requirement.

Risk-based regulation. Part 107 commits the FAA to risk-based regulation, and its content reflects that approach. Repeatedly, the preamble refers to specific risks in justifying relatively modest requirements for microdrones, such as a knowledge test for DROPs, while explaining that other specific risks foreclose certain types of operation such as BLOS.

It is not the right approach, however, simply to ratchet up airframe, operator, and operating rules closer and closer to those for manned aircraft, as the weight of the UAS increases. Instead, any requirement for operator qualification should be related explicitly to more demanding decisions and control inputs that he would be required to make for the larger vehicles and the different types of emergencies he may encounter for different flight profiles.

Airframe requirements should be focused on specific aspects of flight profiles that necessitate greater automation in navigation and onboard control systems and on managing energy dissipation—as in the case of the proposed frangibility requirement.

Operating rules for UAS operating in airspace where manned aircraft typically operate, on the other hand, should resemble those for manned aircraft. The risks of collision are the same whether the aircraft is manned or unmanned.

Pressure to deviate from a concrete, risk-based approach will be continuous, fueled by broad political perceptions of the risks that concern various constituencies, and the absence of data that permit other risks to be quantified. What happens when autonomous safety systems malfunction is an example. The data do not exist until drones are flown frequently in operational environments, but such flight cannot be permitted if the regulators wait for good data. Regulatory flexibility is necessary to permit data to be obtained, which can, in turn support rule revisions that reflect reality.

Performance standards rather than engineering prescriptions. The NPRM invites comment on whether there are additional requirements that could be specified in ways that are more performance-oriented in order to minimize any disincentives to develop new technologies that achieve the regulatory objectives at lower cost.

The need to avoid traditional airworthiness and type certification is manifest, as the FAA recognizes. Not only would delays in the process render candidate vehicles obsolete by the time they are certificated, the burdens would pose a disincentive to technological innovation that would enhance safety, and would encourage operation of uncertificated vehicles. The question is how to address the technologies that have such significant capability to promote safety without imposing traditional airworthiness certification requirements.

Five models are worth considering and adapting, two from the FAA's experience, and three from other regulatory regimes. First, the FAA has a streamlined approval process for equipment installed in experimental, homebuilt, and sailplane aircraft. Its results are apparent in the much lower prices for ADS-B equipment now reaching the market for these categories of aircraft, compared with the much higher prices for ADS-B equipment that has completed the traditional certification process. The characteristics of these equipment approval procedures should be the model for sUAS equipment, rather than the traditional procedures.

Second, the FAA's authority to issue airworthiness directives arguably is not limited to aircraft with airworthiness and type certification. If the FAA, as chapter 9 recommends, establishes performance requirements for sUAS automation, leaving it to the manufacturer to work out the details of the specifications, it can subsequently issue airworthiness directives. If problems develop through experience, the FAA can take a data-based approach to its response, issuing ADs that require appropriate modifications to fix the problems actually experienced.

The third example follows the model in the FCC's "verification" and "declaration of conformity" procedures for standards applicable to unlicensed wireless transmitters under 47 C.F.R. part 15.

The fourth example uses the model adopted by the National Highway Traffic Safety Administration (NHTSA) for certain passenger automobile equipment requirements, such as seatbelts. Manufacturers must certify complains with NHTSA vehicle safety standards prescribed in 49 C.F.R. part 571. Preapproval by NHTSA is not required.

The fifth example uses the model employed by the Consumer Product Safety Commission (CPSC) for certain consumer devices such as walk-behind power lawnmowers in 16 C.F.R. part 1205. The CPSC safety standard specifies performance standards but leaves it to manufacturers to conduct appropriate tests and certify compliance.

Under this approach, the FAA would prescribe certain performance requirements, as in the examples given in chapter 9. Manufacturer or vendors would certify compliance with the performance requirements, and part 107 would limit operations to vehicles having that certification. Chapter 9 elaborates on this approach.

Willingness to modify traditional approaches to certification. Part 107 reflects appropriate flexibility to develop new channels for certifying DROPs rather than imposing poorly suited traditional pilots' license requirements. Likewise, Part 107 wisely avoids imposition of airworthiness and type certification requirements on the vehicles and their systems, recognizing that the years-long delays associated with obtaining such certification are likely to render the vehicles obsolete by the time they are certified and thus represent a disincentive to safety-enhancing technological innovation.

Interest-group familiarity with traditional certification approaches, however, will lead to calls to return to the familiar rather than trying new approaches, which may or may not provide equivalent safety.

8.7 The big issues

The regulatory requirements for drones are not going to crystallize all at once for the full range of vehicles. As the statutory framework and the content of Part 107 explicitly make clear, only the first stage of a multistage policymaking process is underway. Even for microdrones, for which the regulatory environment will become final first, key questions have to be resolved through development of additional data and the interplay of political forces and the interests of affected groups. To a considerable extent, the same issues will require more intense scrutiny as the regulatory envelope expands to include macrodrones.

8.7.1 Aircraft certification

The most intractable regulatory problems relate to airworthiness certification. Airman qualification may be controversial, but it is straightforward to decide what level of existing pilot certificate is required for a particular operation, or to define a new category of DROP and write the knowledge test and practical test requirements for it. A market-based infrastructure will develop quickly around whatever requirements are prescribed. Operating rules likewise will be pretty straightforward. The basic operating rules for manned aircraft will be applied to big drones, with cross-references to pilot qualifications and aircraft type.

Airworthiness and type certification is a different matter. Engineering judgment must be made about the flight vehicle characteristics necessary to mitigate specific risks. Then, designs resulting in those characteristics must be developed and evaluated. For example, a frangibility requirement might be required to absorb kinetic energy in a crash. Such a requirement would rule out certain materials, necessitate certain structural features like breakaway joints, and compel the use of multiple batteries rather than one.

The basic alternatives for configurations of airplanes and helicopters are well understood, as are the behavior of control services like such as pitch links and control rods. Various kinds of rotor hub designs are still evolving. Airbus has been particularly innovative in this area of rotorcraft design. To the extent that multicopter configurations appear in the big drone group, airframe and power plant certification my loom larger in the design, testing, and certification process, closely integrated with the automated control systems.

The challenges will be greatest, however, not with airframe and powerplant design, but with automated systems design and certification. Fault analysis for complex computerized systems is still in a state of flux, especially when it involves the human interface with computerized systems.

When it comes to automated systems for sense and avoid, defining criteria for airworthiness is even more difficult. Technologies for collision avoidance by manned aircraft are still in a stage of rapid development, as ADS-B is integrated into systems designed originally to rely only on transponder signals. Moreover, greater latency in the control loop for big drones raises questions about the safety of systems that rely on operator response to traffic conflict warnings. Safe systems may require greater reliance on automated evasive maneuvers. Automating evasive responses to collision threats requires designers to anticipate maneuvers by the other aircraft that may increase the threat if the system anticipates the wrong maneuvers.

Additionally, bigger drones, if they are to be flown at low levels, must not only sense and avoid other aircraft; they must sense and avoid other types of obstacles. For big drones flying conventional airplane and helicopter flight paths, the same automation advances underway for manned aircraft will be suitable for them as well, including better, cheaper and lighter weight autopilot-coupled approaches, terrain and obstacle avoidance and ground proximity warning systems.

To the extent they fly different flight paths, for example, orbiting or following a grid at altitudes congested by manned aircraft, additional features may be necessary with respect to flight plans and collision avoidance.

8.7.2 Automation requirements

One common issue relates to automated control navigation systems, widely available now on mass-market microdrones but of uncertain reliability. Who should have the burden of demonstrating that a land-immediately, automatic-return-to-home, or preplanned-flightplan-navigation will not go awry? What should the standards be for failure? Should the regulations establish a presumption of adequacy based on a manufacturer's representation of compliance with performance standards, or should pre-certification demonstration and testing be required?

Drone proponents and the FAA itself wisely want to do everything they can to avoid traditional airworthiness and type certification. Traditional certification processes typically lead to more and more detailed engineering specifications, running up costs and delays, as more and more components must be designed from scratch and undergo extensive testing.

Relying on performance standards and shifting the responsibility for implementing them to private sector actors raises questions about effective incentives for compliance and about the soundness of self-interested engineering judgments about how to design a system to meet the performance requirements.

Many of these concerns about system design and rely can be brushed off, as they were in the final rule, as long as vehicle characteristics—especially weight—and limited flight profiles mitigate risks without predominantly relying on automation. Heavier vehicles, operating beyond the line of sight, and in proximity to manned aircraft and structures and people on the ground present commensurately greater risk, and the only obvious way to mitigate the risk is greater reliance on sophisticated automated systems. The pressure will be enormous to ensure that the safety systems as well as basic navigation and control systems perform reliably and fail safely when they fail. It is going to be hard to resist something that looks a lot like airworthiness type certification for larger and more capable drones.

Chapter 9 advocates an approach to drone automation that establishes a presumption of adequacy of automated land-immediately, return-to-home, and preplanned-flightplan-navigation based on a manufacturer's representation of compliance with performance standards. If post sale data collected from users demonstrates that self-certification was unwarranted, the FAA can issue airworthiness directives requiring modification or recall. Pre-sale approval of technology by the FAA should not be required and no specific pre-sale testing would be mandated.

8.7.3 Airman certification

A second major battleground will involve DROP qualification. It remains to be seen whether the NPRM's light hand for microdrone DROPs will survive the current process. The final regulations do not require any kind of practical testing and aeronautical experience, in addition to a basic knowledge test.

Any training and testing likely will be done by a mixture of private sector, voluntary, and government instructors and examiners.

It is unlikely, however, that the FAA will be induced by ALPA or anyone else to return to a requirement that DROPs hold conventional pilots licenses. The lack of fit between pilot requirements and DROP requirements is simply too obvious.

8.7.4 Operating rules

The battle over operating rules will turn on two considerations, one policy driven; the other technology driven.

For microdrones, the FAA got itself off the hook in Part 107 by *segregating,* rather than integrating, them into the National Airspace System. The core strategy for mitigating risk is to confine microdrones to places where manned aircraft are unlikely to be flying—low heights above the ground and clear of controlled airspace.

The attractive capabilities of macrodrones cannot be realized if they are similarly segregated; they will have to be integrated into airspace full of manned aircraft order to be economically productive. That will put stress on operating rules, beyond merely subjecting macrodrones to the same operating rules applicable to manned aircraft. The key default principle of see-and-avoid is not available—or at least not as robust for remotely controlled aircraft as for aircraft with pilots in the cockpit.

That then leads to technology development. Trustworthy collision avoidance systems must be available at affordable prices. The weight and requirements must not only permit macrodrones and manned aircraft to detect and avoid each other, but must also keep macrodrones clear of static obstacles such as terrain, antennas, powerlines, wind turbines, foliage, and buildings. The technologies for fully automated collision avoidance systems with this range of target avoidance are barely beyond their infancy.

And once adequate capability and reliability have been demonstrated in the lab, it is still a long road to commercialization and FAA certification.

8.7.5 Mitigating crash damage

Energy dissipation techniques such as frangibility are part of the equation. Frangibility is a mainstay of airport design regulation, to ensure that taxiway lights and radio antennas break away if an aircraft collides with them, minimizing damage to the aircraft. The NPRM explicitly invites comments on frangibility as a requirement for microdrones, but frangibility may be even more desirable for macrodrones, which have much greater kinetic energy because of their heavier weights and faster speeds.

Emergency measures such as parachutes, crash bags, and small explosives to destroy a distressed drone, though largely untried, should be part of the discussion. See chapter 5 for a fuller exploration of these techniques.

Changes in manned aircraft design also may be desirable, to mitigate the risk of a collision with a drone. Designing helicopter windshield bubbles for greater resistance to penetration might be an example.

8.7.6 Operators

The current regulatory regime relies on a variety of private-sector actors to internalize federal regulatory requirements: Part 141 flight schools, Part 119 commercial operators, including those operating under Part 135 and Part 121.

A similar approach may emerge for drone operators. It is unlikely that the existing operator categories, distinguishing among different kinds of passenger-carrying operations, are suitable.

8.7.7 Designing new airspace management system

Logistics applications such as package delivery are especially daunting from a regulatory standpoint. As chapter 3 suggests, it likely will be necessary to develop a completely new airspace management system for low-level flight and interaction of a couple of hundred drones flying in roughly the same airspace at the same time, close among people, structures, and utility infrastructure. Engineering such systems has barely begun; technologists still are deciding principles on how such a system should work—by peer-to-peer detection and avoidance? By exclusive permission to operate in blocks of airspace? And how should the risk of a child or customer rushing out to greet the drone be handled?

8.8 State and local regulation

The Commerce Clause and federal preemption doctrine interact in determining the legality of state regulation of drones.[152] Under its commerce power, the Congress retains the authority explicitly to preempt state and local regulation, as it has done with respect to economic regulation of airlines.[153] It has explicitly forborne to do so with respect to state law *remedies*.[154] As to the more general realm of aviation safety regulation, it has not spoken explicitly about state power, but it has granted broad authority to the FAA and specified some details as to how the FAA should exercise that authority, supporting the many judicial findings of implied preemption.[155]

Under the Commerce Clause, the Congress could decide in the future to adopt explicit statutory preemption of state regulation of drones, partially or completely. It could not do so, however, beyond the limits of interstate commerce.

Although the aviation preemption caselaw frequently articulates a conclusion that the entire field of aviation regulation is preempted, closer examination of the cases shows, not field preemption, but preemption turning on whether the FAA has exercised its statutory authority with respect to a particular aspect of safety.

As long as it acts within its statutory authority, the FAA could adopt a new rule that not only regulates some aspect of drone operations that theretofore had been unregulated, but it also could explicitly preempt state regulation, either as to the subject of the new FAA rule, or expressing its conclusion that that an aspect of safety should go unregulated. As long it is as it has done neither, states have a plausible argument that they are free to regulate the subject matter.

Derivative of this power to define the boundary between federal and state regulation, the FAA—or the Congress itself—could define a system for cooperative and concurrent state and federal regulation of drones.

In December 2015, the FAA chief counsel issued a "fact sheet"[156] stating that state and local regulation of drones is likely to be preempted. It explicitly says that state and local registration requirements are preempted, and suggests the same preemption conclusion is likely for most DROP qualification requirements, vehicle design mandates, and operating rules. It provides for consultation between the FAA state and local governmental entities as they consider regulatory proposals.

8.8.1 Space for states?

States are free to regulate drone operations when a statute explicitly saves room for state regulation as in tort remedies, or when the FAA has not exercised its authority on a particular subject.

States may not regulate subjects explicitly addressed by the FAA in its final rules. Nor may they regulate activities carried on within section 333 exemptions. That means that states may not impose different weight limits, height limits, preflight inspection requirements, accident reporting requirements, or periodic reporting requirements on operations. It means they may not impose different DROP qualification, training, certification, or experience requirements. They may not impose vehicle design requirements.

It is unlikely that states have the power to enforce FARs directly. States have no inherent power to enforce federal law.[157] As a general matter, judicial enforcement of FARs is reserved to the Secretary of Transportation and the Attorney General.[158] But if a state incorporates federal regulatory standards into its tort law[159] and provides its own remedies when a plaintiff can prove violation of the standards, proximate causation, and injury, preemption is unlikely. What matters is whether the state directly charges a violation of federal law, or whether it charges a violation of parallel state law, incorporating federal standards and requirements by reference.

States, however, retain their authority to enforce generally applicable state and local law against disorderly conduct,[160] public endangerment,[161] refusal to obey the lawful command of a police officer,[162] or refusal to disperse.[163] The FAA has published guidance for local law enforcement personnel confronted with what they believe to be impermissible microdrone operations.[164]

The language of the section 333 exemptions itself does not address state and local regulation. The blanket COAs accompanying the section 333 exemptions however do. A note on the first page says:

> Note—This certificate constitutes a waiver of those Federal rules or regulations specifically referred to above. It does not constitute a waiver of any State law or local ordinance.[165]

State courts remain open to adjudicate claims of invasion of privacy, trespass to land, or negligence so long as the elements of each tort applied in a particular drone case do not conflict with FAA rules.

As aviation matured through the twentieth century, landowners periodically sued aircraft operators for trespass and nuisance.[166] Most of the trespass cases confronted questions about how high above the ground the property owner's rights extend.[167] The caselaw does not provide a quantitative answer; the most that can be said with any confidence is that flights above 400–1,000 ft above the ground are not trespasses to private property. Macrodrone flight is unlikely to engender difficulty with an upper boundary of private property at 400–1,000 ft, but microdrones flying within the section 333 exemption limits and under the Part 107 restrictions will. Then the preemption argument will be that the FAA permits such microdrone flights at that level, thus preempting inconsistent state trespass law. Defenders of such a state or local rule establishing a minimum height would argue against preemption because there is no FAA-established minimum height, and because of traditional police power to regulate land use. The cases involving claims of trespass to land by aircraft can be mobilized by both sides of the argument.

State or municipal regulations *increasing* the height at which drones can be flown, on the other hand, would be preempted, because of the FAA prescription of a maximum height above ground level, justified by reducing interference between drones and higher flying manned aircraft.

When states and municipalities adopt legislation or rules that target drones, it is more likely to be preempted than a state statute or regulation of general effect, as relating to noise, taxation, or environmental protection. When state legislation and regulation specifies limits on flight profiles, crew qualifications, or aircraft design, it is more likely to be preempted than initiatives that address matters not directly related to flight, such as business financial reserves, employee vacation or sick leave, minimum wages, or employment discrimination.

Preemption is straightforward if an existing or proposed FAA rule exists on a particular subject, but if the if the FAA leaves a gap in its regulations on the particular subject, room may exist for states and municipalities to fill the gap, even if the FAA regulates the general area.

Moreover, if a state narrowly targets a particular highly localized area of drone operations, and relates it to matters of traditional state concern, such as personal privacy, security of property occupancy, preemption is less likely. Deference usually given to matters of traditional state concern, and the argument is stronger that the activity is outside the Commerce Clause.

The caselaw validating state regulation of airport siting supports the proposition that states and municipalities have the power to specify where drones may take off and land, effectively limiting where microdrones may fly, given their short range.

Recognizing traditional state power to preserve public order, states should have the power to establish tort liability or to criminalize reckless conduct,[168] although this also is the subject of an FAA rule prohibiting reckless flight operations.[169]

In Alabama, "A person commits the crime of reckless endangerment if he recklessly engages in conduct which creates a substantial risk of serious physical injury to another person."[170] New York has both a first degree and second degree reckless endangerment statute. Reckless Endangerment in the Second Degree occurs, "when the person recklessly engages in conduct which creates a substantial risk of serious physical injury to another person."[171] Reckless Endangerment in the First Degree occurs, "when, under circumstances evincing a depraved indifference to human life, he recklessly engages in conduct which creates a grave risk of death to another person."[172]

In Illinois,

> (a) A person commits reckless conduct when he or she, by any means lawful or unlawful, recklessly performs an act or acts that:
>
> (1) cause bodily harm to or endanger the safety of another person; or
> (2) cause great bodily harm or permanent disability or disfigurement to another person.
>
> (b) Sentence.
>
> Reckless conduct under subdivision (a)(1) is a Class A misdemeanor. Reckless conduct under subdivision (a)(2) is a Class 4 felony.[173]

Limiting the purposes for which drones may be flown, for example, prohibiting flights for surveillance or to capture imagery of a particular individual might be permissible, because the FARs, while imposing different airman and aircraft certification and different flight rules for different purposes such as banner towing, med-evac, and tourism in certain areas do this because of differing types of safety threats. A state or local law limiting purposes would be aimed instead at exercising traditional police power over privacy or land use. A recent student note[174] concludes that state and municipal laws focused on drone safety, such as measures limiting flight altitudes or flights over populated areas, are likely to be preempted.[175] Conversely, the author concludes that state and local regulation of surveillance, justified by protection of personal privacy, may survive preemption challenges, at least if they apply the same limitations to manned aircraft as to drones.[176]

States should be able to regulate data collection, to limit liability for accidents, and to require liability insurance, because there is no federal aviation law on these subjects, and because of traditional—and statutory[177]—state prerogatives over insurance.

The airport regulation preemption decisions suggest that states and municipalities have more non-preempted power over facilities they own and manage than over facilities in private hands. In the drone context, that means that states and municipalities likely have more authority to regulate conduct in public spaces than they do over private property, especially over public parks. They already regulate access to public parks, charge fees, and determine what activities are permissible. It is a relatively simple matter, as the Chicago Park District Commission proposes to do, simply to add microdrone flight to the list of activities that are prohibited unless one obtains a permit.

8.8.2 Model aircraft and "consumer drones"

Hobbyist and recreational drone uses present problems distinct from commercial uses. Moreover, the ready availability of easy-to-fly consumer-oriented drones at modest prices bifurcates the hobbyist and recreation category. Most owners of DJI Phantoms are not members of model aircraft clubs.

§ 8.6.5.5 explains now 336 of the 2012 Act[178] prohibits the FAA from promulgating any rule or regulation applicable to model aircraft weighing less than 55 pounds. Section 336 says nothing about state or local regulations, and its withholding of authority for the FAA leaves a relatively clear field for states to regulate model aircraft operations.

Even if section 336 is amended, or if the FAA decides to impose automation performance requirements on microdrones as a prerequisite for sale, as chapter 9 proposes, anarchy will be the norm for consumer drones unless states and municipalities supplement FAA enforcement resources. If states and municipalities decide to step in, the preemption barriers are modest. States have a long history of regulating recreational activity that may pose safety problems: hunting, archery, recreational boating, all-terrain vehicles. Requiring consumer drones to fly down low, stay within line of sight of the operator, not to fly over people, and to fly in public parks only if they have a permit is not likely to interfere with commercial microdrone operations or to interfere with the operation of the National Airspace System.

8.8.3 State and local initiatives

Only a handful of states have enacted statutes limiting the operation of drones (Unmanned Aerial Vehicles or UAVs). Most of these laws prevent law enforcement use of drones for evidence gathering without appropriate search warrants. Some of these limitations prohibit law enforcement and citizens from weaponizing drones. Some heighten privacy protection by prohibiting aerial surveillance without consent. A few statutes limit drone involvement

Table 8.1 State drone statutes

State	*Citation*	*Date Approved*	*Date Effective*	*Summary*
Florida	C.S.C.S.S.S.B 766 Freedom from Unwarranted Surveillance Act	May 14, 2015	(i) July 1, 2015	Prohibits law enforcement use to gather evidence; prohibits recording an image of a privately owned property or of the owner (tenant, occupant, invitee, etc.) violating a reasonable expectation of privacy. Exceptions: police get warrant, "performance of reasonable tasks" within the scope of one's license, property appraisals, utility inspection, mapping, delivering cargo (if FAA compliant)
Idaho	IC 21-213	(ii) April 11, 2013	(iii) July 13, 2013	No law enforcement searches without a warrant; no aerial photography without prior consent
Illinois	720 Ill. Comp. Stat 5/48-3 Freedom from Drone Surveillance Act			Prohibits law enforcement use of drones without a warrant; no drone interference with hunters
Indiana	IC 35-33-5-9	July 1, 2014		No law enforcement use without a warrant
Iowa	HF 2289	May 23, 2014		No drones for traffic law enforcement; evidence without search warrant is inadmissible
Maryland	SB 370	May 12, 2015	July 1, 2015	Only the state can make drone laws (preempts counties and local ordinances)
Mississippi	SB 2022	April 23, 2015		Defines felonious trespass to include peeping through a window, hole, or opening with a drone; prohibits photographs and video of people without consent
Montana	HB 330	April 23, 2015	October 1, 2015	No weaponized or armored drones for law enforcement

(*continued*)

Table 8.1 (cont.)

State	*Citation*	*Date Approved*	*Date Effective*	*Summary*
Nevada	AB 236			No person shall weaponize a drone or operate a weaponized drone; no operation within 400 ft or 250ft vertically from a "critical facility" and 5 miles from airport without consent; right of action (trespass) if drone less than 250 ft over property and property owner notifies DROP that the flight is unauthorized; no use for law enforcement to collect evidenceCreates a public registry of all state operated drones
New Hampshire	SB 222 Fish and Game – Animal – Harassment	May 7, 2015	January 1, 2016	No activity that disturbs animals with intent to prevent their lawful taking; no drone use with intent to conduct video surveillance of citizen lawfully hunting, finishing, or trapping, without prior consent
North Dakota	HB 1328	April 15, 2015		Evidence obtained by a drone not admissible as evidence without a search warrant nor can law enforcement use drone footage as a basis for probable cause; no lethal weapons on a drone. Does not prohibit drone usage for research and development by educational institution
Oregon	HB 2534 Fish and Game – Fish and Wildlife Commission – Drone Regulation HB 2354: only definition of drone changed to "unmanned aircraft system"	May 12, 2015		Prohibits use of drones related to pursuit of wildlife (angling, hunting, trapping) or aiding through use of drones to harass, track, locate, or scout wildlife; and interfere with angling, hunting, and trapping. The definition of drone includes unmanned water-based vehicles

Table 8.1 (*cont.*)

State	*Citation*	*Date Approved*	*Date Effective*	*Summary*
Tennessee	HB 153 Crimes and Offenses – Drones – Photography and Pictures	April 20, 2015	July 1, 2015	No operation over events with 100+ attendees for a ticketed event; no flight around fireworks without event organizer's consent
Texas	423.008			Law enforcement must submit a report of drone use to the governor
Utah	HB 296	March 27, 2015		Evidence obtained by a drone not admissible as evidence without a search warrant
West Virginia	HB 2515 Wildlife – Animals – Weapons	April 2, 2015		Prohibits hunting with drone
Wisconsin	WSA 941.292WSA 175.55	April 10, 2014		No weaponized drones; no law enforcement use without a warrant

in hunting. Table 8.1 lists the statutes. The discussion considers specific statutory provisions and evaluates the likelihood of federal preemption.

Law Enforcement. The drone statutes reinforce the Constitutional limitation on unreasonable searches and seizures. Some states prohibit law enforcement from gathering evidence without a search warrant. In Illinois, law enforcement "may not use a drone to gather information"[179] unless it obtains a search warrant prior to the search.[180] Law enforcement agents may use drones in certain circumstances like crime scene and traffic investigation.[181] Wisconsin and Indiana, like Illinois, prohibit the use of drones to gather evidence without a search warrant.[182] Violation results in inadmissibility of the evidence. In addition, North Dakota prohibits use of drone imagery to establish probable cause to obtain a search warrant that would lead to drone captured evidence.[183]

Some states prohibit law enforcement from weaponizing a drone.[184] In addition to weaponizing a drone, Wisconsin prohibits law enforcement from equipping a drone with armor.[185] Other states extend this prohibition to civilian drone operations.[186]

Texas addresses concerns about law enforcement abuse but does not limit drone use. The Texas statute does not explicitly require a search warrant when law enforcement conducts an aerial search to gather evidence using a drone.[187] It merely requires that the law enforcement agency must, every two years, submit a written report to the governor, the lieutenant governor, and each member of the state legislature with a list of drone missions, costs of operating and maintaining a drone, and a list of non-criminal drone investigations.[188]

The measures restricting what state or local law enforcement may do with drones are not preempted, because of the traditionally strong state interest in regulating its own law enforcement bodies and the limited effect on air commerce.

Privacy. States with drone privacy statutes address the fear of citizens using drones as "prying eyes" to collect information about their neighbors from an aerial vantage point. The statutes prohibit aerial imagery capture without consent.[189] Florida, for example, prohibits any surveillance of a privately owned property, its owner and anyone legally occupying the premise (landlord, tenant, or licensee).[190] The Idaho statute prohibits capturing imagery of land and occupants without prior consent of the owner or the occupant.[191] Despite these prohibitions, some states allow drones to conduct property appraisals, utility inspections, and mapping if the DROP performs the "reasonable task" under a state occupational license.[192]

Flying a drone over private property without consent can lead to a trespass claim against the DROP or imposition of a statutory penalty. Some states allow a trespass claim after the land owner notifies the DROP about an unauthorized flight over the land owner's land lower than 250 ft.[193] Texas, for example, creates a civil right of action against a violating DROP and allows a land owner to recover a penalty for every captured image, or for distributing images.[194] In Mississippi, a drone trespass is a "felonious trespass" when a DROP uses a drone to peep through a "window, hole, or opening."[195] Tennessee prohibits unauthorized use over events with more than 100 ticketed guests attending.[196]

The Tennessee, Florida, Idaho, and Mississippi statutes present interesting preemption questions. Tennessee's prohibition of flying over major events can be justified by the state's interest in public safety. Limitations on what people can do in connection with large public events are a traditional mainstay of state and local regulation. The Tennessee crowd overflight prohibition is congruent with the section 333 exemption and (probable) eventual final-rule prohibition on flying over crowds. Tennessee could further justify its involvement as simply providing additional enforcement mechanisms for a federally established standard, similar to what happens when state law provides remedies for conduct that violates federal standard. On the other hand, a state crowd overflight restriction that goes well beyond the federal standard is more vulnerable to a preemption challenge.

The prohibition against aerial imagery over property without the owner's consent and of human subjects without their consent can be justified as an extension of traditional state measures to protect private property and personal privacy interests. These matters are generally left to the states and covered by extensive state regulation already. Many states already prohibit capturing or publishing images of persons without their consent.[197] State overflight rules on this subject, however, are more vulnerable to preemption challenges when they extend the height below which permission is required. A limit of 250 ft places half of the FAA's allowable height under off-limits, especially if it is accompanied by restrictions or overflight of public spaces. Such inconsistent height limits interfere with the federal regulatory regime and burden air commerce.

These state privacy measures would fare better under preemption analysis if they simply extended existing state limitations on photographing individuals. Such measures do not single out drones or other aircraft for special restrictions. The case law is more hospitable to state regulation of general application.

Hunting. Aside from privacy concerns, some states worry about the role of drones in the outdoors, in conjunction with hunting, fishing, and trapping. Drone wildlife statutes prevent DROPs from interfering with others' enjoyment of wildlife sports and from taking advantage of drone to gain an upper hand in outdoor sport. New Hampshire prohibits drone use with the intent to prevent lawful taking by hunters.[198] Oregon, for example, prohibits drone use to interfere with hunting, trapping, and fishing.[199] State statutes also prohibit drone use to facilitate hunting. It is illegal in Virginia to track, locate, and scout for wild animals[200] and to use a drone to herd animals.[201] The measures related to hunting are not preempted, because of the traditional state interest in that subject.

8.8.4 Space for municipalities?

The sovereignties in the United States constitutional structure are the federal government and the states; not municipalities. It was the states that met at the Constitutional convention and ceded some of their sovereign power to the United States; counties, towns, and cities were not at the table.

States started out with sovereignty and gave some of it up—part of it upward, to the national sovereign, and part of it downward to counties, cities, and towns. Counties and other municipalities enjoy only such powers as are granted by the sovereign state. The American Civil War established that states, having ratified the United States Constitution, are not entitled to take back any of the sovereignty they ceded to the federal government, but the sovereignty they ceded downward, to local units of government, they can take back at any time. The 2015 Maryland statute preempting municipal drone regulation is an example.

Of course, to the extent that state local government prerogatives are codified in state constitutions, the process for taking it back may be more arduous than simply passing a bill in one session of the Legislature. In some states, local government enjoys only those governmental authorities explicitly granted to them by state statute or constitutional provision. That was the case in Alabama for many years.[202] The trend, however, is for states to adopt home rule legislation that grants general governmental power—roughly equivalent to that exercised by the state—to municipalities unless a specific power is withheld in the home rule statute or by subsequent legislation.[203]

8.8.5 Taking photographs or making video recordings as constitutionally protected expression

Apart from federal preemption, many of the enacted or proposed state statutes and municipal ordinances violate the First Amendment, because they prohibit expressive conduct. Little doubt exists that taking photographs or making video recordings constitutes expression protected by the First Amendment.[204] First Amendment protection does not depend on the particular tool being used. Drone imagery is protected just as much as imagery captured from the ground.

Notes

1 The process for certificating aircraft is to determine the airworthiness of a particular type, which results in a type certificate.
2 5 U.S.C. §§ 551–559, 701–706.
3 49 U.S.C. § 106(a).
4 49 U.S.C. § 106(b).
5 49 U.S.C. § 106(f)(1).
6 49 U.S.C. § 106(f)(2)(A)(iii).
7 49 U.S.C. § 106(f)(3)(A).
8 49 U.S.C. § 106(f)(3)(B)(i)(II).
9 49 U.S.C. § 40104(a).
10 49 U.S.C. § 40103 (b).
11 49 U.S.C. § 44701(a)(1).
12 49 U.S.C. § 44701(a)(2).
13 49 U.S.C. § 44702(a); 49 U.S. Code § 44704.
14 49 U.S.C. § 44711(a).
15 49 U.S.C. § 44101.
16 49 U.S.C. § 106(p)(4)(A)(5)
17 AC 20–27G ¶ 8(j)(3).
18 Ibid. ¶ 9(a).
19 Ibid. ¶ 12.
20 Ibid. ¶ 12(b).
21 Ibid. ¶ 12(a)(2).
22 Ibid. at p. 24 (Figure 3).
23 Ibid. at ¶ 12(c).
24 AC 20–27G Fig. 4 at p. 26.
25 Ibid. Table 7 at p. 27.
26 Ibid. at p. 27 (referring to AC 90–89).
27 Ibid. ¶ 14(b).
28 49 U.S.C. §§ 40102(a)(41), 40125.
29 FAA, AC No: 00-1.1A, *Public Aircraft Operations*, www.faa.gov/documentLibrary/media/Advisory_Circular/AC_00-1_1A.pdf (Feb. 12, 2014) (exploring application to governmental contractors).
30 14 C.F.R. § 91.1.
31 Curtis W. Copeland, *The Role of the Office of Information and Regulatory Affairs in Federal Rulemaking*, 33 *Fordham Urb. L.J.* 1257 (2006) (comprehensively reviewing history and activities of OIRA) [hereinafter "Copeland"].

32 www.reginfo.gov/public/jsp/EO/letters.jsp Under Executive Order 12866 and the Paperwork Reduction Act.
33 Copeland, 33 Fordham Urb. L.J. at 1291–1292 (describing steps to improve transparency).
34 49 U.S.C. § 46301 (imposing civil penalties for violations).
35 49 U.S.C. sec. 44709(b) (authorizing revocation, suspension, or modification of airman or airworthiness/type certificate if "the Administrator decides after conducting a reinspection, reexamination, or other investigation that safety in air commerce or air transportation and the public interest require that action").
36 49 U.S.C. § 46301.
37 14 C.F.R. § 13.19 (setting forth procedures for certificate actions).
38 The term *airman* refers to a class of certificate holders that includes pilots, mechanics, flight engineers, flight attendants, ground instructors, and flight dispatchers.
39 49 U.S.C. § 46301(d)(5)(B) (granting pilots, flight engineers, mechanics, and repairmen right to appeal civil penalty to National Transportation Safety Board). "After notice and an opportunity for a hearing on the record, the Board shall affirm, modify, or reverse the order. The Board may modify a civil penalty imposed to a suspension or revocation of a certificate."
40 49 U.S.C. § 46301(d)(6) (entitling respondent or Administrator to judicial review of NTSB civil penalty order, and making NTSB findings of fact conclusive if supported by substantial evidence); 49 U.S.C. § 44709(f) (providing for judicial review under section 46110 of order suspending, revoking, or modifying certificate).
41 49 U.S.C. § 46110; 49 U.S.C. § 46301 (g) (authorizing judicial review of order imposing civil penalties by Secretary or Administrator, but only under section 46110).
42 49 U.S.C. § 46101 (requiring investigation of written complaint filed by "a person" if reasonable grounds exist).
43 14 C.F.R. §§ 13.15–13.16 (setting forth procedures for assessment and appeal of civil penalties).
44 49 U.S.C. § 46301(f) (granting authority to compromise civil penalties).
45 5 U.S.C. § 554(d) (guaranteeing independence of persons presiding at formal administrative hearings); Federal Maritime Com'n v. South Carolina State Ports Authority 535 U.S. 743, 756 (2002) (concluding that ALJs possess independence nearly equivalent to that of Article III judges, sufficient to make Eleventh Amendment applicable to exercise of their power).
46 14 C.F.R. § 13.16(h)–(k).
47 49 U.S.C. § 46301(d)(5) (A) (entitling pilots, flight engineers, mechanics, and repairmen to opportunity to answer and be heard); 49 U.S.C. § 46301(d)(5)(B); 14 C.F.R. § 13.18(g).
48 49 U.S.C. § 46301(d)(7)(A) (entitling person other than pilot, flight, engineer, mechanic, or repairman to notice and opportunity for "a hearing on the record").
49 Ibid. subsection (d)(7).
50 14 C.F.R. §§ 13.31–13.63 (rules of practice for FAA hearings); 49 C.F.R. § 821.5 (applying Fed.R.Civ.P. to NTSB hearings "to the extent practicable"); Ibid. § 821.38 (applying Fed.R.Evid. "to the extent practicable"). 49 C.F.R. § 821.35 (specifying assignment, duties, and powers of NTSB ALJs). An ALJ decision becomes final if no appeal is filed with the full board. 49 C.F.R. § 821.42. On appeals of an ALJ decision, the Board determines if findings of fact are supported by evidence, and if conclusions comport with "law, precedent, and policy." 49 C.F.R. § 821.49.
51 49 U.S.C. § 46301(d)(5)(C) (providing that NTSB is not bound by findings of fact, but is bound by FAA interpretations of statute and regulations unless arbitrary, capricious or not in accordance with law).
52 49 U.S.C. § 46110.

53 Ibid., subsection (d)(6).
54 49 U.S.C. § 46301(g).
55 49 U.S.C. § 46110(c).
56 5 U.S.C. § 706.
57 Chevron, U.S.A. v. Natural Resources Defense Council, 467 U.S. 837, 845 (1984).
58 49 U.S.C. § 46305 (authorizing civil actions to recover civil penalties and actions in rem on lien against an aircraft for a penalty); 49 U.S.C. § 46304 (subjecting aircraft involved in a violation to lien and summary seizure).
59 United States v. Caribbean Ventures, Ltd., 387 F. Supp. 1256, 1257 (D. N.J. 1974) (holding that preliminary injunction is available against violation of federal regulations without showing of irreparable injury; granting injunction against uncertificated air carrier operations).
60 United States Attorney's Manual, Civil Resource Manual § 4-6.200, www.justice.gov/usao/eousa/foia_reading_room/usam/title4/6mciv.htm#4–6.380 (summarizing procedure for agency to obtain authorization for civil action); Ibid. § 4-1.450 (precribing contents of agency referrals); Ibid. §§ 4-1.500 to 4-1.511 (explaining relationship between Civil Division of Justice Department and US attorneys).
61 Air Line Pilots Ass'n v. Quesada, 182 F. Supp. 595, 599 (S.D. N.Y. 1960) (accepting general principle that injunction against enforcement of invalid regulations is available but finding that mandatory pilot retirement requirement was valid), aff'd 276 F.2d 892 (2d Cir. 1960) (affirming on ground that regulation was not arbitrary and capricious).
62 FAA Modernization and Reform Act of 2012, Pub. L. No. 112-95, 126 Stat. 11 (Feb. 14, 2012).
63 FAA Modernization and Reform Act, § 332(a)(1) (2012). (requiring plan by 270 days from the date of enactment).
64 Ibid. at § 332(a)(5).
65 Ibid. at §§ 331–336.
66 Ibid. at § 332(a)(3).
67 Ibid. at § 332(b)).
68 Section 333, Pub.L. 112-95, 126 Stat. 11 (based on size, weight, speed, operational capability, proximity to airports and populated areas, and operation within visual line of sight).
69 FAA Modernization and Reform Act of 2012, Pub.L. 112-95, 126 Stat. 11 (Feb. 14, 2012).
70 Ibid. sec. 332(a)(1). Section 332 of the 2012 Act obligates the FAA, in consultation with the aviation industry, to develop a plan to integrate drones into the National Airspace System. Sec. 332(a)(1), 126 Stat. 73.
71 *Unmanned Aircraft Systems (UAS) Comprehensive Plan* (Sep. 2013), available at www.faa.gov/about/office_org/headquarters_offices/agi/reports/media/UAS_Comprehensive_Plan.pdf.
72 *Integration of Civil Unmanned Aircraft Systems (UAS) in the National Airspace System (NAS) Roadmap,* (2013), available at www.faa.gov/uas/media/uas_roadmap_2013.pdf.
73 Comprehensive Plan at p. 8.
74 Roadmap at p. 58 (section C.6, discussing goal of accelerated regulatory safe harbor for sUAS).
75 81 Fed. Reg. 42063 (June 28, 2016).
76 14 C.F.R. §§ 107.1, 107.3.
77 80 Fed. Reg. at 9556.
78 80 Fed. Reg. at 9557 (comparing micro UAS category with proposed rules for the entire category).
79 80 Fed. Reg. at 9565.
80 80 Fed. Reg. at 9565.

81 80 Fed. Reg. at 9549 (noting that advances in battery technology may allow lightweight transponder solutions that are presently unavailable).
82 80 Fed. Reg. at 9549–9550.
83 14 C.F.R. § 107.13.
84 80 Fed. Reg. at 9550.
85 80 Fed. Reg. at 9550 (describing lack of congruence between manned-aircraft pilot certification requirements and safe microdrone operation).
86 14 C.F.R. § 107.2.
87 14 C.F.R. § 107.63 (summarizing requirement).
88 14 C.F.R. § 107.63.
89 80 Fed. Reg. at 9568.
90 80 Fed. Reg at 9568.
91 80 Fed. Reg. at 9568 (noting option to sacrifice microdrone).
92 80 Fed. Reg. at 9569.
93 80 Fed. Reg. at 9569 (enumerating and justifying knowledge-test components).
94 80 Fed. Reg. at 9569.
95 80 Fed. Reg. at 9570 (citing 14 C.F.R. § 61.56).
96 80 Fed. Reg. at 9570.
97 80 Fed. Reg. at 9572.
98 80 Fed. Reg. at 9573.
99 14 C.F.R. § 107.61(c).
100 14 C.F.R. § 107.51(b).
101 80 Fed. Reg. at 9562.
102 14 C.F.R. § 107.25.
103 14 C.F.R. § 107.49.
104 14 C.F.R. § 107.41.
105 14 C.F.R. § 107.51.
106 14 C.F.R. § 107.49(d).
107 80 Fed. Reg. at 9577.
108 80 Fed. Reg. at 9577.
109 80 Fed. Reg. at 9561–9562, 9563.
110 80 Fed. Reg. at 9562.
111 80 Fed. Reg. at 9562.
112 80 Fed. Reg. at 9563.
113 80 Fed. Reg. at 9562.
114 14 C.F.R. § 107.23.
115 80 Fed. Reg. at 9558 (soliciting comment on establishment of micro UAS sub-category).
116 80 Fed. Reg. at 9562.
117 www.faa.gov/uas/legislative_programs/section_333/ (summarizing section 333 process).
118 Brooks Barns, *Drone Exemptions for Hollywood Pave the Way for Widespread Use*, www.nytimes.com/2014/09/26/business/media/drone-exemptions-for-hollywood-pave-the-way-for-widespread-use.html?_r=0.
119 www.faa.gov/news/updates/?newsId=82485 (announcing "summary" section 333 exemption grants).
120 www.faa.gov/uas/legislative_programs/section_333/333_faqs/ (explaining relationship between section 333 exemption and COA).
121 FAA, Registration and Marking Requirements for Small Unmanned Aircraft, 80 Fed. Reg. 78594 (Dec. 16, 2015).
122 FAA Order 8130.34C, rgl.faa.gov/regulatory_and_guidance_library/rgOrders.nsf /0/10947cee0052205886257bbe0057bd76/$FILE/8130.34C.pdf.
123 Ibid., section 2, paragraph 4(a)(2), at p. 2–4.
124 Chap. 3, sec. 1, para. 3, at page 3–1; Appendix A, para. 3.

125 Ibid. at p. A–4 (requiring PIC with at least a private pilot certificate)
126 Ibid. at p. 3–3 and 3–4;
127 Ibid. at p. A–7.
128 Ibid. at p. A–8.
129 Ibid. at p. A–9.
130 *FAA Makes Progress with UAS Integration*, www.faa.gov/news/updates/?newsId=68004.
131 Ibid. § 334.
132 Pub. L. 112–95, 126 Stat. 11.
133 Pub. L. 112–95 § 334(c)(2).
134 www.faa.gov/news/updates/?newsId=68004 (May 14, 2012).
135 www.faa.gov/news/updates/?newsId=68004.
136 Ibid.
137 Ibid.
138 Sec. 336(c).
139 Sec. 336(a), 2012 Act. The 1981 Circular adds: "No flight higher than 400 ft above the surface."
140 AC91-57 (Jun. 8, 1981), www.modelaircraft.org/files/540-c.pdf; NPRM at 45 (discussing statutory exemption for model aircraft).
141 www.faa.gov/documentLibrary/media/Advisory_Circular/AC_91-57A.pdf.
142 Example, David Windestal, *FPV FunJet – Beautiful flight over the clouds – RCExplorer.se*, YouTube (Nov. 12, 2009), www.youtube.com/watch?v=iI_UKiFqRsA (showing model airplane flight into and above clouds).
143 NPRM at 47.
144 Michael D. Shear and Michael S. Schmidt, *White House Drone Crash Described as a US Worker's Drunken Lark*, N.Y. Times, Jan. 27, 2015, www.nytimes.com/2015/01/28/us/white-house-drone.html?_r=0.
145 *FAA investigating drone flying near news helicopters* (Mar. 17, 2015), www.kirotv.com/news/news/faa-investigating-drone-flying-near-news-helicopte/nkYk7/ (reporting on near miss between drone and news helicopters).
146 2012 Act § 336(a)(2).
147 FAA Modernization and Reform Act of 2012, Conference Report to Accompany H.R. 658, Rep. No. 112-381, 112th Cong., 2d Sess. 199 (Feb. 1, 2012), www.gpo.gov/fdsys/pkg/CRPT-112hrpt381/pdf/CRPT-112hrpt381.pdf.
148 161 Cong. Rec. S2849 (May 13, 2015); referred to the Committee on Commerce, Science & Technology.
149 161 Cong. Rec. S4300 (Jun. 18, 2015); referred to the Committee on Commerce, Science & Technology.
150 80 Fed. Reg. at 9556 (discussing possibility of Groups A–E).
151 See14 C.F.R. § 33.76(b) (requiring test with "large single bird" aimed at the most critical exposed location on the first stage rotor blades at a bird speed of 200 knots; requiring bird weights of 4–8 pounds, depending on engine inlet throat area); FAA Advisory Circular: Bird Ingestion Certification Standards, AC No. 33.76-1A (Aug. 7, 2009).
152 Henry H. Perritt, Jr. & Albert J. Plawinski, *One centimeter over my back yard: where does federal preemption of state drone regulation start?*, 17 *N.C.J. Law & Tech.* 307 (2015).
153 Montalvo v. Spirit Airlines, 508 F.3d 464 (9th Cir. 2007) (distinguishing field preemption, conflict preemption, and express preemption; holding that federal law preempted state negligence claims for an airline's failure to warn about the danger of developing deep vein thrombosis, because such a state-imposed duty would conflict with federal safety standards for pre-flight passenger briefings); compare Morales v. Trans World Airlines, Inc., 504 U.S. 374, 391 (1992) (state deceptive advertising guidelines preempted as applied to airline fares) with

American Airlines v. Wolens, 513 U.S. 219, 233 (1995) (state breach of contract action for violating terms of frequent flying program not preempted).

154 49 U.S.C. § 40120(c): "A remedy under this part is in addition to any other remedies provided by law."; 49 U.S.C. § 40116(c) (allowing landing fees for commercial aircraft landing or taking off within a state).

155 Compare In re Air Crash Near Clarence Center, New York, 798 F. Supp. 2d 481 (W.D. N.Y. 2011) (holding that occupation of the field of aviation safety by the Federal Aviation Act leaves no room for state safety standards) with Martin v. Midwest Express Holdings, Inc. 555 F.3d 806, 809 (9th Cir.2009) (holding that state standards of care for airplane stairs were not preempted because the FARs established no requirements for airplane stairs); Gilstrap v. United Airlines, Inc., 709 F.3d 995 (9th Cir. 2013) (embracing both Martin and Abdullah in finding disability claim not preempted).

156 Federal Aviation Administration, Office of the Chief Counsel, State and Local Regulation of Unmanned Aircraft Systems (UAS) Fact Sheet (Dec. 17, 2015) (explaining preemption of state and local drone regulation), www.faa.gov/uas/regulations_policies/media/UAS_Fact_Sheet_Final.pdf.

157 Margaret H. Lemos, *State Enforcement of Federal Law*, 86 *N. Y. Univ. L. Rev.* 698, 708 (2011) (asserting that states have no inherent power to enforce federal law); Hawaii v. Standard Oil Co., 405 U.S. 251, 263–64 (1972) (affirming dismissal of state *parens patriae* suit for damages under Clayton antitrust act); Connecticut v. Health Net, Inc., 383 F.3d 1258, 1262 (11th Cir. 2004) (affirming dismissal of action by state to enforce ERISA; no evidence of Congressional intent to give states enforcement standing).

158 Bonano v. East Caribbean Airline Corp., 365 F.3d 81, 84–85 (1st Cir. 2004) (holding that Congress meant to reserve enforcement of aviation regulations to the FAA); Schmelling v. NORDAM, 97 F.3d 1336 (10th Cir. 1996) (interpreting 49 U.S.C. section 46108 and holding that Federal Aviation Act does not grant private right of action to enforce FAA rules; affirming dismissal of action by former maintenance employer challenging dismissal for failing drug test).

159 The common law doctrine of negligence per se is an example of such incorporation.

160 720 Ill. Comp Stat. 5/26-1 (2013) (disorderly conduct).

161 Mont. Code. Ann. § 45-5-207 (1987) (criminal endangerment).

162 720 Ill. Comp Stat. 5/31-1 (2014) (interference with public officers).

163 City of Chicago v. Morales, 527 U.S. 41, 57–58 (1999) (affirming conclusion that gang-dispersal ordinance was unconstitutionally vague; explaining that laws criminalizing disobedience of police order are similarly questionable because of the possibility of arbitrary police orders); CA Penal Code §§ 409, 416 (refusal to disperse).

164 FAA, *Law Enforcement Guidance for Suspected Unauthorized UAS Operations*, www.faa.gov/uas/regulations_policies/media/FAA_UAS-PO_LEA_Guidance.pdf.

165 FAA FORM 7711-1 UAS COA Attachment accompanying section 333 exemption No. Exemption No. 11310 (Colin Hinkle), docket no. FAA-2014-0608 at p. 1 (Apr. 9, 2015).

166 Example. Hinman v. Pac. Air Lines Transp. Corp., 94 F.2d 755 (9th Cir. 1936) (rejecting trespass liability for aircraft overflying private property).

167 Compare United States v. Causby, 328 U.S. 256, 264 (1946) (holding that military flights at 83 ft over plaintiff's property constituted a compensable "taking" because it encroached on plaintiff's property rights), with Laird v. Nelms, 406 U.S. 797 (1972) (holding that high-altitude flight creating sonic booms did not constitute a trespass); Pueblo of Sandia ex rel. Chaves v. Smith, 497 F.2d 1043, 1045 (10th Cir. 1974) (rejecting trespass action against aircraft operator because no proof of actual injury to concrete uses of land). "The landowner owns at least

as much of the space above the ground as he can occupy or use in connection with the land." *Causby*, 328 U.S. at 264.

168 www.nydailynews.com/new-york/manhattan/drone-hits-nypd-helicopter-2-men-arrested-article-1.1858159; www.collegespun.com/sec/alabama-sec/someone-got-struck-by-a-drone-outside-bryant-denny-stadium-saturday-afternoon; www.7online.com/archive/9292217/; www.rt.com/usa/185480-new-york-tennis-drone/ (news reports of drone flights resulting in charges of reckless endangerment).
169 14 C.F.R. § 91.13 (prohibiting careless or reckless operation).
170 Ala. Code § 13A-6-24.
171 N.Y. Pen. Law § 120.20.
172 N.Y. Pen. Law § 120.25.
173 720 ILCS 5/12-5.
174 Ray Carver, *State Drone Laws: A Legitimate Answer to State Concerns or a Violation of Federal Sovereignty*, 31 Ga. St. U. L. Rev. 377 (2015).
175 31 Ga. St. U. L. Rev. at 404.
176 31 Ga. St. U. L. Rev. at 404–405.
177 15 U.S.C. §§ 1011–1015 (saving state insurance regulation from federal preemption).
178 2012 Act § 336.
179 725 Ill. Comp. Stat. Ann. 167/10.
180 725 Ill. Comp. Stat. Ann. 167/15 (2).
181 725 Ill. Comp. Stat. Ann. 167/15 (5) The Illinois statute confines law enforcement drone operation to the geographic location and imposes a time limit on investigation.
182 Wis. Stat. Ann. § 175.55 (West); Ind. Code Ann. § 35-33-5-9 (West); Fla. Stat. Ann. § 934.50 (West); Idaho Code Ann. § 21-213 (West); Iowa Code Ann. § 808.15 (West); Aircraft – Unmanned Aerial Vehicles – Rules And Regulations, 2015 Nevada Laws Ch. 327 (A.B. 239); Surveillance – Unmanned Aerial Vehicle – Limitations, 2015 North Dakota Laws Ch. 239 (H.B. 1328) (prohibiting drone use to gather evidence without a search warrant).
183 Surveillance – Unmanned Aerial Vehicle – Limitations, 2015 North Dakota Laws Ch. 239 (H.B. 1328).
184 Mont. Code Ann. § 46-5-109.
185 Wis. Stat. Ann. § 175.55 (West).
186 Surveillance – Unmanned Aerial Vehicle – Limitations, 2015 North Dakota Laws Ch. 239 (H.B. 1328).
187 Tex. Gov't Code Ann. § 423.008 (West).
188 Ibid.
189 See notes for text in this paragraph.
190 Fla. Stat. Ann. § 934.50 (West).
191 Idaho Code Ann. § 21-213 (West).
192 Fla. Stat. Ann. § 934.50 (West).
193 Aircraft – Unmanned Aerial Vehicles – Rules And Regulations, 2015 Nevada Laws Ch. 327 (A.B. 239).
194 Tex. Gov't Code Ann. § 423.006 (West).
195 Crimes And Offenses – Trespass – Photography And Pictures, 2015 Miss. Laws Ch. 489 (S.B. 2022).
196 Crimes And Offenses – Drones – Photography And Pictures, 2015 Tennessee Laws Pub. Ch. 240 (H.B. 153).
197 Cal. Civil Code sec. 3344 (prohibiting commercial use of another's name, voice, signature, photograph, or likeness without permission); 765 ILCS 1075/30 (prohibiting commercial use of person's identity); the statute defines "identity" as defining identity as "any attribute of an individual that serves to identify that

individual to an ordinary, reasonable viewer or listener, including but not limited to (i) name, (ii) signature, (iii) photograph, (iv) image, (v) likeness, or (vi) voice. 65 ILCS 1075/5.

198 Fish And Game – Animals – Harassment, 2015 New Hampshire Laws Ch. 38 (S.B. 222).

199 Fish And Game – Fish And Wildlife Commission – Drone Regulation, 2015 Oregon Laws Ch. 61 (H.B. 2534).

200 Ibid.

201 Va. Code Ann. § 19.2-60.1 (West).

202 Paul Diller, *Intrastate Preemption*, 87 *B.U. L.Rev.* 1113, 1127 n.64 (2007) (characterizing Alabama's lack of meaningful home rule).

203 David J. Barron, *Reclaiming Home Rule*, 116 *Harv. L. Rev.* 2255, 2277–2322 (2003) (analyzing history and competing philosophies of home rule); City of Commerce City v. State, 40 P.3d 1273, 1279 (Colo. 2002) (describing home rule authority).

204 Compare Ex parte Thompson, 442 S.W.3d 325, 336–337 (Tex. Crim. Ct. App. 2014) (analyzing First Amendment caselaw and holding that statute prohibiting video recording without subject's consent facially violated First Amendment) with State v. Lashinsky, 404 A.2d 1121, 1124–1125 (N.J. 1979) (affirming conviction of news photographer for disorderly conduct; restrictions on close-in photography of accident victim did not violate First Amendments, because photographer was interfering with public-safety personnel).

9 Allowing only "law-abiding drones"

9.1 Introduction

Multicopters—the dominant configuration for microdrones—require sophisticated automatic control systems to fly at all. Even the most skilled DROP cannot hand fly a multicopter, in the same sense that he can fly an airplane or helicopter—model or manned. A human cannot adjust the differential thrust and torque on each motor quickly enough to hover and to fly forward, backwards, and sideways. Instead of requiring the DROP to make decisions and adjustment on the current delivered to each motor, all practical control systems allow the DROP to command speed, direction of flight, climb or descent. The onboard electronics computes motor current to result in that flight condition. To do this, the onboard electronics must receive signals representing the DROP control inputs, typically movements of a cyclic and collective stick on the DROPCON, configured so that yaw inputs also are possible by moving the collective sideways. Then computers embedded in the electronics—the *flight computer* (FC)—must determine the drone's current orientation in space and rates of change in that orientation along each of the vehicle's three axes. The FCB computes the difference between what the vehicle is doing and what the DROP commands, and then calculates the current required by each motor to conform the vehicle's behavior to what the DROP wants. Finally, the systems must translate the commanded current through the power supply regulator to send the necessary current to each motor. All of this is built into a small onboard computer, linked to the control-link radio receiver, to the onboard sensors, and to electronic speed control (ESC) chips for each motor. Some microdrone systems include a small computer on the DROPCON as well, capable of running custom applications and storing flight information.

These automatic control systems, packaged onto integrated circuit boards weighing only an ounce or two and costing less than $100 enabled the technological revolution associated with microdrones.

This capability, necessary for basic flight, it can be extended relatively easily to automate flight control further. The sensors and flight computer already know where the vehicle is and how it is moving, and they already know how to adjust its position. Coupling these capabilities with rules and additional data permit regulatory requirements to be built into the vehicle.

For example, the vehicle can record the GPS coordinates where it began its flight or alternative safe landing GPS coordinates, and be programmed to return to the launch point or to the closest safe-landing point if the control link is lost, or if the DROP commands it. That capability is known as *return to home*. If it has access to data defining the details of the national airspace system, such as the GPS coordinates of airports, it can be programmed to avoid airports. That capability is known as *geo-fencing*. Such airspace data is available for download on ordinary flash memory cards, which typically drive GPS navigation systems on manned aircraft.

Most off-the-shelf microdrones have at least some of these flight autonomy capabilities built into the basic package. The techno-legal question is: how such autonomy should be combined with regulatory requirements without subjecting microdrone design to detailed engineering prescriptions that would burden designers and manufacturers, increase costs, and delay deployment of safety-enhancing technologies.

This chapter explores that question, drawing on the mission profiles in chapter 1, the explanations of technology in chapters 2 and 3, the human factor considerations in chapter 4, and the regulatory analysis in chapters 6 and 7.

9.2 Making microdrones law-abiding through vendor self-certification

A straightforward way for safety of flight to be autonomous is for the FAA simply to prohibit sale and distribution of drones unless they contain certain capabilities. The problem with this approach is that it invites application of the traditional regulatory approach for manned aircraft: they cannot be flown unless they have airworthiness and type certification. The FAA issues a type certificate for a model of aircraft; it issues airworthiness certificates for each aircraft of that model. Airworthiness requirements condition the issuance of a type certificate.

As the FAA itself recognized in its NPRM, airworthiness and type certification takes years and costs millions of dollars. An overview of the airworthiness requirements for normal category helicopters shows why that is so. The FAA imposes standards for flight, strength of components, design and construction, powerplant, equipment, operating limitations and information disclosed to customers.[1] The helicopter must have main rotor high- and low-pitch stops, and low-rotor RPM warning.[2] Airspeeds for minimum rate of descent and minimum angle of descent in autorotation must be determined.[3] "Control system forces and free play may not inhibit a smooth, direct rotorcraft response to control system input."[4] Inflight vibration may not be excessive.[5] Instruments must be easily visible under various cockpit lighting conditions and must be located so as to prevent confusion.[6] Safety belts must be equipped with metal to metal latching devices.[7] And hundreds of other equally specific requirements exist.

The applicant must prove compliance at each combination of weight and center of gravity for which it requests certification by flight test data or engineering calculations.[8] These requirements are not unreasonable for manned helicopters; they are perfectly logical to assure safe flight. Some of them, such as the cockpit instrumentation and the seatbelt latch requirement, are manifestly irrelevant to drones, but that is not the main problem. The problem is that establishing compliance with the requirements is arduous, expensive, and time consuming. The FAA must approve design documents before the applicant can begin fabrication; it must approve flight test plans before flight tests can begin; its inspectors scrutinize fabrication and flight testing. Once the FAA gets data that the applicant believes is sufficient, the FAA frequently requires redesign or retesting or both. An aircraft intended to be sold for $50 million can tolerate the cost; a microdrone for which no market exists at prices higher than $5,000 cannot.

Moreover, by the time a type achieves certification, its technology often is obsolete. So a traditional certification approach would discourage and delay technological enhancements that would make the vehicles safer. A better approach is for manufacturers to self-certify their products through a declaration of conformity (DoC) to performance standards.

The challenge is to develop a performance based regulatory approach that mandates what the vehicle must be able to do without prescribing the details of how it does it.

9.3 Regulating microdrones as consumer products

The desirable regulatory strategy would recognize that microdrones are essentially consumer products (even when they are used for commercial purposes), because of their low cost. Thousands of people buy them more or less casually and are willing to tolerate their loss.

The price the customers pay, the disposability of the product, and the lack of attachment between purchasers and the safety oriented aviation community means that traditional FAA regulatory approaches will not work. Traditional vehicle certification would drive the cost up by orders of magnitude. Purchasers are unlikely to spend the money or invest the time to earn traditional pilots' licenses, and their lack of attachment to a traditional training infrastructure makes them unlikely to know about or to comply with detailed operating rules. As with consumer products such as automobiles and lawnmowers, the only feasible regulatory approach is to regulate the design of the products and set a framework to ensure compliance with the design requirements.

9.4 Requiring reliable built-in rule compliance

The FAA can promulgate performance requirements that reflect the limitations developed in this chapter and prohibit the sale or operation of drones that do not have those capabilities.[9]

Basic enforcement would be achieved by having designers and manufacturers certify compliance, much as is done with motor vehicles, wireless devices, and certain potentially hazardous consumer products like lawnmowers. Then, the FAA would spot check promotional literature and user guides to make sure the required capabilities are advertised and request test measurements from manufacturers to make sure their products comply with the safety standards.

But a return-to-home feature or geo fencing capability does not protect safety very well if it only works some of the time. Reliable operation under all anticipated flight conditions is at the heart of traditional airworthiness certification. Manufacturers must make available to the FAA detailed engineering calculations on the strength of structures, conduct elaborate destructive and non-destructive testing to verify the engineering calculations, and perform extensive flight testing. All of this is intended to assure reliability: to make sure that systems actually perform as they are intended. The challenge is to achieve the desired reliability of microdrone safety systems without following this burdensome regulatory approach.

One attractive answer is that problems in the operation of systems can be determined through their actual use rather than through engineers predicting behavior by extrapolating from laboratory or test flight environments.

That approach is tolerable in the microdrone world because of the relatively small risks they present. If something goes wrong with the return-to-home feature on a DJI Phantom, it is not going to kill the crew or a load of passengers.

But for such a post-hoc, experience-based strategy to work, the FAA and the manufacturers must know about malfunctions. A procedure for doing this already is visible in the NPRM in the section 333 exemptions. Operators must report anomalies in navigation systems and malfunctions to the FAA. Compliance with these reporting requirements is made more likely because of the natural desire of drone purchasers to have their vehicles perform as advertised.

A viable self-certification approach requires that the reporting be made automatic.

9.5 Enforcement

If the FAA prescribes autonomous safety capabilities and collects data on malfunctions, it can adapt its own system of airworthiness directives and borrow the systems of recalls under the highway safety regulatory regime to require the refinement of imperfect systems.

While the number of microdrones in operation already exceeds the number of manned aircraft, their numbers will always remain smaller than the number of automobiles and trucks on the road. Although the supply side of the market is fragmented now, with many small producers and user-built aircraft, most microdrones are designed and manufactured by a relatively small number of vendors. Economies of scale will result in further consolidation. The concentrated structure creates channels of communication between vendor and customer that can facilitate distribution recall notices and notification of airworthiness directive is pertinent to a particular model.

Chapter 8 explains the system of penalties under the Federal Aviation Act and FAA rules. That system is applicable to drone vendors who fraudulently certify compliance with performance standards and to drone operators who fly drones that do not have valid compliance certification, as, for example, when a certified drone has been sold, but subsequently does not comply with an airworthiness directive. False statements to the FAA are crimes, punishable by up to five years imprisonment.[10] DROPs holding FAA operator certificates could have them revoked or suspended for flying drones without certification or in violation of airworthiness directives.[11]

And, of course, misrepresentations to the FAA and non-compliance with airworthiness directives would be powerful evidence for the plaintiff in a private lawsuit.

9.6 Required limitations

The following subsections describe the necessary autonomy requirements and justify each. They explain that autonomous safety capabilities must address at least height, weight, and operating radii.

9.6.1 Height

Limiting the maximum height above the ground for microdrone flights promotes safety by reducing the likelihood of encountering conflicting traffic. Lower heights also mean less kinetic energy stored in a drone falling out of the sky, but a few calculations about terminal velocity will show this is not very important.

Height restrictions for microdrones do not take care of all safety problems, however. Seaplanes, for example, regularly fly enroute only 300 ft above the water, and helicopters often fly below 500 ft for certain types of missions, such as boat photography, law enforcement, and heavy lift operations.

There are maybe a half-dozen parts of the country where low-level seaplane or helicopter flight is concentrated. Seattle is one example of dense seaplane operations. The routes of pipelines and major powerlines provide other examples. One way to deal with this is to prohibit either microdrone or macrodrone flights in these areas.

9.6.2 Weight

Imposing a height restriction on microdrone operation is not enough; manned aircraft sometimes fly below 500 ft, and a microdrone mishap can endanger persons or property on the ground regardless of the height at which the drone was operating. A weight restriction also is appropriate. If a small drone weighing less than, say, ten pounds were to collide with another aircraft, survivability would be much greater than if a much larger drone is involved in a similar collision. Bird strikes illustrate the problem. Bird strikes are the number two accident cause with helicopters.[12] In one case, the pilot was incapacitated when a bird penetrated the windshield bubble. In another, a larger bird impacted the main rotor causing it to separate, killing the occupants of the helicopter. Larger birds cause more damage and worse injuries.[13] Bigger birds have a mass similar to microdrones now on the market. The Phantom microdrone, for example, has a mass roughly equivalent to that of a mallard duck or a seagull.[14] Microdrones also have batteries constructed of denser material than bird skeletons.

A weight limitation promotes safety by limiting the kinetic energy in the drone and therefore reduces the damage to another aircraft, ground objects, or persons on the grounds if a collision or crash occurs. Engineering ballistics analysis backs this up and shows the relationship between drone weight and expected damage.

Weight matters if there is a collision with another airborne object or a crash. Damage done when two objects collide depends on the kinetic energy of each when the collision occurs. As chapter 5 explains, kinetic energy of a moving object is half of its mass times its velocity squared. If an aircraft collides with a hailstone, likely damage to the aircraft is modest. A small hailstone weighing one gram (0.002 pounds) has kinetic energy of 20.5 foot-pounds relative to a helicopter moving at 60 knots.[15] A 2.5-pound bird or microdrone has kinetic energy of 795 foot-pound. As mass increases, the kinetic energy increases linearly.

Even modest deformation of the aircraft skin or windshield can absorb all of that energy. Aluminum absorbs more energy before it fractures under impact than Plexiglas.[16] Energy absorption of either is linearly proportional to thickness, for relatively thin structures.[17] On the other hand, a 55 pound drone with a similar velocity would have 17,490 foot-pounds of energy, and a windshield or skin designed to withstand impacts with smaller objects cannot absorb that much energy without suffering penetration.

A similar relationship between mass and damage to persons or objects on the ground exists: the heavier the object, the more damage and the greater the hazard. A 9.2 pound Canadian goose can do serious damage. Collision or crash of a 50-pound drone could have even more dramatic consequences.

As weight increases, the need for more capable collision avoidance systems increases commensurately.

9.6.3 Operating radii, line of sight, and visibility requirements

When the microdrone remains within line of sight of its DROP, two risks are reduced. First, the DROP can control it by looking at it; he just does not need more sophisticated FPV, graphical display, or collision detection and avoidance telemetry and logic; nor does he need ATC communications that would be necessary to mitigate the same risks for a drone flown beyond the line of sight.[18]

Second, keeping the drone in sight is a rough proxy for keeping it within range of relatively low powered control links, usually implemented on currently available microdrones through Wi-Fi connectivity.[19]

Keeping a microdrone in sight involves limiting height and horizontal distance from the DROP. A microdrone at 500 ft directly over the DROP is easier to see in detail than one 1,000 ft away at 20 ft AGL; what matters is the actual range and not simply the height.

Moreover, seeing the microdrone in sufficient detail to determine its orientation, flight path, and distance from obstacles depends on visibility—the visual range is much less on a foggy day than on a clear day. It also depends on the DROP's visual acuity and on the possibility that other objects might interpose themselves between the DROP and the microdrone.

The purpose of the line of sight restriction is to increase controllability. So long as the DROP can see the microdrone, he can fly it so as to avoid collisions with other aircraft and ground objects. He can keep it under control and navigate by watching it, without reliance on more sophisticated video systems installed on the drone and in his console.

Minimum visibility requirements are a mainstay of operating rules; the basic three-statute-mile minimum visibility for VFR operations at most altitudes[20] is probably too great for microdrones, especially when they are flown below treetop level, where no manned aircraft fly. At greater heights, however, some visibility limitation, say ½ mile, is appropriate, if only to allow other aircraft to see the drone. Visual acuity requirements are mainstays of airmen certification.[21]

And, as with the visual contact between DROP and drone, what matters is range, not simply height. A microdrone directly over the DROP's head at 500 ft is likely to be well within control-link range; one at 500 ft AGL at a 2,000-foot horizontal distance is more likely to be out of range. This, as with maintaining visual contact, invites consideration of limits on operational radii as well as height.

Line of sight restrictions can be problematic, however, if they are literally interpreted. It is hard to fly a microdrone in any realistic mission environment without occasionally flying behind a tree or a building. Indeed, some missions require flying a drone behind an obstacle, such as searching a backyard or an alley in a law enforcement operation. The NPRM acknowledges that momentary loss of line of sight is not a violation.[22]

Autonomous drone systems can enforce line of sight restrictions by enforcing control-link range limitations. Limiting transmitter power on the drone and DROP limit the range of the control link, and the flight computer can be programmed to respond safely to loss of control link.

9.6.4 Performance requirement for autonomous operations

The FAA should express performance requirements in regulations that leave maximum flexibility for engineering design, while ensuring that drones obey the most basic safety requirements. One way to mitigate risks is for microdrones to behave in a certain way when they detect certain conditions. They can cut their engines and simply fall to the ground. They can reduce power and enter a controlled descent to the ground, wherever they are. Or, they can return to their launching point. None of these is foolproof. Cutting the engines simply results in a crash with whatever consequences result from weight, terminal velocity, and what is below. A controlled descent to the ground reduces impact speed, but the consequences depend on what is underneath when the controlled descent ends.

The best response is usually autonomous return to home or to the closest preprogrammed alternative safe landing position. The crucial requirements are that the drone stay out of certain airspace, that it respond appropriately to certain potentially dangerous flight conditions, and that its automatic systems be reliable. The most difficult to ensure is reliability.

9.6.4.1 Exclusion from controlled airspace

As the introduction to this chapter explains, many off-the-shelf microdrones have geo-fencing, making use of widely available geographic databases and their built-in GPS positioning to prevent the drone from flying into the controlled airspace. Consequently, there is no particular technological challenge to require such capability in new products.

9.6.4.2 Loss of control links, battery exhaustion, obstacles

The vehicle knows when the control link has been lost; any digital receiver can detect loss of signal. When that occurs, or if the vehicle has received no DROP inputs for a predetermined period of time, it automatically should enter return-to-home mode.

Ordinarily, that will result in a safe conclusion of the flight, although risks still remain if there is insufficient battery charge to fly to the launching point, if the winds are greater than the speed at which the drone executes return-to-home functions, or if obstacles exist between the point at which the failure occurs and the launching point and the altitude when the control link does not allow it to clear obstacles on its return flight.

The battery capacity problem can be mitigated by triggering return to home as soon as the battery charge reaches a certain minimum level. The wind problem can be mitigated by careful preflight, taking into account weather conditions.

The obstacle problem is harder; collision avoidance technologies are just beginning to be available, as explained in chapter 3.

Alternatively, the DROP can program a list of safe landing GPS coordinates, as part of his preflight preparation. When the drone loses a control link, the drone will immediately perform a computation and land at the nearest safe landing coordinates. There, the drone will attempt to reestablish a link with the DROP.

9.6.4.3 System reliability

As § 9.1 notes, most mass-marketed microdrones come with basic autonomous safety features: return to home, land immediately, automatic hover, and automatic takeoff and landing. Many also come with geo-fencing. So to require these features and to place them beyond the control of the DROP imposes relatively trivial burdens on vendors. The regulatory challenge is how to assure that they deliver what they promise. The reliability of the currently delivered systems is less than is needed to ensure safety. Flyaways are a frequent problem for virtually every popular microdrone model. A flyaway means, at a minimum, that the return-to-home feature did not work as intended.

So in addition to assuring that the autonomous safety functions exist and are installed on the drone, the regulatory regime must require that the critical ones be reliable. Reliability engineering is a well-defined field of expertise, discussed in chapter 3. Its norms would require that a return-to-home feature—the most basic of all safety functionality—be tested under all reasonably expected flight conditions to collect enough data to permit statistically valid assessment of its reliability, expressed as a percentage of the time that it functions correctly.

The problem is that testing to ensure compliance with any performance standards can be very expensive. The testing requirements easily can dwarf whatever benefits are obtained by self-certification and engineering standards. NHTSA's recent standard for electronic stability control on busses and trucks is a good example. The final rule published in the federal register has more than a dozen pages devoted to arguments over test standards in the proposed rule.[23]

In the microdrone context, executing even rudimentary reliability testing protocols can impose significant cost. The challenge, and the main driver of cost and duration, is not only that many—probably thousands—of flights are necessary to collect the necessary data, but also that multiple causes of return-to-home failure exist—even as a theoretical matter; never mind real-world complications. To function successfully, any return-to-home subsystem: (1) must know where the vehicle is when the feature is triggered; (2) must know where home is; (3) must be able to calculate a path from its present position to home; (4) must communicate that path to a navigation system capable of causing the drone to fly the path; (5) the path must be one that the drone's thrust, climb and descent capabilities permit it to fly; (6) the path must not be interrupted by obstacles; and (7) the drone's return speed must be greater than opposing wind.

Failure of steps (1) and (2) results from not having GPS lock at the points when coordinates are recorded. Failure of step (3) can result from a misdesigned algorithm, data errors in the coordinates input to it, or a hardware fault as the algorithm is being executed. Failure of step (4) can result from a poor physical connection, data errors, or misalignment of data-structure frames. Failure of step (5) could result from the commanded path requiring altitudes, speeds, or turn rates exceeding the drone's performance capabilities. Failure of step (6) results if the drop has flown around or above a tree, pole or building on the outbound flight. Failure at step (7) can result if the drone flew downwind on its outbound flight, or if the wind speed has increased during the flight or is greater at a higher altitude at which the drone is flying.

A comprehensive test protocol must collect failure rate data under each of these conditions, many of which must be simulated for the test. Room for argument always exists as to whether a simulation adequately models reality. Some of the testing, such as that for steps (3) and (5) may not require actual flight, however. Requirements for any kind compliance testing are controversial, even among engineers skilled on the subject matter.[24] The same room for argument exists with respect to drone autonomous safety system testing.

The cost of all this is considerable. Suppose 1,000 flights or other test cycles for each condition are necessary to collect the data necessary for statistical robustness. The actual number may be much larger. Suppose a DROP, a reliability engineer, and a data analyst are necessary for each series of tests. Suppose further that the replacement cost of the test vehicle is $1,200, and that the vehicle loss rate during the tests is 10 percent. Finally, suppose that the duration of each test flight is 20 minutes, and that return to home can be triggered every 5 minutes on each flight.

Those assumptions result in total test-flight time of 8,333 hours.[25] Assuming personnel compensation of $30,000 annually for the DROP, $50,000 annually for the reliability engineer, and $25,000 annually for the data analysis, labor cost for the testing totals $125,000.[26]

This is just one part of a comprehensive test protocol. Tests also must be designed to determine how much return-to-home capability is achievable without a GPS lock by reliance on the onboard IMU, or with onboard magnetometer and altimeter alone. An IMU can record spatial movements from the launch point and therefore enable the drone to retrace the path to return to home. A magnetometer and altimeter alone can allow a drone to fly directly toward the launching point—assuming it knows where it is—but is incapable of compensating for wind.

On the other hand, not every component has to be subjected to reliability testing if the return-to-home subsystem includes particular component designs or off-the-shelf components have passed reliability testing with specified failure rates.

Even if a vendor elects to do minimal testing before placing a product on sale, performance standards nevertheless can be useful. They can represent a trigger for product recalls as in the motor vehicle safety standard context. They represent a standard of care for courts to apply when litigation occurs to compensate victims of drone accidents.

9.7 Regulatory language

The FAA can ensure that drones have autonomous safety features by following the regulatory models used to regulate motor vehicle safety, to protect against interference by radiofrequency devices such as cellphones and Wi-Fi points of presence, and to assure the safety of consumer products.

The agency would promulgate a requirement with the following substance:

14 C.F.R. Part 107

Subpart E-Small Unmanned Aircraft Autonomous Safety Systems

§ 107.90

Small unmanned aircraft systems are deemed airworthy under Part 21 if, and only if, they are equipped with the following autonomous safety features:

- *a. GPS-enabled automatic land-immediately and return-to-home capability*
- *b. Interlock that does not permit takeoff unless a GPS lock has been obtained*
- *c. Backup land-immediately and return-to-home capability not dependent on GPS lock*
- *d. Live collection of flight data and malfunction indications, automatically recorded to non-volatile memory permanently installed on the vehicle and the operator's console as it is collected*
- *e. Capability to upload flight data and malfunction reports to the vendor or to an Internet server accessible by the vendor and*
- *f. Vendor certification of compliance with these requirements.*

The requirement in § 107.90(c) would be satisfied by software that collects the necessary data from an onboard IMU, magnetometer, and barometric altimeter so as to enable the drone to retrace its outbound flight path. Section 107.90(d) requires recording on *both* the vehicle and on the operator's console, because if the control link is lost, the console will not receive what may be the most critical data about the malfunction in the autonomous safety features. If the vehicle is lost in the mishap, the only recorded data would be on the operator's console.

This data collecting, recording, and evaluation is exactly what happens now with state-of-the art airplanes and helicopters. The aircraft lands, and the mechanic plugs in her laptop and downloads system performance data to facilitate troubleshooting. The warranty requires the mechanic to upload certain data to the manufacturer. Similarly, emissions testing of modern automobiles relies mostly on data downloads from the vehicle self-monitoring system.

Self-certification by manufactures and vendors would allow vehicles to be sold and distributed in interstate commerce. The FAA would not, in the ordinary case, do any pre-sale certification. This is the approach followed by NHTSA for automobile safety standards and by CPSC for consumer products. No particular testing protocol would be required, but prudent designers and manufactures would conduct presale testing as a way to back up their self-certification. This approach would avoid the enormous costs and delays associated with traditional airworthiness certificate, a fact the FAA itself acknowledged in the NPRM. It also avoids negating the benefits of performance standards when the FAA inserts itself in a pre-sale approval process. Recognizing that even rudimentary testing for reliability of return-to-home features can cost more than $100,000, the approach avoids prescribing a test program, and instead allows vendors to choose what mix of presale testing, exposure to recalls, and after-the-fact liability for mishaps their business model accommodates.

The most aggressive part of the proposal involves the mandatory backup for the return-to-home system, one that does not rely on GPS. Many different factors can interrupt reception of the necessary satellite GPS signals. It is unlikely that any design and any practicable testing program could assure GPS reliability at the necessary level; GPS signals are lost, not because of any malfunction of drone systems, but because of poor reception. Paragraph (b) mitigates this risk partially by requiring an interlock that does not permit the drone to take off unless it has GPS lock.

But a GPS lock can be lost once the drone is in flight. The only way to automate a response to this eventuality is to require a backup navigation system not reliant on GPS. Virtually all microdrones now marketed have inertial measurement units, and they can collect the necessary data to permit the drone to retrace its flightpath and return to home—as long as it knows where home is.

Notes

1 14 C.F.R. Pt. 27.
2 14 C.F.R. § 27.33.
3 14 C.F.R. § 27.71.
4 14 C.F.R. § 27.151.
5 14 C.F.R. § 27.251.
6 14 C.F.R. § 27.1321.
7 14 C.F.R. § 27.1413.
8 14 C.F.R. § 27.21.
9 Henry H. Perritt, Jr. & Eliot O. Sprague, *Law Abiding Drones*, 16 *Colum. Sci. & Tech. L. Rev.* 385 (2015) Henry H. Perritt, Jr. & Albert J. Plawinski, *Making civilian drones safe: performance standards, self-certification, and post-sale data collection*, 14 *Northwestern J. Tech. & Intel. Prop.* (forthcoming 2015).
10 18 U.S.C. § 1001.

11 Armstrong v. FAA, 296 Fed. Appx. 88, 89 (D.C. Cir. 2008) (affirming revocation of pilot's license for operating Mooney aircraft in violation of airworthiness directive).

12 blogs.usda.gov/2013/06/06/helicopters-and-bird-strikes-results-from-first-analysis-available-online/; Washburn *et al.*, *Bird Strike Hazards and Mitigation Strategies for Military Rotary-wing Aircraft* (Dec. 31, 2012) (Project No. 11-944), www.denix.osd.mil/nr/upload/FINAL-OSD-Legacy-Report-Joint-RW-Bird-Strike-Hazardswithcomments-FINAL.pdf [hereinafter "Bird Strike Study"]. Helicopters are much more likely to be damaged by bird strikes than fixed-wing aircraft. Bird Strike Study at 2.

13 Bird Strike Study at 17, 47. 200 bird strikes with US civil helicopters were reported in 2011. Bird Strike Study at 41.

14 Roger Nicholson, *Kinetic Energy of Bird Strikes & Damage to Aircraft* (Boeing technical paper).

15 The hailstone is not likely to have much horizontal velocity on its own.

16 Figure 16, www.kazuli.com/UW/4A/ME534/lexan%20VS%20Acrylic4.htm.

17 David Roylance, *Introduction to Fracture Mechanics*, (Jun. 14, 2001), p. 11 ocw.mit.edu/courses/materials-science-and-engineering/3-11-mechanics-of-materials-fall-1999/modules/frac.pdf [hereinafter "Roylance"].

18 NPRM, *supra* note 9 at 66–72 (discussing need for LOS restriction as a substitute for traditional see-and-avoid requirement).

19 Use of Wi-Fi for control links has significant advantages. It uses unlicensed spectrum and spread spectrum technologies, which greatly reduce radio interference problems.

20 14 C.F.R. § 91.155 (2015) (imposing minimum visibility requirements at different altitudes in different kinds of airspace).

21 14 C.F.R. §§ 67.103; 67.203; 67.303 (2015) (prescribing vision requirements for first-, second-, and third-class medical certificates).

22 NPRM, *supra* note 9, at 67–68 (acknowledging possibility of momentary loss of sight).

23 Federal Motor Vehicle Safety Standards; Electronic Stability Control Systems for Heavy Vehicles, 80 Fed.Reg. 36049 (Jun. 23, 2015), www.gpo.gov/fdsys/pkg/FR-2015-06-23/pdf/2015-14127.pdf (final rule).

24 NHTSA's recent standard for electronic stability control on busses and trucks is a good example. The final rule published in the federal register has more than a dozen pages devoted to arguments over test standards in the proposed rule. Federal Motor Vehicle Safety Standards; Electronic Stability Control Systems for Heavy Vehicles, 80 Fed.Reg. 36049 (Jun. 23, 2015), www.gpo.gov/fdsys/pkg/FR-2015-06-23/pdf/2015-14127.pdf (final rule).

25 1,000 test cycles, divided by 4 cycles per flight, multiplied by 20 minutes per flight, multiplied by 5 scenarios (excluding tests for steps (3) and (5)).

26 Total test time of 8,333, divided by annual work hours of 2,000, multiplied by the sum of the annual salaries for the three test professionals.

10 Economics

10.1 Introduction

The performance of aviation enterprises is determined by an ongoing war between thermodynamics and economics. The laws of thermodynamics limit how much speed and payload can be delivered by an air vehicle to a fraction of the energy supplied by fuel or electricity. Increasing the fraction of energy turned into useful work requires advanced technologies, which often cost more.

Private enterprises, including drone enterprises, must earn enough revenue to cover their costs and to generate a return on invested capital. To identify the range of viable business requires projecting revenues and costs under various conditions. This chapter uses techniques well known to the field of operations research for fifty years to develop its various cost and revenue estimates.

Assessing the economics of civilian drone use requires a separate analysis of the economics of demand and the economics of supply. With respect to both supply and demand, the questions, and therefore the answers, are considerably clearer for microdrones than for macrodrones. Everyone knows now what microdrones look like, what they can do, and how much they cost. Vendors offer new, upgraded, models several times a year, and it is reasonably certain that they will get better—automation will be extended to new functions and get more reliable and accurate; battery technology will improve, increasing endurance. Prices will continue to fall, as the scale of production ramps up.

The designs of macrodrones, their capabilities, and their costs, are more difficult to predict with confidence. The ones available now are, for the most part, straightforward adaptations of manned aircraft designs. Macrodrone development is likely to continue to be led by the defense contractors. National security decision-makers are already committing themselves to grow drone procurement at the expense of manned aircraft, as they put their budgets together for the next 5–10 years. It is clear that there will be substantial growth in macrodrone R&D, regardless of how active actual procurement and operations are.

Much of the defense-oriented R&D will benefit civilian macrodrone development. At the same time, manufacturers are unlikely to commit themselves to aggressive civilian macrodrone projects until the size of the market is clearer. The size and nature of the civilian market for macrodrones will remain murky until potential users and operators know more about what macrodrones will be able to do and how much they will cost to acquire and operate.

About the only thing that is certain now is that macrodrones will not require a pilot or copilot, and that at least for the first few generations, they will not carry passengers. So, a potential purchaser of macrodrones or of contract macrodrone services knows that it will not incur the costs for pilot and copilot, but cannot know much else. The purchaser will have to pay DROPs and mechanics who must maintain aircraft with more complex control systems. The cost of development, in turn, depends on what kind of DROP and vehicle certification requirements and operating requirements the FAA imposes.

The world of macrodrones will not be like the world of microdrones. As this book advocates, microdrones may be regulated with a very light touch, with many existing autonomy features simply mandated. If the law-abiding-drone approach advocated in chapter 9 shapes regulatory philosophy, regulation will impose relatively modest costs and limit operational utility relatively little. The demand for microdrones will be driven by perceptions of how much aerial photography and other sensor-based analysis will enhance existing business activities, and whether their benefit is worth paying the quite modest cost of using them.

A light regulatory touch is unlikely for macrodrones. They weigh much more, have greater range, and will be flown outside the line of sight of the operator. All these characteristics make a much stronger case for traditional aviation regulation, supplemented by new forms of airspace management.

10.2 Quantifying demand

The level of demand for any product depends on its price and on what it can do—its productivity—relative to the price for substitute products and their productivity. Of course, price is not everything. Bragging rights count, too. Adding a drone point of view to the wedding video is something the newlyweds can brag about and thus often is worth a high price. Productivity is not often the issue for wedding arrangements. Nevertheless, relative price and productivity govern most purchase decisions and thus shape demand.

In evaluating price and productivity, one must consider price elasticity of demand and must carefully define the product and its function.

10.2.1 Price elasticity

Microeconomics teaches that both supply and demand for particular product vary with price. A higher price means less demand, but more supply. A lower price means greater demand, but less supply. How much demand varies with price is the *price elasticity of demand*. How much supply varies with the price is *price elasticity of supply*.

For a given product at a given time, supply and demand are in equilibrium at only one price. Market exchanges permit sellers and buyers to find each other and to discover the equilibrium price.

This simple view of how a market economy works is useful, but seriously incomplete. This model assumes there are many identical buyers and sellers, each one of which is too small to affect price. This assumption leads, in turn, to the assumption that there are many different sources offering identical products. And it further explicitly assumes that the only determinant of quantity supplied or demanded is price.

In the real marketplace for aviation services—and anything else—these assumptions are, at least to some extent, invalid. More than price determines demand; a buyer may be willing to pay a higher price for products from a particularly trusted seller. A seller may be willing to sell at a lower price in order to obtain a new customer, with the expectation that future revenue from that customer will make up for early losses.

Furthermore, the products offered by competing sellers rarely are perfect substitutes for each other. Different sellers offer different drone models; they offer different pricing packages; they offer different kind of support. And often, the relevant competition is not between two different sellers of drone services, but between a seller of drone services and a seller of manned aircraft support, often helicopters.

Nevertheless, price matters, and the price for microdrone services is a small fraction of the cost for competing services using manned aircraft. That price difference means that the demand for microdrones will be much greater than the demand for manned aircraft in those industries where microdrones can do almost as good a job.

The attractiveness of prices for microdrone services is premised on dramatically lower cost associated with much more modest capabilities. Their modest capabilities nevertheless can fill important gaps in air support now unable to be filled by manned aircraft. But no customer will pay as much for microdrone support as it is willing to pay for manned-aircraft or macrodrone support. Why pay more for less capability?

As the market matures for microdrones, the point at which price resistance occurs can be estimated with greater precision. The same is true for macrodrones, but their comparable capability to that of manned aircraft makes it likely that customers will be willing to pay roughly equivalent prices.

They may not be willing to pay higher prices. Operators were skeptical, for example, of an unmanned K-Max equipped for wildfire suppression. "The bottom line, will the government and business community be willing to pay for the new technology?" one operator said. "Why would I make the significant investment to be unable to compete at a competitive price?" said another. "I don't see how it would work without the government basically subsidizing the investment."[1]

10.2.2 Defining the product

Quantifying the size of the market for civilian drones requires being as precise as possible in defining the service the drones can provide. That can be tricky when new technology is involved, because new technologies often make users aware of "needs" they had not known about before. The (perhaps apocryphal) story of the introduction of the Xerox machine provides an example. After Chester Carlson proved that his technology for photocopying was workable, he set out to sell it to major enterprises with a foothold in the office-supply and office-equipment markets. They were uninterested after they calculated potential based on data about the demand for carbon paper—the prevailing technology at the time for making copies. When the Xerox machine was finally introduced, the market for copying exploded, because Xeroxing was so much more convenient and produced better copies than using carbon paper.

The relevant market was copies, not carbon copies.

One needs to be careful not to make the same mistake, to define the market too narrowly, for drones. The analog of the demand for carbon paper in the Xerox story is the demand for airplane and helicopter support in the aviation context. Unscheduled air transportation in the US (SIC 4812) was a $6.8 billion industry in 2012, comprising 213,805 establishments,[2]

The wide availability of drones is likely to stimulate new demands, beyond what is apparent for airplanes and helicopters. Only the most expensive weddings justify helicopter involvement, but almost any wedding can justify flying a drone to take aerial video. So, once drones are taken into account, the market for wedding photography will explode. Some imagery in high-risk contexts may be worth putting the unmanned aircraft at risk, although it is not worth risking an aircrew. The result will be expansion of the market for aerial inspection of hazardous operations.

As with any market analysis, segmentation is helpful. The reasons for using air vehicles to transport people are different from those for transporting things, and those, in turn, are different from reasons inviting aerial sensing such as photography. Aerial insertion is another category of use, often referred to in the industry as *heavy lift* or *longline* operations. In these operations, a helicopter inserts or removes a load usually attached to the helicopter by a long line cable. The load may be equipment or cargo, including animals such as whales and elephants, or it may be one or more human beings.

Demand depends, not only on price, but also on how much performance a drone can offer. Autonomy expands the possibilities. Drone autonomy means that the question is no longer whether a DROP on the ground can ever match the performance of a human pilot on board the aircraft—whether he can control and direct his vehicle as precisely, only by means of a (probably) flat screen display and radio controls that have at least some lag in them. Automation systems on board the drone may do a *better* job than a human pilot in finding particular locations on the ground, monitoring groundspeed for landings and computing the complex dynamics of a load at the end of a 50 ft line under a helicopter.

Over the long run, it is likely that the flexible, low-cost photography available through microdrones and other kinds of imagery and mapping data about the ground will cause potential users to discover a pressing need for aerial sensing that had not had occurred to them before. Likewise, if highly automated macrodrones can do better than manned helicopters in certain dangerous activities like long line operations, or repetitive operations like aerial survey, the demand for them may far exceed whatever has been experienced with helicopters.

The question always will be whether it is worth paying for extensive autonomous capability. That question pops up regularly in other contexts, and many purchasers pay the necessary price. Farmers install GPS auto-steering systems on their tractors and combines. They buy GPS tracking systems to improve production, save fuel and soil compaction. They could do the job with simpler equipment but the benefits of automation justify the cost.

10.2.3 Competitive assessment

Relative price and productivity of aircraft and the people who fly them are the starting points for evaluating the competitive relationship between drones and their manned counterparts. Drone aircraft, at least microdrones, cost much less than manned aircraft that perform the same missions. A typical microdrone is likely to cost $1,000–$20,000. A typical helicopter for similar missions costs $250,000–$2.5 million. A typical microdrone DROP is likely to get paid $50,000. A typical commercial helicopter or airplane pilot outside the airline industry is paid $50,000–$150,000, depending on the difficulty and risk of the type of flying.

Physical productivity is measured by the ratio of output to input. A transport airplane that can carry 300 passengers 1,000 miles in 2.5 hours is more productive than one that can carry 100 passengers 1,000 miles in 5 hours. A microdrone that can capture 100 acres of crop imagery in 2 hours is four times as productive as one that takes 8 hours to cover the same 100 acres.

Economic productivity comes closer to measuring what a business enterprise cares about. Economic productivity measures how much revenue an asset produces per unit of cost: how much each $9,000 microdrone earns per week. Physical productivity and economic productivity are related by the price for (or economic value of) each unit of output, and by the cost of the inputs.

Both measures of productivity depend on how much an aircraft flies—its utilization. Utilization is the number of hours per unit of work time, e.g. per day, that an asset is available to do work. Productivity is the amount of work it does per unit of input, e.g. per workday. For an asset like a DROP or a pilot, the two values may be the same: flight hours per workday. Drone operations open up the possibility that one DROP could fly more than one drone at a time, which would increase DROP productivity. For a vehicle, on the other hand, the calculation is likely to be quite different. Utilization would be expressed in flight hours per day, while productivity would also include the number of passengers or tons of cargo carried: ton miles per day or passenger miles per day or simply passengers per day.

Airline economics is of limited utility in evaluating drone operations. Airline aircraft acquisition costs are in the millions to tens of millions of dollars; drone acquisition costs are in the thousands to tens of thousands. Fuel costs are similarly disproportionate.

Focusing on aircraft utilization is appropriate for an airline and for many other aircraft operations because aircraft are the most expensive physical assets those enterprises must acquire and operate. This is not necessarily the case for a microdrone operator. Personnel costs, such as those for the DROPs are likely to be more significant economically than the vehicle cost, at any reasonable level of utilization. While microdrone utilization is important for sizing the fleet necessary to satisfy demand, it is not a particularly strong determinant of overall enterprise productivity or the return that it can generate on its invested capital. Likely, the utilization of other, more expensive assets, such as DROPs, is more important. DROP productivity and utilization is considered in section § 10.9.1.

Nevertheless, some basic characteristics are common to all commercial aviation. The success of low-cost airlines illustrates some conclusions that are pertinent to drone operations.[3] Southwest Airlines, for example, has significantly higher labor productivity than legacy airlines, although it pays competitive wages to its pilots. A significant reason for the higher productivity is that Southwest Airline pilots perform collateral activities such as cleaning the aircraft and carrying bags.[4] Another low-cost airline, Jet Blue, has considerably higher aircraft utilization, approaching 15 hours per day instead of the 9–10 hours a day typical of United, Delta, and American.[5]

Aircraft utilization is less likely to drive the overall productivity of a drone enterprise than it is for manned-aircraft enterprises. Drone productivity, however, is a measure of the basic competitive strengths of an enterprise centered on flying drones, as compared to one flying manned aircraft for accomplishing the mission, or using ground cameras or satellites.

For manned aircraft, the typical measure of physical productivity is expressed in terms of how many passengers or how many tons of freight it can carry, and at what speed. Microdrones and early generations of macrodrones are not likely to transport either people or cargo from one place to another. Instead, they are likely to do surveillance, capturing video or other data from above and transmitting it consumers of the data on the ground. So, a more plausible measure of drone productivity than passenger miles or ton miles is surveillance imagery produced per flight hour.

Similarly, a plausible measure of drone economic productivity is paid surveillance hours per day per dollar invested in the platform. Properly calculated, this takes into account the capability of faster aircraft of longer endurance, such as a typical ENG or public-safety helicopter, to cover multiple assignments on one flight by moving quickly from one location to another.

It also should reflect the effect of greater heights at which manned aircraft can be flown. A manned aircraft flying higher with larger sensors and faster processors may be able to capture the same imagery in much shorter periods of time than a drone flying lower and making more passes over the same targets. For certain types of surveillance, however, height may be unimportant. For example, an ENG helicopter might fly at 400 ft AGL for safety reasons, while the photog uses a zoom lens with long telephoto focal length to give the illusion of being lower. A drone, flying at 25 ft AGL could use a standard focal-length lens and get the same picture.

As is the case for almost any economic parameter, simplifying assumptions are necessary to develop values for aircraft productivity. One basic assumption for much productivity analysis, for example, is that the operator enjoys as much demand for aerial photography as it can satisfy. The reality is that both aircrews and aircraft spend much of their time waiting, or in the case of aircrews, spend much of their day performing non-flying tasks. Sometimes aircrews and aircraft must be available for a particular event, the timing of which is impossible to predict. There is no reason that there would be any difference between that standby time for manned aircraft and drones, and so it does not distort the comparative analysis to omit that reality.

The maximum vehicle productivity would be achieved by a vehicle flying continuously, collecting imagery 24 hours per day. Commercially available vehicles are unable to do that. Helicopters must refuel every three hours or so; microdrone battery life in 2015 does not exceed 30 minutes. Macrodrone endurance may exceed that of manned aircraft, but macrodrones costing comparable amounts are likely to mimic their manned aircraft equivalents in terms of endurance.

Then one must consider turnaround time—to refill to refuel a manned aircraft or a macrodrone or to swap batteries for a microdrone. Any of these types of aircraft will have to go off station to get fuel or fresh batteries: a certain amount of ferry time will be necessary on each flight.

Twenty minutes seems like a good figure for a helicopter and three minutes for a microdrone, which is flown within the line of sight of its DROP. In some helicopter operations, the turnaround figure is much lower: helicopters often land next to a fuel truck on location and refuel while the engine and rotor are turning—a process known as *hot fueling*. Otherwise, refueling involves time required to shut down and start up the engine, and the time required for the fuel to be pumped into the tank of the aircraft. Ten minutes seems like a good figure for helicopter or airplane, taking into account hot fueling when it's necessary. Three minutes seems like a good figure for changing the batteries on microdrones. Both times of aircraft also must travel from mission location to the refueling or battery-change point. A helicopter might take 5 minutes out and back; a microdrone 3 minutes out and back.

Aggregating these numbers for a 24 hour workday, one can began with *sortie* time, defined as on station time, plus ferry time, plus refueling or battery-change time. For an ENG or public-safety helicopter, that is 3 hours on station, plus refueling time of 10 minutes, plus ferry time of 10 minutes out and back to the refueling point, or a total 3.33 hours. For a microdrone it is 30 minutes on station, plus a total of 6 minutes ferry time, plus 3 minutes of battery change time, or a total of 39 minutes. Dividing each figure into a 24 hour day means that the helicopter can fly 7.2 sorties per day, producing 21.6 surveillance hours per day, and that the microdrone can fly 36.9 sorties per day producing 18.5 surveillance hours per day. Based on these estimates, a microdrone is less productive than a helicopter.

One can extend this analysis to take into account nonproductive time, such as that required for maintenance. One also can extend it to include operating costs as well as acquisition costs. But without more, one can conclude that a light helicopter such as the popular AS350 can deliver 7,892 surveillance hours per year for an acquisition cost of $2.1 million (0.003 hours per dollar), while a microdrone can deliver 6,738 surveillance hours per year for a $9,000 acquisition cost (0.7 hours per dollar).

But that is misleading. The economic life of helicopters is considerably greater than that of microdrones. Economic life refers to the length of time an asset remains economically productive. Because of that, one can acquire a used helicopter for much less than a new one, but its productivity is the same. One reason that is so is that helicopters receive extensive maintenance, which typically is reflected in their operating costs. So, using depreciation schedules acceptable to accountants for helicopters, one can divide total acquisition cost by the economic lifetime of 20 years for the helicopter, and the total acquisition cost by economic lifetime of, say, one year, for the microdrone. Little data is available from which to estimate the economic life of a microdrone. It is likely, however, that their useful lives will be limited more by the rapid pace of innovation, and by aircraft losses and damage, than by their wearing out mechanically. One year seems a reasonable figure, all things considered. Many will be lost or damaged well within a year, and others might have luckier lives and remain in service for several years. The residual value of helicopters, in contrast, often is almost the same as their acquisition cost.

That produces 0.075 surveillance hours per dollar of investment for helicopters and 0.749 surveillance hours per investment dollar for microdrones. That is still a ten-to-one advantage for drones.

Quantitative comparisons do not tell the full story, however. The aphorism, "You get what you pay for," is apt in this regard. Helicopters cost more than drones because they provide much more flexibility. They fly faster and further. They can stay in the air for two to three hours, compared to less than an hour for microdrones. Pilots aboard helicopters have more complete visual perceptions than DROPs watching a drone from the ground or looking at a two-dimensional video display with a limited field of view. This permits them to capture imagery with details or artistic characteristics that might escape a DROP's notice.

As macrodrones with capabilities matching those of helicopters and airplanes become more widely available, the quantitative difference between them and manned airplanes is likely to vanish or reverse itself.

10.3 Development costs

However great demand might be for theoretical drones, it cannot materialize unless the drones are available for purchase. Whether they are available, and at what price, depend largely on their development costs. Of course, whether suppliers are willing to incur the development costs depends on the demand they project. More demand means that the development costs can be absorbed by more revenue.

Aircraft development costs have skyrocketed. One cause is increased levels of detail in FAA airworthiness certification requirements, a subject that has received congressional attention because of its impact on low-end general aviation aircraft. The general aviation community is concerned that ordinary aircraft owners and would-be owners are progressively getting priced out of aviation.

Another force driving up development costs is concern about tort liability. Several large judgments and a larger number of ultimately unsuccessful lawsuits asking for large amounts of damages have made manufactures risk-averse. They invest more effort in evaluating alternatives during the design process and documenting what they have considered in order to reduce the likelihood of a successful lawsuit. A perception exists that manufacturers avoid, whenever they can, the need to get a new type certificate for a new aircraft, instead trying to design the aircraft so it can come with an existing type certificate.

Few civilian type certificates now exist for macrodrones. So most developers are confronted with the need to design a macrodrone on a clean sheet of paper (metaphorically speaking) and to get it type certificated under the full range of FAA requirements. The process will almost certainly be even more burdensome than for a new model helicopter or airplane, because macrodrones with equivalent performance will have subsystems that never before have been approved by the FAA for vehicles because they have pilots on board. The developer will be burdened to demonstrate that the onboard systems combined with the remote DROP can handle all of the requisite flight conditions without onboard human intervention. That is a far different test environment from one that assumes the possibility of pilot intervention. Chapter 9 provides more detail on the airworthiness and type certification process.

It is hard to get reliable figures for the exact development costs of new light aircraft and helicopters. The only thing it is safe to assume is that macrodrone development costs will rival or exceed these. That means, in turn, that the price will be higher or that the production run to satisfy demand will be greater to spread the development costs over a larger volume of aircraft. In any event, the higher price and the longer production necessarily raise the threshold for anyone confronted with making a decision about whether to proceed in offering a macrodrone.

The same thing is not true for microdrones. Microdrones already in the market have most of the features likely to be required by regulation or insisted on by customers. If the FAA requires that some of these be placed beyond the reach of the DROP, as in chapter 9's law-abiding drone proposal, some redesign, recalls, and remanufacture may be necessary, but most of what is like to be required can be implemented by software changes rather than necessitating hardware changes.

It is not necessary to quantify exactly the total development costs, price, or production run for macrodrones. To assess their competitiveness with manned aircraft, it is sufficient to look at recently introduced aircraft, determine their pricing and their projected production runs. That alone is enough to conclude that macrodrones will not offer competitive advantages in terms of their price. It will have to be something about them that makes them attractive to customers for other reasons. That might be superior sensor systems. But then why wouldn't the customer wanting the superior sensors simply insist that they be installed on a manned aircraft?

It is unlikely that competitive advantage could be obtained by superior endurance, range, or speed because all three of these involve the same kind of design tradeoffs, engineering decisions, and fabrication techniques regardless of whether the ultimate aircraft possessing those characteristics is manned or unmanned.

The magnitude of the development costs for macrodrones requires deep pockets. The major defense contractors involved in supplying military drones are the most likely enterprises to move first. They already have technologies from their military drone R&D and manufacturing, and they already are offering civilian adaptations to purchasers. Eventually, more specialized developers may enter the market, but the likelihood of significant overlap between microdrone suppliers and macrodrone suppliers is low.

10.4 Supply—the production function for drone operators

Economists analyze firms that supply goods or services through a concept known as a *production function*. A production function expresses output as a function of various inputs. The traditional inputs for the economists who developed the idea in the nineteenth century were *labor*, *capital*, and *land*. Modern economics often adds a fourth factor: *technology*. The technology input in production functions is broader than its ordinary meaning in that it includes managerial inputs as well as and knowledge. Inanimate machines are subsumed by the traditional factor of *land*, because land is a proxy for physical assets.

The production-function ideas are a good starting point for evaluating the economics of drone operation, which does indeed require labor, capital, and land input—and, of course, technology as well. The point is not to develop a detailed equation for the drone operations production function, let alone to quantify it; the point is to use it as an intellectual structure for analysis.

10.5 Equipment acquisition

A drone operator can acquire drone equipment in one of three basic ways: it can purchase it outright and pay cash, it can purchase and finance the purchase, or it can lease the equipment.

In a perfect market, the cost of each of these options would be the same. They are close substitutes, and if the price of one drops relative to the others, increased demand for that option will bid up its price, while decreased demand for the other options will cause their price to weaken until equilibrium is restored.

But no market is perfect, and neither is the market for microdrones or macrodrones. Information about price changes is not immediately known by every participant in the market. People may calculate each element of price differently, and other factors, such as overall demand levels in the macroeconomy and changes in the tax law, may disturb the equilibrium from time to time. Accordingly, a particular drone operator is likely to prefer one of these options compared to the others for reasons other than price.

The following sections evaluate each alternative.

10.5.1 Outright purchase for cash

A drone operator can buy one or more drones—including kit components—from the manufacturer or on the used market and pay cash for the equipment. The principal advantages of this approach are:

- The operator avoids interest payments
- The operator gets the benefit of tax deductions for depreciation of the equipment, and may be in a better position to deduct other costs such as maintenance
- Management of the asset may be somewhat simpler, because there is no lienholder.

The principal disadvantage of a cash purchase is that it has a bigger impact on short-term cash flow. The operator must come up with the full purchase price immediately, rather than spreading payments over time as is the case for a lease or a finance contract.

The decision should be driven by the operator's cost of equity capital. If investors are willing to contribute enough capital for the purchase and expect a rate of return equal to or less than lease or finance charges, then the operator is better off accepting the equity capital and making a cash purchase rather than choosing either leasing or financing. Of course, there also may be issues of maintaining control of the enterprise; equity investors often want a measure of control of early-stage ventures, as § 10.8 explains.

10.5.2 Financing an outright purchase

Instead of paying the full cash amount at the time of purchase, the operator may borrow money to purchase the equipment. The simplest finance arrangement involves the lender coming up with enough money to pay for the purchase in exchange for the operator's signing a promissory note and, most likely, a security agreement (called a *chattel mortgage* at common law). The note obligates the operator to repay the loan over time. The security agreement creates a property interest called a *lien* in the financed equipment. The lien, under the terms of the financing agreement, allows the lender to take possession and achieve full ownership of the equipment if the operator defaults on its obligation to repay the loan. Most aircraft financing arrangements, like most other equipment financing deals, do not distinguish very sharply between the note and the financing agreement. The negotiated terms may be in one or the other document.

Variation from the simple arrangement described is limited only by the imagination of the parties and their shared interests. For example, the agreement may shift the cash flow required of the borrower sooner or later, through a bigger down payment, or conversely, with a balloon payment at the end.

A financing arrangement is distinguished from a lease in that the operator acquires ownership of the asset at the beginning, subject to the lien in favor of the lender, while a lease leaves ownership in the lessor, with the operator having a right only to use the asset during the term of the lease.

A number of well-established aircraft financing companies have been in business for decades. Although their primary business is financing manned aircraft, they are willing to finance micro- and macrodrones as well. Banks and other general purpose lenders also are possible sources of financing; although their lack of familiarity with aviation technology and markets may cause them to be more reluctant to lend, or to insist upon more costly terms.

10.5.3 Leasing

In a simple lease arrangement, the operator does not have to come up with the money to buy the equipment; the lessor buys it. Then the operator obligates itself to make periodic payments over the term of the lease, and, in exchange, has a right to the exclusive use of the leased equipment. Leasing and financing resemble each other in that the operator makes periodic payments instead of needing enough capital for an outright cash purchase.

Equipment leasing is attractive when the operator's cash flow is insufficient to fund cash purchases, when the lease costs are lower than finance costs, or when the operator cannot take full advantage of depreciation deductions for the equipment because of its low level of profitability. Under most leases, the lessor's ownership entitles it to take the deduction under the Internal Revenue Code. Equipment leasing long has been common in the transportation industry. Depending on macroeconomic conditions and overall railroad industry profitability, most of American railroad rolling stock is leased rather than owned by the railroads. Likewise, aircraft leasing is at the forefront of any negotiation over the purchase of new or used air carrier aircraft. A similar phenomenon is prominent in truck transportation.

A number of large aircraft leasing firms are active in all segments of the aircraft market and are willing to negotiate lease deals for macrodrones. There is no reason to think that they will not also get involved microdrone leasing, although the smaller size of the likely equipment packages may cause them to be less interested and to cede the market to start up enterprises like ArchAerial in Houston, Texas. The business model for leasing companies depends on the present value of the flow of lease payments exceeding the cost of capital to purchase equipment to be leased.

10.6 Fixed and variable costs

Any income producing activity incurs costs. Some costs vary directly and immediately with the level of activity. They are *variable costs*. Fuel costs for airplanes and helicopters are an example. Any vehicle with an internal combustion engine burns fuel for each mile that it travels and each hour that it flies. Other costs are incurred regardless of the level of activity. They are *fixed costs*. Rent for hangar space is an example.

All costs are variable in the long run, as economists like to point out. One can, after all, not renew the lease on the hangar or rent more or less hangar space, depending on how the business is doing. How frequently a fixed cost can be changed and the minimum size of positive or negative increments is referred to as the *lumpiness* of the cost. Lumpier resources can be changed less frequently and only in larger increments. Less lumpy resources can be changed more frequently and in smaller increments.

Lumpiness also refers to the smallest size of a resource that can be purchased in the marketplace. Can one rent only 300 square ft of hangar space, or must one rent an entire bay or an entire hangar? Can one buy six hours a week of mechanic time, or must one hire a full-time mechanic?

The relationship between fixed and variable costs is of central importance to any enterprise. If fixed costs are high in proportion to variable costs, the enterprise must achieve a minimum scale of operations to produce enough revenue to cover the fixed costs. If, on the other hand, fixed costs are small, relative to variable costs, one can operate at almost any scale and still make money. For example, if a commercial pilot can rent a helicopter for $300 an hour, he can take a passenger on a tour in it for $400 an hour and make a profit of $100, even if he flies only one tour per year.

The cost structure for a microdrone operation is dramatically different from that for a macrodrone or manned aircraft operation. It is less lumpy. Fixed costs and equipment are much lower for common civilian microdrones. Each one costs on the order of $1,000–$20,000, compared with the typical cost for a commercial airplane or helicopter of $300,000 upwards.

Several results flow from this fact. First, the price of services performed by the aircraft and the necessary utilization of the aircraft are much less to cover equipment cost and debt service.

Second, the feasibility of offering overnight replacement of a malfunctioning vehicle is much greater.

10.6.1 Fixed costs

Some items of fixed costs, such as office space, marketing expenses, travel, utilities, and Internet connectivity, are likely to be the same for similarly sized enterprises in the same geographic area regardless of whether they fly microdrones, macrodrones, or manned aircraft.

Other items of fixed cost, however, are likely to be lower for microdrones. Premiums for both liability and hull insurance should be lower because the value of the hull is a couple of orders of magnitude lower, and the smaller size means that the amount of damage for which the operator might be liable is lower. A counterintuitive example is lower premiums for motorcycle insurance than automobile insurance. A smaller motorcycle can do less damage to third parties than an automobile, and it typically carries only one rider rather than a driver and multiple passengers.

Hangar rental for microdrones will be dramatically lower than for macrodrones or manned aircraft, simply because microdrones are smaller. In fact, microdrone operators do not need a hangar; they can store the vehicles in ordinary office space.

10.6.2 Variable costs

Variable costs in aviation enterprises are those that increase as flight time increases. Fuel costs are the most obvious example. Crew costs are another example, although many crew members are paid regardless of how much they fly. Maintenance is similar to crew costs in that it can be paid for on an hourly basis or the mechanics performing maintenance can be salaried—paid regardless of how many hours they work on aircraft. Moreover, some maintenance requirements increase in direct proportion to how much the aircraft is flown. Others must be satisfied regardless of flight time.

10.6.3 Estimating total operating costs

Estimating operating cost for macrodrones could proceed from a zero-based calculation of fuel requirements, and so on, but that would require making assumptions without much data supporting them about a specific macrodrone designs. A more useful approach is to take operating cost calculations for a particular aircraft that now exists and to consider how different a macrodrone of similar size and ability might be. That can lead to a more credible estimate as to whether its direct operating costs will be higher, lower, or the same.

The Robinson R66 is a good starting point. Its small turbine engine represents the state of the art of propulsion systems and has better specific thrust, compared to reciprocating engine alternatives. The aircraft also is at the low-end of what is acceptable for law enforcement and ENG support. It is a good proxy for what an early civil macrodrone entrant will look like. Robinson also offers police and ENG configurations of its piston-engine R44, but the turbine-engine offers more advanced propulsion technology.

The Robinson list for the R66 has the right categories for a basic estimate for macrodrones. Insurance cost should be roughly similar, because insurers care about hull value that limits their exposure under hull insurance coverage, and they care about how much damage the aircraft might do if it crashes. That is a function of gross weight, speed, the number of people on board, and the probability of a crash. A macrodrone intended to do the same job as an R66 would have an empty weight about the same, maybe a bit higher to accommodate more sophisticated electronics. It would have a comparable speed. It would not, however, carry people. Insurers might offset this lower-risk factor by factoring in a somewhat higher probability of an accident, resulting in liability insurance for a macrodrone at about 15 percent less than that for the R66.

The hull-insurance risk, derived from the purchase price, is likely to be higher, because of the development cost factors discussed in 10.3. So the hull insurance should be proportionately higher, say 15 percent higher.

Fuel costs should be a bit lower. The propulsion systems are likely to be similar—or the same, with equivalent specific thrust, but macrodrones will operate without an onboard pilot or photog, lightening its payload by, say, 450 pounds. The relationship between gross weight and feel consumption is not linear, and fuel consumption by turbine engines tends not to be sensitive to gross weight variations. Fuel consumption might be 5–10 percent less if the payload is 450 pounds lighter.

Electronic maintenance costs will be higher, and mechanical maintenance costs should be the same, because the drive train and rotor control mechanisms are the same. As a result, the periodic and unscheduled maintenance figures are adjusted upwards by 20 percent.

Whether macrodrones would have life limited parts like the R66 is a design decision, but to simplify the way this section is calculating operating costs for macrodrones, that aspect of the cost calculation is left intact. The results from these assumed differences in the elements of operating cost macrodrones results in a direct operating cost and total operating cost almost identical to that of the R66—only one dollar per hour higher.

The zero depreciation figure for the R 66 is not useful, however. It may be warranted based on projections about residual value of R66s, but the market for macrodrones all just too immature to justify any such assumptions. Anyway, amortization of purchase price is more properly considered in the fixed cost section. That figure should be omitted from the direct cost calculation.

10.6.4 Fuel costs

Fuel costs dominate direct operating costs for fixed wing aircraft and rival maintenance cost for helicopters. A light helicopter such as the Robinson R44 burns 16 gallons of gasoline per hour. At a price of $6.50 per gallon, the result is fuel costs of $104 per hour. An AS350 burns about 45 gallons of jet fuel per hour. At a price of $5.50 per gallon, its fuel costs are $248 per hour. It is likely that macrodrones will have similar propulsion systems, and therefore their fuel costs will be roughly the same as those for manned aircraft.

Microdrones have electric propulsion systems and do not require fuel. They do require electricity. The price per kilowatt hour of electricity in the United States was about fifteen cents in 2014.[6] A high-end microdrone consumes about 20,000 mAH for a 20 minute flight. Combining these figures, the electricity cost for a one-hour microdrone flight is about twenty cents.

10.6.5 Crew costs

The defining characteristic of drones is that they do not have onboard crew. Accordingly the direct operating costs associated with pilot, copilot and observer compensation is zero.

On the other hand, drones do have DROPs, and DROP compensation costs are as directly associated with flight hours as are those for onboard pilots. In some highly automated missions, such as surveying, one DROP might operate multiple drones simultaneously, reducing labor costs.

The question in quantifying drone direct labor costs relative to those for manned aircraft depends on:

- The hourly compensation rate for pilots, copilots, observers, DROPs, and DROSOPs, the amount of collateral-duty time required, compared to actual operational flight time and billable flight time
- Utilization of each crew member
- Whether DROPs and DROSOPs are paid on a periodic basis as opposed to a strictly hourly basis

Many uncertainties exist in quantifying these variables. Uncertainty, however, creates opportunity. Drone operators may structure labor costs differently from the way they traditionally are structured in commercial aircraft operations. For example, it is reasonable to believe that the total compensation packages for DROPs can be significantly less than those for pilots. Supply and demand is not well-established yet in the labor market, but dramatically lower labor costs for DROPs may result from dramatically lower training costs.

Training for a DROP to fly a sophisticated microdrone similar to a DJI Inspire or S1,000 requires 5–8 hours of instruction, at a price of $50–$80 per hour, resulting in total training costs significantly less than $1,000. Commercial pilot training costs approximate $100,000. Macrodrone DROP training may cost more, because of the additional proficiency the DROP needs on sophisticated remote navigation and collision avoidance systems.

A DROP who pays for his own training will enter the labor force with much less invested in training, whether that represents actual cash expenditures or student loan debt.

Training costs incurred by the airman before employment typically are reflected by student loan debt. A DROP with lower student loan obligations can accept lower compensation than a pilot with higher debt and still have the same amount of discretionary income.

10.6.6 Maintenance

Any company or operator must have plans in place to maintain its drones. This requires a combination of skilled personnel to diagnose problems and to repair or replace parts and subsystems that have malfunctioned. The economics of maintenance requires estimating the labor cost of ad-hoc diagnosis, actual repair and replacement, and the cost associated with periodic maintenance, such as hundred-hour and annual inspections on manned aircraft. Some of these are best treated as fixed costs, while others are inherently variable with flight time. Annual inspections, for example, are fixed costs because they must be performed each year, regardless of how many hours the aircraft has flown. Similarly, the compensation of a full-time mechanic is a fixed cost, even though it may be allocated to items of variable cost based on a *maintenance-hours-per-flight hour* factor.

The cost of maintenance on microdrones is much less than that for macrodrones and manned aircraft. Manned aircraft and macrodrones require, under the FARs, 100-hour and annual inspection. Some kind of periodic and annual inspection is likely to be necessary for microdrones, as well, but their simplicity reduces the number of things that could go wrong, and their small size necessitates many fewer maintenance man hours for a thorough inspection. Additionally, the complexity in microdrones is mostly in the electronics, not in mechanical devices. If a chip or circuit board malfunctions, it simply can be replaced. The possibilities for mechanical malfunction are minimal. If a rotor blade breaks, it can be replaced in seconds. If a motor goes bad, it similarly can be replaced and a matter of minutes. There are no pitch control links, gearboxes, or sprag clutches, no control surfaces or links connecting them to the cockpit that need to be adjusted.

Macrodrones, on the other hand, are likely to resemble manned aircraft in their basic airframes, propulsion systems, and control linkages. Their electronics will be more complicated than those for manned aircraft, because of the need for a robust control links and backup systems. Their sensor equipment will likely be about the same. Macrodrones will not be much in demand if their mission performance capabilities are weaker than those of their manned competitors.

Maintenance costs include, in addition to labor costs, the cost of parts. Microdrones are simpler mechanically than manned aircraft, but they have control and mission systems that are, in many ways, more sophisticated than those on typical manned aircraft. Accordingly, a microdrone operator can expect much lower maintenance costs associated with mechanical inspection, diagnosis, and repair and relatively higher costs associated with electronic systems.

The skills required for a traditional mechanic are different from those required in a good electronics technician. Typical compensation for a good aircraft mechanic is $80 per hour billed to the customer and $60,000 annual salary for full-time employment. Electronics technicians earn somewhat more, maybe up to $90,000 on average and bill at $90 per hour.

Macrodrones are comparable in mechanical complexity to their manned aircraft counterparts, and they have more complex and sophisticated electronic systems for their control links. Accordingly, one can expect comparable maintenance costs for mechanical systems and higher costs for electronics systems, compared with those for manned aircraft.

The simplicity of microdrones compared with manned aircraft and macrodrones also impacts a user's philosophy toward parts inventories—*spares*. It is difficult—sometimes impossible—to repair an integrated circuit. Instead, when one malfunctions, the maintenance task is simply to replace it. In contrast, a malfunction in the landing gear often can be repaired.

The choice between repairing and replacing determines whether an operator that wants to minimize downtime for aircraft concentrates on assuring the availability of sufficient skilled personnel to perform repairs promptly and a sufficient parts inventory to replace defective subsystems. In most cases, it is more efficient simply to swap out the offending component and take care of the repairs later—or just to discard the defective one.

Calculating how many spares should be an operator's inventory starts with estimating failure rates for each component. The estimated failure rate multiplied by the number of microdrones in the fleet determines how many spares of each type of component should be in the inventory. Part of the analysis is recognizing that many spare parts have a shelf life. A rarely needed spare part may become obsolete before it is needed, obviating the advantage of keeping it on hand. Parts manufacturers often can ship a part overnight, reducing the time advantage of having the same part in inventory.

Obviously, if a microdrone operator standardizes its fleet—limiting it to only one model of drone—the number of spares and the total cost of spares is less than if it must keep spares for multiple types of aircraft.

The lower price for microdrones also opens up the option for more operators to have spare aircraft, in addition to spare parts. The choice between spare components and spare aircraft should be a straightforward cost comparison.

10.7 Uncertainty

The economics of real world commercial operations do not depend upon averages. The operating cost for manned aircraft or drones can be computed based on average maintenance costs, fuel prices, and hours flown per year, but the actual hourly operating cost is like to vary considerably from projections based on averages. Weather may be worse than normal, cutting back on flight hours; a particular model of microdrone may develop mysterious problems in the interactions inside its autonomous electronic systems; an airworthiness directive may come out for a helicopter or airplane part that is notoriously difficult for a mechanic to get to and inspect. A pipeline or powerline mission may take twice as long as expected because it requires more time to get detailed imagery of a particular fault, or to put skilled personnel on the ground to perform tests. And, of course, whenever mission time substantially exceeds that planned, a refueling stop may be necessary, adding further to the variance above planned hours. Post-maintenance ground checks are about to be completed, and the generator drops off line for an unknown reason, requiring more troubleshooting and more ground tests, resulting in three more hours of downtime for the helicopter than estimated. Adverse variances in flight hours ripple through crew costs and maintenance costs.

Pilots get sick, aircraft break, and back up aircraft and relief crews are not always readily available, necessitating either losing a business opportunity and displeasing a customer, or paying the cost of expensive short-term temporary hires, or short-term leases of equipment.

All this means that, while a basic business plan may start with averages and assumptions based on economic and technological benchmarks, an enterprise is likely to go out of business if it goes no further than that. It needs to know variations around the average—standard deviations—and make a managerial policy decision about whether it wants to plan for the 90th percentile or 60th percentile. The 90th percentile for aircraft availability would capture 90 percent of the actual hours per day of aircraft availability, and the 60th percentile would capture 60 percent. The average (actually the median) is the 50th percentile.

Planning for the worst case, however, does not provide a free lunch. Planning for the worst case drives up the costs for which a business plans. That, in turn, drives up its price. Being too conservative will price a firm out of the market.

10.8 Sources of capital

Business enterprises require money to cover the gaps between revenues and expenditures. When an enterprise starts up, it almost always will have to make expenditures before it receives revenues from customers. Even established enterprises experience seasonal or cyclical revenue slumps, but cannot cut expenditures proportionally.

In some enterprises, especially very small ones, the entrepreneur provides all the capital from his personal resources. Or, he teams up with a handful of partners who contribute capital. Most enterprises, however, require some form of outside capital as they grow.

Outside capital takes one of two basic forms: *equity* or *debt*. Equity represents an ownership share in the enterprise. The rights of an equity investor are defined by a subscription agreement, which may be incorporated into a US Securities and Exchange Commission (SEC) registration statement and prospectus. While the details very enormously and are limited only by the imagination of the issuer, the investor, and their lawyers, the basic idea is that the investor is entitled to earn a return determined by the performance of the enterprise. If the enterprise just not profitable, the investor gets nothing. If the enterprise fails and is liquidated or reorganized in a bankruptcy proceeding, the equity investor has the lowest priority for distribution of assets, after employees and other creditors. Most equity shares confer the right to participate in the management of the enterprise, in other words, to exercise a measure of control by electing members of the board of directors. Boards of directors have ultimate responsibility for management. Boards of directors select the officers and usually delegate to them substantial day-to-day responsibility, subject to oversight and removal by the board.

It is not uncommon for an enterprise to issue different classes of equity shares, with varying degrees of control and claims to profits and assets in the event the business is terminated. For example, an entrepreneur might need to raise outside capital but be unwilling to give up control of his business. He can arrange that by holding one class of equity himself which has all the control rights, and issuing to outside investors only another class of shares which include no control rights. Whether he can persuade investors to accept this—to put their money into an enterprise over which they exercise no control—is a matter for negotiation in the marketplace for capital.

Debt entitles the creditor to repayment and some form of interest payments. The creditors' rights and the debtor's obligations do not depend on the success of the enterprise.

Debt obligations, like obligations to equity shareholders, can be structured in many ways, causing each to resemble the other. For example, preferred equity shares may obligate the enterprise to make certain payments, independent of profitability. Debt obligations may be convertible into equity at the option of the creditor, or give the creditor certain rights to oversee management.

Financing an enterprise with debt has the advantage that the entrepreneur need not give up or share control with the creditors, while he would with equity investors. On the other hand, obligations to service the debt and eventually to repay it impose a continuing cost and cash-flow burden, unlike equity investment which need not be repaid and earns dividends or appreciates in value only as the enterprise thrives.

An alternative source of capital is early customer payment. Real estate development and many profit-seeking Kickstarter projects are financed entirely or mostly this way. Customers pay all or part of their obligations before they receive anything. The payments often are characterized as "deposits."

When customers make advance payments, the risk of business failure shifts to them and away from the owners of the startup enterprise.

Enterprises can raise any of these types of capital through private offerings, public offerings on the securities exchanges or the over the counter market, or through crowd sourcing. Chapter 11 evaluates the alternatives in the context of starting a new drone business.

10.9 Organization design

Management of an enterprise involves adapting economic theory to practical realities of assembling relevant assets, maximizing their productivity, acquiring customers and keeping them. No management, no matter how skilled, can achieve the results predicted by economic theory, any more than an engineer can design a system to approach the limits predicted by thermodynamics.

10.9.1 Productivity and its limits

Any enterprise maximizes its economic efficiency when it increases productivity of its assets. An asset's cost per unit of output depends on how fully it is utilized, as § 10.2.3 explains. A drone in which an enterprise has invested $10,000 produces a rate of return directly proportional to the number of revenue hours it flies. It can produce 260 revenue flight hours per year if it is flown for one hour every weekday, or 1560 revenue hours per year if it is flown six hours per day. The cost per revenue hour is $5.77 in the first case and $0.96 in the second case, assuming a cost of capital of 15 percent.

The same thing is true of human assets. If a DROP is paid a salary of $40,000 per year and flies one revenue hour per day, the labor cost per revenue hour is $153.85. If he flies six revenue hours per day, the cost is $25.64. USAF's crew ratio—the number of qualified DROPs to operate a drone 24 hours per day—is 10:1.[7]

A certain amount of nonproductive time is associated with almost any asset—aside, perhaps, for real estate. Any aircraft, manned or unmanned, must be refueled, or to have its batteries swapped or recharged. It must be inspected and repaired periodically. A pilot or a DROP must have time to eat, relieve himself, and inspect and warm up his equipment. But an efficient aircraft operator is not going to be comfortable if its pilots fly one or two missions per day for a total of three hours and spend the rest of the day waiting for the next mission. A good manager will seek collateral assignments for the pilots: conducting training, participating in staff meetings with management, holding performance improvement meetings with customers, making sales and marketing calls, and assisting with maintenance of aircraft.

The reality is, however, for many operations, pilots must for remain ready to be called out on short notice. That limits the kinds of collateral activities that can be assigned. Tasking a pilot to attend a meeting with a customer in the downtown office building makes him practically unavailable for a call, when the aircraft is at an airport in the suburbs. If he flies microdrones, however, he can take the microdrone to the meeting in a case, ready to go.

Workforce efficiency militates toward split shifts in many operations; an aircrew member is in pay status only when he flies or is actively preparing to fly. But qualified personnel naturally oppose this; they prefer more regular and predictable employment. Even a day laborer can expect a full day's work, once he gets hired.

Demand is not constant. Almost all demands for aviation services are seasonal, cyclical through the day, and, in many cases, episodic. These demand fluctuations distinguish aviation from industries whose assets can be scheduled so that their utilization approaches 100 percent. A factory assembly line is an example, both in terms of asset scheduling and crew scheduling. Such activities are good candidates for the scientific management approach. Some other, non-manufacturing, activities are, as well. United Parcel Service is famous for micro-dissection of its delivery drivers' activities and its constant effort to redesign their work to shave a few seconds off the time required to deliver a package.

It is very difficult to make this approach work in the aviation flight operations context, however. Two realities make flight operations unlike assembly lines or ground package delivery. First and most important, demand is episodic. Tourists want to take helicopter tours when it fits with their other sightseeing and shopping plans. Parents and couples want wedding imagery while the wedding is taking place. Neither aerial tours nor weddings are likely to be at 2:00 AM. Power lines and pipelines generally are not inspected at night. TV stations want aerial imagery of traffic during rush hour, which usually occurs only twice a day; and of breaking news events, which may occur at almost any time. When stations have a choice, they want imagery proximate to or during scheduled news broadcasts when it can be labeled "live." Some aerial imagery can be stockpiled, but only so much B-roll is useful. EMS flights occur when patients need to be transported. Law enforcement needs helicopter patrol support during high crime times. It needs helicopter support for tactical events when they occur, which is impossible to predict.

The time and place of principal photography for movies is discretionary with the producer and director, of course. But anyone who has juggled actor, cinematographer, and sound personnel schedules, and sought the best natural light for a movie shoot, knows that a shoot is not going to be scheduled around the time that a helicopter or drone is not otherwise engaged.

In all these examples, an operator seeking to meet the demand needs aircrews on duty and close to the aircraft. Customers will not buy a service instead of building in-house capability if a contractor cannot provide prompt aerial support. Calling in crew members from home or from other jobs simply takes too much time. Crewmembers could be required to be on call for prolonged periods and prohibited from venturing more than, say, thirty minutes from the worksite, but few job candidates would be willing to adhere to such limitations without getting paid for call time.

Matching labor inputs to fluctuating demand is even more difficult when the customers are spread out geographically. Then DROP and mechanic—and perhaps other employee—duties are scattered around the country. Larger markets have multiple TV stations, but few have more than two or three. The enterprise might succeed in clustering its customers regionally, but most of the DROPs will work in locations widely separated from each other. An enterprise cannot deliver an acceptable level of service based on the premise that its DROPS will roam around the country, flying microdrones for one station for a couple of hours, and then moving to another city and flying for another station there. Moreover, TV news collection tends to follow the same pattern, regardless of the market: covering traffic in morning and afternoon rush hours and on-call for breaking news and background coverage only proximate in time to scheduled news programs. The stations in different locations would need DROPs at the same time.

This puts the operator in the position of paying for crew-time that is almost certain to be idle. A single-minded efficiency seeker might make collateral assignments, but that will increase the likelihood that the crew member, who now is essentially engaged in active work for the employer, will demand additional pay. Moreover, the operator is likely to run out of productive collateral assignments. Any practicable collateral assignment for an on-call crew member must be one that can be interrupted on a few seconds notice so the worker can go flying.

FAA regulations limit duty time for Part 121 (airline) and Part 135 (air taxi, charter, EMS, and other non-airline commercial) pilots. Not allowing for reasonable amounts of leisure time can jeopardize pilot safety.

Furthermore, aviators are highly trained professionals. Much of their training emphasizes independent decision-making. They are proud of their craft, and they value their autonomy. Aircrew members who work together value the opportunity to interact with each other when they are not flying.

Subjecting them to a work environment in which the boss hovers over them with a stopwatch, insisting that every moment be occupied with work activity is not one where any aviator will be happy. Morale is important for any workforce, especially one that needs energetic professionals motivated by genuine internalization of employer goals; who exercise independent judgment in the employer's interest; at places remote from the physical, on the ground, worksite.

An employer is not going to get those workforce traits, by regularly pulling a pilot aside and saying, "You can't go to lunch with the others. I have a little job for you to do now."

The second challenge for introducing scientific management to aviation flight operations is that flight operations are dispersed geographically. The geographic dispersal is obvious in the case of airlines and long-distance charter flights or executive flight. But even for those activities that are thought of as being essentially local—law enforcement support, ENG, EMS, and power line and pipeline patrol—are also dispersed, although over a smaller area. None of the flights for any of these activities occur in the hangar next to the base manager's office. They occur in the field, with flight personnel making decisions about what the aircraft does. The base manager is back at the base, not well-positioned to provide close supervision of every decision that involves allocation of time. The pilot, not the base manager, must make the judgment whether to abort the mission to return to base because of a low fuel state or a warning light. The pilot, not the base manager, must make the decision, perhaps in conjunction with customers, whether to hover, orbit, or stop nearby for fuel, rather than returning to base, only to called out again for some new development at the same incident site.

For operations in which the aircraft is not returned to its launching point after every mission, geographic dispersion presents an additional challenge: getting the crews to the same place as their aircraft. That gives rise to the necessity for crews to deadhead, maybe by air if air transportation is available, but often by land. That adds additional nonproductive time, and it is even more difficult to find collateral assignments that can be performed by someone who is actively traveling.

Many of the same realities exist in the drone context, limiting effective pursuit of the abstract goal of improving productivity by scientific management techniques. Important differences exist in the drone context, however.

Macrodrones are likely to be operated by DROPs seated in front of DROPCONs at a fixed location. That means that they are more likely than airplane or helicopter pilots to be close to their supervisors, who therefore can monitor their activities more closely and assign collateral work activities to fill in idle time. Also, even long-range macrodrones are likely to return to their launching points after each mission. That means that neither aircraft nor crew is likely to be out of place to fly the next mission.

Microdrones potentially have greater utilization than manned aircraft. They are simpler, and require less inspection and maintenance. Microdrone DROPs on the other hand must be near their microdrones when they fly them. Sometimes it may be feasible to transport drone and DROP by air to a job location; more often, the DROP must pack up the drone and travel by ground. That introduces an element of non-productive time indistinguishable from airline crew deadheading in terms of its economics: it is extremely difficulty to conceive of a collateral assignment that can be performed effectively while traveling. Then, once a microdrone is on location, it requires a certain minimal set up and breakdown time before and after each mission. This time is equivalent to, but shorter than, the preflight and postflight times for manned aircraft.

For example, in a congested urban setting, there are practical limits on the utilization of aircraft; utilization of 10–12 hours a day is excellent performance for an average airline, and airlines have more control over their scheduling then do on-demand operators such as air taxi services, EMS, and ENG. The kinds of activities for which drones, especially microdrones, are suitable, have demand patterns that are more within the control of the operator that most passenger carrying operations. For example, aerial photography of real estate, powerline or pipeline patrol, or inspection of agricultural fields can be conducted at almost any time of the day. It is difficult to perform them effectively at night absent substantial investment and additional lighting and imaging equipment. Greater control over drone schedules makes possible more efficient management of their DROPs.

10.9.2 Owning, leasing, and contracting out

The difficulty in evening out the demand for aircraft and pilots may tempt an operator to contract out for the assets—renting aircraft instead of owning them or using them under a long-term dry lease; contracting for pilots to fly only certain times of the day rather than putting them on the payroll. Which approach makes the most sense depends on the total compensation cost for salaried pilots, as compared with the contract price and flexibility in scheduling available from a contractor. Is it better off operating its own ENG helicopter fleet, even though the number of helicopters does not require the full-time services of a mechanic or an avionics technician? Or is it better off hiring an ENG helicopter contractor whose scale of operations warrants full-time mechanics and avionics technicians?

Ronald Coase, the founder of the field of law and economics, offered a theorem in 1938 that has been richly extended by economists thinking about and working on institutional economics. Under this Coase theorem, enterprises define the boundary around their enterprise according to whether they can perform a particular activity more efficiently inside, under a system of bureaucratic rules, or whether they can perform it more efficiently outside, under the forces of the marketplace. If a dispute arises about job performance, is the enterprise better off being able to resort to a breach of contract lawsuit with an outside contractor, or is it better off applying corrective action and discipline to employees performing the same job inside the boundary?

No enterprise can do everything; indeed one of economist Adam Smith's most important insights was that markets become more efficient as their economic components specialize. A firm can specialize narrowly, performing only one function, and enter into contracts for everything else it needs. Or, it can become more fully integrated, making most of what it needs. This *make-or-buy* philosophy, often referred to as *contracting out* or *contracting in*, is a fundamental strategic decision, and it is made, explicitly or implicitly, with respect to every function of a business. Drone businesses are no exception.

Some supporting activities are more central than others, and the interdependencies between those activities and core functions may be such that coordination is too cumbersome when the supporting activities are performed by another firm. For example, in television news, the interaction between the news or assignment desk and the technical operations center (TOC) are so intimate that few TV stations or networks would contract out their TOC function. Even less likely is the contracting out of the news and assignment desk functions, leaving only reporters as employees.

The most important consideration in a make or buy decision for drone services is the likelihood that the user needs drone services that do not have the economies of scale that a specialized drone contractor has. *Economies of scale* exist when larger enterprises can produce their products and services at lower costs than smaller enterprises that compete with them. A rides and tours helicopter operator with $50 million in annual revenue can cover fixed costs for helicopters, hangars, and full-time salaried pilots and mechanics more easily than a competitor with $1 million in annual revenue. A drone manufacturer that sells 250,000 microdrones per year can afford a bigger, more efficient factory, than a competitor that sells only 1,000 per year. Economies of scale typify industries with high fixed costs, compared to variable costs.

Economies of scope is a related concept that refers to efficiencies obtainable when a firm offers a broader line of products. A firm that offers both helicopter and drone services can spread the cost of a full-time salaried electronics technician over both lines of business. A firm that offers only drone services must earn enough revenue from drone operations alone to cover the cost of the technician.

Enterprises that specialize in drones can afford a larger inventory of parts, standby vehicles, standby crew members, more specialized scheduling and operations management software, and more expensive diagnostic machinery.

Drone support is a new operation, and therefore presents more explicitly the make-or-buy decisions that managers may take for granted with respect to other functions that have become habitually performed inside or outside. In other words, the market for drone services need not mimic the market for manned aircraft services in its patterns of contracting out.

In evaluating the cost of performing drone activities in-house, with employees, potential employers must be careful to consider *all* of the costs. Compensation costs for employees must include not only straight salary or wages, but also fringe benefits such as vacation, holiday, and sick time and training costs. Suppose, for example, a drone firm hires a new DROP. The employer's standard contract requires the DROP to work a 40 hour week. He gets one week's vacation in the first year, 10 sick days a year at full pay, and eight standard holidays. The salary is $35,000 per year. The employer's share of Social Security taxes is $2,170 per year. The employer pays half of the health insurance premium, or $330 dollars per month or $3960 per year, and the entire premium for life insurance and long-term disability insurance: $85 dollars per month, or $1,020 per year. Medicare taxes for the employer are $508 per year. State unemployment compensation taxes are $2,852 annually, and Worker's Compensation premiums are $19 per month, or $228 per year.

The result is a total outlay of $45,738 for 237 eight hour workdays per year.

Opportunity cost is also an issue when an enterprise performs drone operations in house. Opportunity cost as an aspect of the make or buy decision is more likely to be in play with respect to microdrones than manned aircraft. *Opportunity cost* refers to what an asset could earn if it engages in alternative operations. If a pilot could earn $85,000 per year flying helicopters, $85,000 is her opportunity cost for flying drones at $50,000 per year.

It is unlikely that an enterprise needing aviation support from helicopters or airplanes will assign flight activities as a collateral function to employees whose main responsibilities do not involve flying. If the firm wants to perform aviation services in house, it will hire full-time pilots.

With the advent of microdrones, however, it is far more likely that being a microdrone DROP would be a collateral assignment. A ground-based field news crew would be given a microdrone and trained to use it along with its other equipment. Police patrol officers would be issued a microdrone and trained to use it in conjunction with their other police activities. Powerline and pipeline inspection personnel similarly would be assigned a microdrone and asked to fly it when it would be helpful. In all of these cases, opportunity cost comes to the forefront of economic analysis.

Opportunity cost is the cost of *not* doing something. If a lawyer decides to become a judge, the opportunity cost of the decision is the value of the legal fees from private practice that she will give up. If a freelance news photographer becomes a salaried reporter for a TV station that prohibits outside employment, she gives up the fees she has been learning from private clients. That is the opportunity cost of going to work for the TV station.

In the microdrone context, if a police officer flies a microdrone at a crime scene, he has less time to perform other patrol functions. If a field reporter for a TV station launches a microdrone, she has less time to conduct on-the-scene interviews.

Several reasons exist, including the ones identified in the preceding paragraphs, for an operator to prefer subcontractors. One motivation—to obtain better utilization of flight personnel despite fluctuating demand—is likely to prove illusory. Contractor pricing will be driven by many of the same limitations on paying personnel only when they are needed that affect in-house operations. Firms that offer contract services participate in the same regional, national, or international labor markets as the firms that contract with them. It is unlikely that the fundamental economics of who will do what work for what wage will differ significantly depending on whether the worker works for a contract services firm, or directly for a firm that performs the same service within the firm. If a subcontractor can find qualified people to work split shift, so can the firm that contracts with it. A contract services firm will have the same difficulty getting aviators at any particular level of experience to accept a constant stream of collateral assignments as a firm that has them directly on its payroll.

On the other hand, short-term exigencies may warrant paying a premium price for a brief period. The purchasing entity may need to supplement its workforce on a short-term basis to cover a surge in demand or a spike in training or maintenance requirements.

So a rational decision whether to contract out or perform services in-house depends on more than a microeconomic assessment of labor markets. It is more likely to turn on the economics of specialization and economies of scale.

The user may want specialized expertise, beyond its own knowledge and experience. This motivation may be temporary; the operator may contract out work until it gets more familiar with the activity and hires and trains its own workforce, intending from the outset eventually to replace the contractor with employees. A firm that specializes in recruiting and supervising pilots and whose main business is pipeline and powerline patrol by air will be better at it than a firm that whose main business is delivery of natural gas, which carries a few pilots on the side to inspect its pipelines.

The user may prefer not to deal with personnel administration associated with DROPs and mechanics, monitoring training and recertification requirements, and helping to shape their career plans. A contractor that specializes in drone services may be much better at performing those functions. The specialized drone firm will have better knowledge about what qualifications matter, what kinds of economic and professional incentives keep pilots happy, and have a better recruitment pipeline.

Moreover DROPs, mechanics, and avionics specialists may prefer to work for an enterprise that specializes in aviation rather than one whose primary business purpose is something else. They will have better opportunities to share professional experiences and they can make more relevant career contacts. If that is the case, an enterprise hiring a drone contractor may get the benefit of better qualified and more enthusiastic DROPs, DROSOPs, and mechanics.

Economies of scale are just as important. Many aviation assets are, as § 10.6 explains, lumpy. A firm with only one light helicopter is unlikely to keep a full-time mechanic busy. The firm with many helicopters will have no difficulty doing that. A firm that provides EMS helicopter services to a variety of hospitals can afford a centralized flight dispatch desk that knows the pertinent regulations, helicopter performance, and meteorology. A hospital with one helicopter could hire someone with the requisite expertise, but could hardly keep her busy. The same thing is true with parts inventories and backup aircraft. A firm specializing in aviation services with many aircraft can support a large spare parts inventory, and keep a backup helicopter reasonably busy. Its customer, with only one aircraft, cannot.

These realities for manned aircraft operations are likely to apply to drone operations as well. Specialized drone operators who sell contract services will know the constantly changing product and labor markets better than any of its customers, who likely employ drones only at the periphery of their main business operations. Economies of scale will be smaller in magnitude, because drones—microdrones, at least—are so much cheaper than manned aircraft. The relative difference between contractors and their customers will be similar, however.

An equally important consideration relates to transaction costs of integrating the pieces that comprise an aviation operation. Does a TV station or a real estate broker want to develop the knowledge necessary to put together the right vehicle, with a suitable maintenance operation, with a lawyer who knows aviation law, with a cadre of skilled pilots and observers, or would it be better off buying a turnkey operation from a specialized contractor? If it contracts on a turnkey basis, it can put pressure on the contractor to improve performance, without having to mediate a finger-pointing game among employees who make different contributions to the whole. The same motivation drives both contractors and in-house operators to prefer ready-to fly drone packages rather than kits.

Often, the most important advantage of contracting for aviation support is predictability. Enterprise managers are held accountable to someone else for operating within their budgets—unless they own the enterprise all by themselves. When a manger contracts out for aviation support, she has a fixed, highly predictable figure to put in her budget. If she performs the same services with employees under her supervision, she can budget, of course, but she may have to hire additional employees, discharge ones that prove unsatisfactory, or pay employees more than the budget reflects to recruit and retain them.

When outside contractors do the work, they take the risk of variations in cost. When the same work is done in-house, whoever is in charge of it takes the risk. Of course, purchasers of aviation services cannot always negotiate a fixed-price contract. Potential contractors may be unwilling to absorb the entire risk. That leaves room for negotiation, diluting what may be one of the chief advantages of contracting out. And of course contracting on the cost plus reimbursement basis leaves all of the risk of unpredictability with the purchaser of services.

10.9.3 Employee or independent contractor?

A firm that decides to operate its own drones must decide between using employees or individual independent contractors. Labor law distinguishes between *employees* and *independent contractors*. Employees are covered by the national Labor Relations Act or the Railway Labor Act and entitled to band together to engage in collective bargaining. They enjoy minimum-wage and overtime protections under the Fair Labor Standards Act. They are protected against discrimination by Title VII of the Civil Rights Act, the Age Discrimination in Employment Act, and the Americans with Disabilities Act. The Internal Revenue Code subjects them to income tax withholding and Social Security and Medicare tax. Independent contractors are not afforded these statutory protections, and they are taxed differently. Employees are entitled to workers compensation; independent contractors are not. While the definition of employee varies somewhat under the different statutes, the basic distinction between the two types of workers depends on the right of the entity paying for their services to control them under the so-called *right to control test*. The following factors determine the right to control:

- If the worker provides his own tools and other instrumentalities of work, he is more likely to be an independent contractor; if the hiring entity provides the tools of work he is more likely an employee. For example, if a DROP uses his own drone, he is more likely to be an independent contractor.

- If the worker determines the details of how the work is done, subject to only general goals and standards set by the hiring entity, he is more likely to be an independent contractor. If the hiring entity exercises close direction over the details of work, the worker more likely to be an employee.
- If the worker works for more than one entity, he is more likely an independent contractor; if he works only for one entity, he is more likely an employee.
- If the worker is paid on a per-job basis, he is more likely to be an independent contractor; if he is paid on an hourly, daily, weekly, monthly, or annual basis, he is more likely an employee.
- If the worker does work that is not a central part of the hiring entity's business, he is likely to be an independent contractor; if he does work that is central to the entity's business, he is more likely to be an employee. For example, a DROP flying drones for a realty firm is more likely to be an independent contractor than a DROP flying drones for a drone operator.
- Whether the entity and the worker explicitly characterize the worker as an independent contractor or employee also is a factor, but it is not determinative.

The courts apply these criteria under seven-factor, 20-factor, and other formulations, but the various tests can be simplified into this list of factors.

Terminology can be confusing. The term independent contractor is a term of art in labor law, but it also refers to businesses to which an enterprise subcontracts certain parts of its activities. For clarity, this chapter uses the term *subcontractor* to refer to a business, and the term *independent contractor* to refer to an individual.

Enterprises often prefer that their workers be independent contractors to reduce employment law obligations and reporting burdens associated with employees. But enterprises also must exercise a measure of control over their workforce, and that often causes them to elect greater control and pay the price of having their workers classified as employees. A common reason for an operator to prefer employees over independent contractors is more detailed control over employees produces better customer service.[8] Indeed, even if the operator classifies a worker as a contractor, but exercises detailed control over her, the law is likely to classify the worker as an employee.[9]

A recent decision by the California Labor Commission finding Uber drivers to be employees rather than independent contractors has created some turmoil in the labor law community. Uber had constructed the responsibilities of Uber drivers to make an almost perfect case for independent contractor status under generally excepted legal principles. But, in finding the workers to be employees, the California Board placed much greater emphasis on the fact that Uber drivers perform a function integral to Uber's business. Uber has appealed the case, and its long run implications are uncertain.

At the least, however, the Uber decision serves as a caution for hiring entities that seek to classify their workers as independent contractors.

10.9.4 Collective bargaining

Organization design for a drone enterprise depends significantly on whether its flight crews and other employees work under a collective bargaining agreement. Collective bargaining agreements prescribe pay, crew assignment, work hours, and training entitlements, matters that otherwise would be determined unilaterally by enterprise management.

Union representation of flight personnel varies greatly among different parts of the aviation industry. Most airline pilots are organized. Few flight instructors are. Typically, unionization is greater in larger enterprises. Air Methods, Bristow, and PHI, Inc. are organized, and their pilots are covered by collective-bargaining agreements (PHI and its union are suing each other in negotiations). It is hard to find a smaller local operator that is unionized, however.

In the United States, and most other developed democracies, all employees below the managerial level have the right to be represented by a union, if they so desire. Section 7 of the National Labor Relations Act (applicable to industry and commerce generally) and section 2, fourth of the Railway Labor Act (applicable to airlines) express that right, and the statutes provide a mechanism through which federal agencies called the National Labor Relations Board and the National Mediation Board can receive petitions for representation. They then hold workplace elections, after which any union receiving majority vote of the employees in a representation unit is entitled to be certified by the board as the exclusive representative.

Thereafter, the employer commits an unfair labor practice (as it is called under the NLRA), for which it may be subject to economic and other penalties if it refuses to bargain with the representative. Neither employer nor union is obligated to reach agreement, and the law does nothing to prescribe the terms of any eventual agreement. It does, however, limit the scope of the bargaining, and therefore any resulting agreements, to matters related to wages, other terms conditions of employment, and working conditions. Pure business matters are outside the scope. An employer is not obligated to bargain, for example, over whether it enters a new market for ENG services. It is, however, required to bargain over the impact on employees of contracting out a function within the scope of their employment. Whether a unionized airplane and helicopter operator would be obligated to bargain over contracting out drone support which might have the effect of reducing its demand for manned aircraft pilots presents just such a question. On the one hand, the work to be performed by the drone contractor does not presently exist. On the other hand, drone operations by the contractor may diminish the amount of work to be performed by pilots. Then it is closer to core employee interests protected by the NLRA and RLA.

10.10 Market structure: fragmentation and consolidation

In the field of microeconomics, the term *market structure* (sometimes called *industry structure*) refers to the number of buyers and sellers buying and selling a particular product or service. In a *monopoly*, only one seller exists. Regional markets for electricity distribution are examples of monopolies, because only one electricity distributor exists in each.

In *perfect competition*, many sellers exist, no one of which has a sufficient share of the market to affect price. The market for funny YouTube videos is an example. The markets for babysitting and dog walking are other examples.

In a *monopsony*, only one buyer exists. The domestic market for F-35 fighter aircraft is a monopsony; only the United States Defense Department is a customer. Intermediate structures exist as well; a market dominated by only a few sellers is an *oligopoly*. Cable TV service and cellular telephone service are examples.

Monopolies often result from intellectual property protections. Apple's patents assure it a monopoly over iPhones and iPads. If one characterizes the product more broadly, however, as the market for smartphones and tablet computers, the market should be characterized as an oligopoly—with more than one, but only a few, sellers.

Market definition is crucial: is it a market for iPhones, or is it a market for smartphones? If one defines the market as one for large airline transports, the structure is an oligopoly; the only sellers are Boeing, Airbus, Bombardier, and Embraer. If one defines the market as one for all types of aircraft including kit-built aircraft, it is much more fragmented.

The economics of production and marketing drive industry structure. In industries with high economies of scale resulting from high fixed costs, larger sellers are able to offer lower prices or earn greater profits. The ultimate example resulting from very high economies of scale is a natural monopoly. An electricity distribution company or cable television enterprise in a particular geographic area is an example. Economies of scale can result from high capital costs for physical assets like transport aircraft or factories. They also can result from high costs for integrated, international marketing. The motion picture industry is an example.

Industry structures often fluctuate. In its earliest days, the airline industry was quite fragmented, approaching perfect competition. Anyone with an airplane, a relatively inexpensive possession at that time, could carry s few passengers from point-to-point, and many did. Gradually, as more costly, higher performing aircraft carrying more passengers became available, economies of scale caused consolidation into a handful of trunk airlines. Then, when the airline industry was deregulated in 1974, eliminating legal barriers to entry, start up airlines proliferated. Now the process of consolidation is reducing their numbers again.

Similarly, the early days of cable television were marked by fragmentation. Every community had its own cable television provider. Monopoly existed at the local level, but it nearly perfect competition existed when viewed nationally. The economies of scale of offering nationwide access to high-cost programming drove consolidation to the point that, now, almost no purely local cable providers exist; a handful of national providers represent an oligopoly.

The drone industry is highly fragmented. Scores, maybe hundreds, of separate enterprises offer basically the same microdrones for sale. Consolidation is likely, however; DJI can afford more efficient manufacturing facilities and spread the costs of elaborate marketing programs over much larger revenue streams. To compete with it, a smaller producer must offer a distinct product, representing a separate market from that for the generic Phantom.

The market for microdrone services, as contrasted with that for microdrone hardware and software, will remain fragmented longer. Because of their limited endurance and range, microdrones are able to serve only local markets. Unless the cost of successful marketing and operations management centers are high enough to produce economies of scale, market fragmentation is likely to continue. Hundreds of microdrone operators will sell their services to hundreds of largely local customers. Microdrones, the central physical assets, are inexpensive. Therefore, in contrast to the airline industry, the capital cost of assets does not produce economies of scale.

Fragmented markets often produce new, secondary markets, for intermediaries who serve matchmaking functions. Economies of scale for matchmaking are significant. Accordingly, the market for microdrone services is likely to involve many buyers and sellers operating in conditions of nearly perfect competition, while a handful of matchmaking services reduces the transaction cost for buyers and sellers to find each other.

Notes

1 Jen Boyer, *Unmanned K-Max Completes Firefighting Demo*, Vertical, Dec. 2015/Jan. 2016 at p. 30, 31–32 (quoting aerial firefighting contractors).
2 www.factfinder2.census.gov/faces/tableservices/jsf/pages/productview.xhtml?pid=ECN_2012_US_48I3&prodType=table.
3 Bijan Vasigh *et al.*, *Introduction to Air Transport Economics* 374 (2nd edn 2013) (providing overview of evolution of low-cost airline industry).
4 Ibid. at 378.
5 Ibid. at 385.
6 www.eia.gov/electricity/monthly/epm_table_grapher.cfm?t=epmt_5_6_a.
7 2012 GAO DROP study at 11.
8 Farhad Manjoo, *When the Best Employees Are Actually on the Payroll*, N.Y. Times, Jun. 25, 2015, at p. B1 (reporting on interviews with startups that prefer employees to independent contractors because they provide better service to customers).
9 Henry H. Perritt, Jr., *Should Some Independent Contractors be Redefined as "Employees" Under Labor Law?*, 33 Vill. L. Rev. 989 (1988).

11 Starting a drone business

11.1 Introduction

This chapter provides concrete guidance to someone who intends to start a microdrone business. Whenever possible, it provides templates and explains actual choices that are plausible with respect to physical and human resources, marketing, and financial management. The number of permutations of all the choices that are legitimate and might lead to a successful business is enormous, however. Any effort to consider all of them would be unwieldy. Accordingly, the chapter supplements analysis of particular pathways to success with checklists of other possibilities an entrepreneur should consider.

The chapter simplifies terminology in several respects. While some microdrone businesses will supply products in a tangible sense—an enterprise that manufactures and sells drones is an obvious example—most will provide aerial support to customers, an activity more precisely termed a *service*. Rather than constantly using the cumbersome phrase *product or service*, the chapter uses *product* to refer to both tangible products and services.

Many types of microdrone businesses exist. An entrepreneur may decide to design and manufacture microdrones. She may decide to buy drones and lease them to operators. She may build certain components, such as inertial measurement units, for final assembly manufacturers. She may, if she is a member of the bar, offer legal services in connection with drones. She may own a television station and start a newsgathering drone operation to build market share for the station.

Another entrepreneur may decide to fly drones for other people, essentially to become a contract operator. This chapter takes the perspective of the drone operator. Picking one type of business affords a specificity and coherence to the discussion about starting a business that would be lacking if it tried to cover the full range of different types of businesses. Moreover, the main issues associated with other types of drone activities, such as manufacturing or providing legal services, have more in common with other types of manufacturing and legal services activities and have fewer unique characteristics associated with drones. Drone operations, on the other hand, revolve around the unique characteristics of drones, as contrasted with some other type of service business.

The chapter covers two scenarios, one in which an entrepreneur starts a new independent microdrone business; the other in which an employee of an existing enterprise starts a new line of business for the enterprise involving microdrones. The basic tasks involved are the same. Section 11.9 explains how the tasks involved in starting independent microdrone business would be carried out in the context of an existing enterprise.

11.2 Entrepreneurial philosophy

Entrepreneurs succeed by embracing opportunity. The essential ingredients are entrepreneurial vision, energy, and courage. Without vision, all of the energetic bustling about in the world is unlikely to produce a successful business—only fruitless attempts to steal market share from real innovators. Without energy to implement it, a vision is nothing more than a fantasy, interesting to talk about perhaps, but incapable of having any real impact on the world. Without courage, the entrepreneur will be talked out of it.

Almost any new business idea worth considering encounters many naysayers. Early reaction to an innovative business idea often cautions that, if it were a good idea, some else would have already done it, or the idea may achieve success in the marketplace and then some big company will come in grab all the customers, or, simply, "that will never work."

That was the early reaction to the Wright Brothers, to Frank Whipple, the inventor of the turbojet engine, to early Internet innovators, and now to more imaginative drone entrepreneurs. It is possible that some of the best ideas for commercializing drones are still sitting inside someone's head because he listened to the pessimists.

The key to evaluating new business ideas is to *evaluate* them, not to abandon them because someone else says they lack merit. The analytical steps presented in this chapter are not especially demanding, and to perform them costs nothing, assuming one has a computer and spreadsheet software. It may, indeed, be the case that a business idea lacks merit. One may find this out in trying to crystallize a credible message as to why potential customers should buy the contemplated service or product. One may discover fatal flaws the first time she compares estimated revenue with estimated costs and cannot come up with realistic assumptions that result in a profit. The numbers may be attractive, but one nevertheless may be unable to get potential investors even to consider financing implementation. But entrepreneurs cannot know that without doing a careful market analysis, running the numbers, and presenting them to potential investors.

The idea may receive financing but run aground in implementation. Most small business startups fail, and most early-stage venture-capital is lost. Indeed, most hugely successful entrepreneurs have failed at least once. Steve Jobs, for example, failed with NEXT, his idea for an innovative microcomputer and operating system to improve on the early Mac. But every attempt, successful or not, improves an entrepreneur's knowledge, enabling him to do a better job the next time around.

It is prudent, of course, to minimize the losses associated with failure: to have an exit strategy that can be executed before too much money has gone down the drain. The desirability of a viable exit strategy should be incorporated into early drafts of the business plan narrative; it always can be removed before the business plan is distributed to investors, although it might be a useful part of the risk factor section of any good narrative.

According to AUVSI's excellent analysis of the first hundred section 333 exemptions,[1] 80 percent of the exemption holders are new businesses. That is not surprising; innovation usually is easier in a startup enterprise than in an established business that has legacy products and individual internal interests to protect against disruption. Innovation is inherently disruptive and enterprises that have a successful track record do not want their trajectories disrupted.

11.3 The five functions of management

The tasks before the entrepreneur fall into the five basic functions of management, topics that comprise the first year curriculum of virtually every MBA program in the United States: marketing, finance, production or operations management, accounting, and organizational behavior.

Some of the activities associated with one function also involve another. For example, deciding on pricing is sensitive to the entrepreneur's sense of the market, but it also drives the revenue side of financial planning. Likewise, operations management relies heavily on assumptions and experience with organizational behavior.

11.4 Marketing

Marketing has to be the starting point for planning any business enterprise. It does no good to have lots of capital and strong management skill, if the entrepreneur has nothing to sell that buyers want. Finance, operations management, accounting, and organizational behavior planning flows from assumptions about demand.

Marketing is a multistage process. It begins with *market research*, assessing what potential customers need. *Need* is the appropriate word rather than *want*, because customers often do not yet know that they may want a new product that has not existed before. This initial aspect of market research, assessing potential customer need, is subtle and sophisticated but not necessarily burdensome analytically; it requires the entrepreneur to have a deep understanding of how potential customers think, and what their marketing and financial situation is.

Once an entrepreneur defines need, *competitive assessment* comes next. Competitive assessment involves identifying everyone else who presently is meeting, or is planning to meet, the customer needs identified in the first step of market research. Once the entrepreneur understands the market need and the competition, step three is to differentiate the entrepreneur's intended product: why will it meet customer needs better than competing products?

An example from a drone business idea illustrates the first three steps. Suppose an entrepreneur wants to sell drone operating services comprising real-time video capture. He could begin with two sources of information: the list of industries in which section 333 exceptions have been issued, and a list of industries that use manned helicopters to support their principal activities. The exemption applications indicate customer interest in drone support sufficient to tolerate the burdens of applying for an exemption. The helicopter customers indicate customer interest in aviation support strong enough to pay the substantial price for helicopters.

From these two sources, he selects real estate marketing, precision agriculture, motion picture and television production, event photography, infrastructure inspection, and newsgathering as likely targets. The resulting set of potential customers will reflect the industries considered in chapter 1.

Refining and understanding customer needs necessarily overlaps the competitive assessment step: what a customer needs depends on how well the need is being satisfied already. Television and motion picture producers, for example, need high-quality video imagery at the level obtainable by the best film cameras. Aerial camera platforms to obtain this must be as flexible as conventional camera tripods, booms, and dollies. So the market research and competitive assessment steps are recursive, in that an initial assumption about customer needs must be revised as more is learned about the competition.

11.4.1 Market research

Substantial amounts of research and analysis may be necessary to understand the target industry, or very little may be required. Entrepreneurs launching a new drone business may already know potential customers; indeed the interest expressed by those customers may have been the incentive for the drone-business idea. In other cases, however, the entrepreneur may know very little about customer industry structure. That lack of knowledge, of course, is a factor to be considered in formulating a strategy: an uninformed belief that demand exists for drone services is inherently less reliable than a similar belief formed from real knowledge of the target industry and from actual conversations and actual expressions of interest by decision-makers within it.

When organized market research on potential customers is necessary, data maintained and made available online by the US Census Bureau is a valuable starting point. It includes data on individual enterprises organized by industry and by geographic area. If the subject is labor markets, the Bureau of Labor Statistics in the US Department of Labor provides similarly detailed data. Industry trade associations often publish reports providing further analysis of industry statistics and friends. Additionally, the Harvard Business School is famous for its *cases* based on actual enterprises, and the cases usually provide sophisticated strategic analysis of the particular industry. They are available for a small fee from Harvard.

Once categories of potential customers have been identified through data analysis, the entrepreneur must explore potential customer interest more rigorously. Focus groups and formal opinion research are ways to do this, but it may be difficult to get potential customers to participate in focus groups, even if they are paid to do so, and it is challenging to get adequate response rates to opinion polling. Moreover, good opinion polling is expensive, though less so with increasingly popular web-based surveys that pop up when someone visits a website.

Informal approaches are more accessible for a low-budget startup. The entrepreneur can simply email or telephone a small set of potential customers and ask them for a brief meeting or telephone call to kick around the idea. Favorable responses to such outreach are more likely if the entrepreneur already has a relationship, or can enlist help from someone else who already has a relationship. The results may identify barriers that the entrepreneur had not anticipated and justify abandoning or modifying the idea. Alternatively, these informal meetings may be encouraging, warranting further, more rigorous research, or they may be sufficient to justify pitching the idea to potential investors.

11.4.2 Competitive assessment

Rarely is a new business idea so unique that no one else has thought of it. Even when it is unique, its success depends on persuading customers that the way they do business now is incomplete or obsolete, and that they should embrace new methods and new technologies. That is the case, for example, for customers who rely on helicopter support and must be persuaded to pay for drones as a supplement or substitute.

Competitive analysis requires careful assessment of how customer needs are being met now, and whether the new business idea will have to displace a competitor's offering or whether the entrepreneur must persuade customers that they need to embrace and pay for something they have never used before. A thorough competitive analysis begins with a complete inventory of who is offering the same product that the entrepreneur contemplates. Then, it must explore why the entrepreneur's product will attract favorable customer attention and eventual commitments. That competitive advantage may depend on more attractive pricing; or it may depend on better capability, compared with that offered by competitors. In any event, the competitive analysis must explain how the new product will rise above the competition.

11.4.3 Market scope

The scope or breadth of the market an entrepreneur intends to serve depends on strategic decisions about where the need is greatest, and where the intended business faces the weakest competition. In the market for aviation support of newsgathering, for example, many television stations in smaller markets want aerial coverage of news for the same reasons that stations in larger markets devote significant parts of their budgets to news helicopters. In the smaller markets, unlike the bigger markets, the need is not already being met by helicopters because the smaller stations cannot afford them. Thus the smaller stations are an attractive market segment for ENG microdrones.

Similar opportunities exist in other markets, for many of the same reasons. Microdrones permit aerial imagery at a price significantly below what helicopter imagery costs. Helicopter imagery is better in many ways because greater payloads permit more sophisticated equipment, and helicopters themselves have much more flexibility than microdrones. But microdrones place aerial imagery within the reach of smaller enterprises that have never been able to afford it before. Infrastructure inspection is a case in point; so are real estate marketing and event photography.

Understanding who those enterprises are, where they are located, and how to get them interested in what the new technology can provide, is an important part of the strategic equation for any new microdrone business—and the novelty of the technology makes every microdrone business, to some extent "new." A successful entrepreneur will focus on those market segments where opportunities are greatest, for these and other reasons.

11.4.4 Pricing

As chapter 10 points out, supply and demand are determined by much more than price. Still, an appropriate starting point for determining a pricing strategy is to recognize that lower-prices ordinarily are more attractive to customers than are higher ones, but that a lower price makes it more difficult for the enterprise to cover costs and produce a return.

Appreciating the factors other than price that shape customer demand draws upon the art of careful market analysis. Any entrepreneur who understands what drives customer decision-making will be in a much better position to obtain a decision favorable to a new drone product, than a competitor who can make only broad assumptions about customer behavior.

Pricing allows for virtually unlimited creativity by seller and buyer. The buyer can pay for services on an hourly basis, on a cost-plus basis, or for a fixed price. Various discounts and incentives can encourage the seller to outperform baseline commitments, or for the buyer to buy more than originally agreed to.

The services can be disaggregated. For example, the buyer can pay a flat fee for a block of services, and a separate fee to rent the equipment provided by the seller. Buyers can pay on a per-mission basis, or be entitled to a certain level of service under one-month, six-month, annual or longer contracts. Payments can be frontloaded, which might provide the seller with capital to buy equipment to support performance under the contract, or they can be backloaded, if that helps the buyer's cash flow, and the seller believes he will eventually get paid.

It is not uncommon for new market entrants to price below cost in the early stages of the business. Sub-market prices can help the new entrant penetrate the market. If that is the pricing strategy, the overall business plan must be clear on when and how the enterprise will generate a profit sufficient to provide a good return on investment.

Any pricing is subject to being revised as financial planning and preparation of a business plan proceeds. It is quite common to start out with one set of price options and then discover that the price is too low to cover costs, or that it is higher than necessary to cover costs and generate a return. In the latter case, the entrepreneur has the opportunity to be more attractive in the marketplace by charging a lower price.

In formulating a pricing strategy, it is important that creativity trump tradition. It is only slightly relevant how competitors or providers of close substitutes price their products. A customer may expect the pricing that she is used to, but also may be intrigued by a variation.

An important part of understanding the customer is to realize that most decision-makers within a large organization care more about meeting their budget than about the objective price level. As long as a decision-maker can budget for a needed service, she wants relative certainty that the service will cost that amount. An overrun obviously is embarrassing, but sometimes underruns are embarrassing as well. They make the decision maker look like a poor manager.

11.4.5 Sales

The marketing analysis so far is conceptual. Once the marketing concept is sound, the entrepreneur must know how it will be translated into actual sales. Just knowing who the potential customers are is not enough. The seller must convince them to buy its services. That implicates the *sales* part of marketing. Sales comprises advertising and direct interaction to close sales. The starting point is advertising.

11.4.5.1 Advertising

Advertising includes defining a message and determining how it should be communicated to potential customers.

11.4.5.1.1 MESSAGE

The content of the sales message proceeds from the competitive assessment. The message must answer the question of how the seller meets the customers' needs better than anything offered by competing products and services. For example, if the marketing strategy focuses on aerial news gathering, its central message would be:

> You need to add or increase your aerial coverage of news in order to enhance the competitive position of your station. We can help you do that for much lower cost and greater flexibility than you can get from news helicopters or from other providers of drone support. Our services cost less than a hundredth of the cost of helicopters, and we understand the television news business better than other drone operators.

That is not a very poetic way of expressing the idea, however. Effective message crafting concentrates on the best way to present the idea, depending on the channel through which it will be communicated. A knack for story telling—or for writing song lyrics—is useful in that regard.

Face-to-face presentations are more effective when they are not scripted. But, print, email, Web, and video messages, should be carefully honed so that their verbal expression is inviting and persuasive, and the graphics and video images are evocative.

11.4.5.1.2 CHANNELS

After she crafts a persuasive message, the entrepreneur must decide on the most cost-effective way to get it to potential customers. Many sales channels are available, including direct mail, booths at conventions and trade shows, television and radio advertising, targeted Web ads, telephone and email contacts, and direct sales calls. Which is most desirable depends on cost, how well it will reach potential customers, the expected conversion ratio—the percentage of the customers to respond to initial contacts—and the ability to target the advertising. It is unlikely to be effective, for example, to buy broadcast television advertising in a large metropolitan area in order to reach a handful of television station customers spread out over a state or a region. Customers for drone services do not constitute a mass market. It is not like selling a new type of soap or deodorant or a new app for an iPhone or Android device. In contrast, potential customers for wedding photography do represent a mass market.

The purpose of communicating the message through the appropriate channels is to find potential customers who are sufficiently interested to warrant a direct conversation.

11.4.5.2 Direct contact

Once potential customers have surfaced, by responding in some fashion to broader advertising, the entrepreneur must decide on the nature of a direct conversation. Should it be preceded by or accompanied by submission of a formal proposal? Should the conversation be structured through PowerPoint slides and video presentations? How prominent a part of the discussion should pricing be, or should price discussions follow a meeting of the minds on what the customer needs?

It is never a bad idea to begin a sales presentation by summarizing the seller's understanding of the customer's needs, so that the rest of the interaction can revolve around the customer's perspective. That can be the case with a formal proposal, in a relatively formal presentation, or in a completely informal and less structured conversation.

Once the initial customer meeting is over, participants should have an explicit understanding of what happens next, whether that is simply for the customer to think about it, whether a subsequent presentation or meeting will involve more specialized personnel or more senior personnel associated with the customer, or whether the customer wants more detail on pricing, design or support.

After the initial contact and the completion of any explicit follow-up steps, the seller should take the initiative in following up with a customer who has been silent. The frequency of such follow-up is a matter of judgment; the seller doesn't want to become a pest. On the other hand, it is not uncommon for a potential buyer to be favorably disposed toward a proposal and yet not get around to taking any action to follow up on it. A polite, timely reminder to attend to it cannot do any harm.

11.5 Operations management

To a considerable degree, the early stages of the marketing and finance functions are theoretical, involving assumptions about costs, prices, revenue, and behavior. But theory is not what matters; reality is what matters. The operations management function is responsible for delivering on a planned product levels at planned for costs.

In order to construct financial projections and the business plan, the entrepreneur needs to know how she will organize and deliver services to customers. That implicates the operations management function. The first step is to define the necessary resources. They include the following:

- Physical assets
    - Drone vehicles
    - Backup parts such as rotor blades, spare batteries, and parts likely to break in service, such as landing gear legs
    - DROPCONs
    - PHOTOCONs
    - Tablet computers or other video displays
    - Base-of-operations facility
    - Cellular bonding devices such as Dejero or LiveU
- Services
    - Cellphone subscriptions
    - Internet access accounts
    - Internet domains and hosting accounts
    - Insurance

- Personnel
  - Dispatchers
  - DROPs
  - Photogs
  - Mechanics
  - Computer electronics technicians
  - Webmasters
  - Administration and financial staff

Each of these resources can be obtained in a variety of ways. DROPs can be employed full- or part-time, or they can be independent contractors dispatched for particular assignments. Mechanics, computer electronics technicians, and webmasters can be individuals employed or acting as independent contractors, or they may be sub-contracting enterprises. The base of operations can be leased office space to which the enterprise has exclusive use, or it can be shared space in the entrepreneur's residence or other business premises.

Chapter 10 explores the range of alternatives for acquiring resources.

A certain amount of industrial engineering or operations analysis, borrowing some of the principles of scientific management discussed in chapter 4, are necessary to develop a concrete operating plan. The entrepreneur must consider questions such as:

- When will a DROP go on pay status if he is an independent contractor?
- Where will he get possession of the drone for a particular assignment?
- What will be the average transit time from where the DROP and drone start out to the mission site?
- How much notice will a customer give in requesting support?
- Will DROPs be obligated to accept callouts, or may they turn them down?
- How many DROPs must be in the pool to assure coverage of customer requests?
- What is the recourse if a drone experiences a malfunction?
- How many backup vehicles should be maintained?
- Where should they be kept?
- How will a backup drone be transported to the mission site if a malfunction of the primary vehicle is detected only during final preflight inspection?
- Who will get it there?
- Is this a 24/7 operation or one with more limited hours?
- Who will be the duty officer, always available to receive customer requests by text, email, or cell phone?
- How will vehicle and DROP time be recorded?
- Who will do the billing? At what frequency?
- When will DROPs be paid?
- What kind of financial reporting to investors is required by law or promised in the investor agreement? Who will prepare them?

11.6 Organization behavior and human resources

Business enterprises are run by people. Recruiting and managing the people effectively implicates the subject of organization behavior, which is a more general aspect of the human resources topic considered in chapter 4.

11.6.1 Delegation versus DIY

In the smallest enterprises the entrepreneur does everything himself. In larger enterprises, one person cannot do everything, and multiple people must be involved in performing different functions. As the number of people rises, more attention must be paid to coordination and management, including designation of coordinators and managers. Almost every activity considered in this chapter can be done by the entrepreneur himself, or delegated to a specialist. Most people are not willing to work for free; at least not for long. So the larger the group of human resources in the enterprise, the higher the cost is. An important part of the operations management planning is to make tough-minded decisions about how much can be performed by one person well with reasonable expenditures of time; everyone has to sleep and eat.

The entrepreneur must make objective decisions about who can and should do what, based on workload and expertise.

11.6.2 DROPs

The entrepreneur should decide on selection criteria for DROPs and other personnel and crystallize them in a written job description. Chapter 4 provides illustrations.

The entrepreneur should know how DROP performance will be evaluated. This is necessary to know whether to retain a DROP employee, and also how often to call a DROP who is an independent contractor. Someone whose performance is acceptable but not exemplary might be called out only under conditions of unusual demand. DROPs whose performance is exemplary would be called out more frequently.

Chapter 10 analyzes various management styles and organizational structure issues. The entrepreneur should think about this material and make choices that are recorded and crystallized in an operations manual. The operations manual will be useful, in any event, once operations begin. Indeed, such a manual might be required by the eventual FAA regulations. Arguably, such a manual is contemplated in the standard section 333 exemptions.

11.6.3 Tight ship or club of evangelists?

One can imagine the same drone enterprise being managed in two dramatically different ways. It can be run as a tight ship—UPS makes an advertising slogan out of that management philosophy. Or, it could be run like a club—the early days of Apple and Google are popular examples. In the tight ship model, workers show up, do their jobs strictly according to the rules, get paid, and go home. Managers correct for any deficiency in performance on a direct and straightforward basis. The advantage of this approach is clear organization, maximum efficiency, and unambiguous responsibilities. It produces acceptable morale for employees that want a job that is just a job.

In the club model, the degree of differentiation between managers and line workers is vague; owners, managers, and line workers not only work together, but they play together, often during working hours. Workers also often participate in strategy formulation. When the philosophy works well, no one says, "It's not my job," when someone else is overwhelmed. Everyone is an evangelist for the enterprise. This kind of environment is likely to attract well-educated workers with broad interests and excitement about the subject matter.

The disadvantages are that everyone may spend so much time playing together that they don't do enough work. Referring every decision to a committee of the whole means that decisions not made promptly, and they reflect the mediocrity that consensus often produces rather than the sharp edges of a creative idea.

11.7 Finance

At its most basic level, the finance function is responsible for knowing where the money will come from to cover costs and generate a profit. Financial planning involves both parts of this equation: what will the costs and revenues be? Where will the money come from?

Anyone starting a drone business can employ one of two basic strategies for beginning operation. It can it can acquire and manage resources so as to minimize costs, and therefore the need for capital, until it has established itself with a critical mass of customers. Or, it can begin operations more grandly, sufficient to handle the level of business expected after becomes profitable. The chosen strategy will determine whether DROPs are employees drawing a salary or freelance independent contractors hired on a per-assignment basis. Similarly, the choice of strategies will determine whether the enterprise rents a facility as a base of operations with ample space for parts, meetings with customers, and offices for its staff; or whether it shares space in a residence or another business.

The grander approach may be justifiable to assure that potential customers find the new enterprise credible, but it requires considerably more capital, which may not be available.

11.7.1 Accounting

The finance function is closely associated with accounting. Anyone deciding whether to invest needs a standard way of evaluating the likely fortunes of the enterprise. Managers of an enterprise need a reliable way to know how the enterprise is performing, so that they can take any necessary corrective action before it is too late.

Projections and reports of financial performance adhering to generally accepted accounting principles are the way to satisfy both needs. Three types of reports portray the condition of the enterprise. The *income statement*, often called *profit and loss statement*, shows whether the firm earned a profit or suffered a loss during a preceding time period—usually one year. It reports income, expenses, and the difference: net income (profit or loss). The *balance sheet* shows the firm's financial situation at a particular point in time—usually the last day of the year for which the income statement was prepared. It reports assets, liabilities, and the difference: equity. The *statement of cash flow*, often called *sources and uses of funds*, shows what cash came into the firm and what it was used for. It reconciles the difference between net income or loss and the cash in the bank. The balance sheet and the statement of cash flow reflect concrete reality: the money actually was received or paid out; the bank balance is indisputable. In comparison, a number of assumptions are necessary to build the income statement.

The difference between cash flow and net income arises from the fact that accounting principles for *accrual* require that costs and revenues be allocated to the time periods with which they are associated, even if money actually was paid or received earlier or later. Accrual follows two somewhat conflicting principles: the *conservatism principle* and the *matching principle*. Under the conservatism principle, expenses and liabilities are recognized (formally posted to the financial statements) as soon as they reasonably can be estimated, despite uncertainty as to their amount; while revenues and assets are recognized as late as possible, until uncertainties about them are resolved. The matching principle requires that expenses be recognized in the same time period as the income they helped to generate.

Insurance is a good example. Often the terms of an insurance policy require that the full annual premium be paid at the commencement of the policy. Yet the costs of the policy actually are associated with each month for which it is in effect. The profit and loss accounts accordingly reflect an insurance cost of one-twelfth of the annual premium in each month; the statement of cash flow shows the full cash premium only in the month in which it is paid. The balance sheet shows the bank balance after payment of the premium and any favorable balance of premium paid, as separate assets—cash and prepaid insurance premium.

Likewise, a drone service that buys its drones must expend the cash for their purchase in one time period, before the enterprise actually offers its services and earns any revenue. The drone-purchase outlay is what shows in the cash flow analysis. But the cost of using the drone fly customer missions is associated with a different time period: the time period in which it is flown. The matching principle requires that a figure associated with the cost of flying the drone be entered for each time period in which it is used, in the profit and loss part of the accounts.

That requires a decision about the best method of allocating that fixed cost, under two closely related concepts: *depreciation* and *depletion*. Depletion is the appropriate concept for an asset that gets used up physically—a stock of coal, or an aircraft that has a life limited by total flight hours. Depreciation is the appropriate concept to spread the cost of an asset that is not physically used up. Microdrones, unlike many light helicopters, do not wear out; they are more likely to become obsolete as innovation occurs and new models are introduced. So they should be depreciated. One way to do it to estimate the economic life of the drone—12 months in the example—and divide it into the total acquisition cost. That assumes in practical terms that the drone loses its value at the end of the 12 months. More formally, at that point, it is *fully depreciated*. That is a reasonable assumption because of the likelihood that the rapid pace of innovation will render any particular model obsolete within a year or two.

If the enterprise leases its drones from someone else, an allocation calculation is a unnecessary—for a simple lease; the monthly expense for drone flight time is simply the lease payment for that time period—the same as the entry for that month on the statement of cash flow.

Accrual deals similarly with revenue and income. If a customer pays only some time after the enterprise performs a service, the revenue received is recognized when it is paid on the statement of cash flow; income is recognized on the income statement for the time period in which it is earned.

All of this assumes accrual-basis accounting, which is required by GAAP and similar systems of accounting principles. In cash-basis accounting, there is no difference between the income statement entries and the statement-of-cash-flow entries for a particular time period.

Few investors are simply willing to accept an entrepreneur's financial statements. Usually they want assurances from an independent accountant. Assurances may take the form of a formal audit, or they may be less formal.

11.7.2 Spreadsheets and pro-forma projections

The only way to do a satisfactory job of basic financial planning is to construct financial projections based on good estimates of revenues and expenditures, called a *pro-forma* financial analysis. The pro-forma financial analysis is a future looking, planning document; it would not be misleading to think of it as a budget. It contains projections rather than representing actual financial experience. Though dependent on assumptions about the future, it is necessary to provide investors and managers with a sense of the likely prospects for enterprise success. A startup business has no actual financial experience, and pro-forma projections are the only financial information available to assess the likelihood that the enterprise will produce an adequate return on investment.

Losses in the early periods are common, and negative cash flow is inevitable. Whether the enterprise is viable depends, not on the timing of profitability or positive cash flow, but on whether the enterprise ever becomes profitable and whether cash flow ever turns positive.

The only sensible way of working up pro-forma projections is to use spreadsheet software. Using a spreadsheet to construct the financial projections is advantageous for three reasons. First, it automatically performs arithmetic calculations that otherwise would be tedious. Second, it makes assumptions explicit and makes it easy to see the effect on profitability and cash flow of changing a particular assumption. The user simply overtypes the number representing a particular assumption, and the spreadsheet automatically recalculates everything else. Third, spreadsheets also format everything in a useful structure, which can be printed or emailed. Tables 11.1 to 11.3 provide examples.

Imagine a startup enterprise that intends to provide microdrone services for electronic newsgathering for TV stations. It intends to offer its customers drone support in four-hour blocks of time referred to as *callouts*. To conserve cash, it will contract with DROPs on a per-callout basis and pay them an hourly rate. The enterprise assumes that customers will want HD video captured by a gimbaled camera with zoom capability.

To deliver the intended product to its customers, it needs the following resources:

- The drones themselves
- Associated equipment to allow the DROP to fly them
- Equipment to capture video good video and send it to the customer
- A physical base of operation
- DROPs, mechanics, and electronics technicians
- Internet connectivity, cell phone service, a web domain, and web hosting
- Advertising and trade show appearances
- Insurance
- Legal and accounting services
- Electricity and gas

The management is committed to the leanest possible operation, having concluded that it must aggressively keep its costs down in order to offer a price that is attractive to stations that have limited news budgets. It must, however, provide for upgrading of its systems to keep pace with rapidly developing technologies.

Some of these resources can be obtained on a highly variable basis. Legal services are a likely example. The enterprise almost certainly can arrange for legal services to be bought only when needed on a per-hour basis; it is unlikely to need a full-time general counsel. On the other hand, insurance is certain to be a fixed cost. Buying coverage on an as-needed basis would be a very bad bargain for the insurer. Other items are fixed, but not necessarily lumpy—administrative support, for example—answering the mail, and the bills, keeping payroll records, tracking track records, and maintaining inventory can be provided by a full-time office manager, but the needs of the enterprise might be such that they could be met by a part timer, working half days or fewer than five days per week.

Spreadsheets are generally familiar, but a brief review of their structure and nomenclature may be helpful. Spreadsheets comprise *rows* and *columns*, which in turn define *cells*. Rows run across the page and are named according to the numbers that appear down the left-hand side; columns run down the page and are named according to the alphabetic characters that appear across the top. A cell is referred to by the concatenation of its column and row number. The cell of the upper left hand corner of Table 11.1 is cell *A1*. The cell in Table 11.2 indicating month number 1 is cell *D49*.

The starting point in constructing the spreadsheet is to list the values for critical assumptions. Reality, of course, constrains the assumptions. An entrepreneur is not likely to find a DROP who will work for five dollars an hour, although he may be able to find one who will work without immediate payment in exchange for an ownership share in the enterprise. A customer is unlikely to pay more for a drone mission then he would pay for a helicopter to fly essentially the same mission.

Moreover, assumptions about different values have implications for strategy. In the sample scenarios, profitability can be improved by ratcheting up the number of customers in each time period. At least that will help cover fixed costs; it will not do anything if variable costs are such that each mission results in a loss. But that will require more resources to put into sales, marketing, and customer support.

The actual figures used reflect the authors' business experience in in proposing drone operations to potential customers and in consulting with other entrepreneurs.

Table 11.1 shows the following key assumptions:

The flight system comprises a DJI S 1000 with a sophisticated gimbal, a Sony Handycam with an optical zoom lens, and an iPad as a video monitor. The total cost for the flight system is $7,712 (cell *C2*), representing the sum of the entries for the S1000, the gimbal, the iPad, and the camera.

Cell *D16* shows the assumed hourly pay rate for DROPs ($50), and cell *D17* shows the assumed hourly pay rate for mechanics (also $50). Rows 27 through 35 contain assumptions for various fixed costs.

The revenue assumptions begin in row 37. Hours flown on each callout are shown in cell *B37*, frequency of callouts per week is in cell *B38*, and price of $850 per callout appears in cell *B42*. They proceed from the pricing decision to offer drone support in blocks of four hours. Whenever a customer calls out the drone, the customer is obligated to pay for a four hour block.

If the entrepreneur expects the number of customers to be fixed, he might also include a number-of-customers assumption in the assumptions section. But in the early stages, at least, the business will grow. Accordingly, the number of customers is reflected by a different number in the revenue and cost section of the spreadsheet, which grows over time.

Table 11.1 Business plan assumptions

	A	B	C	D
1	Assumptions			
2	Flight vehicles	Total	7,712	
3		DJI S1,000	4,000	
4	Economic life (months)		12	
5	Control equipment			
6		DROPCON		
7	Video capture, display, and transfer equipment	Gimbal	1,500	
8		iPad	700	
9		Dejero		
10		Camera	1,512	
11	Physical facilities			
12		Rent (per month)	600	
13			Annual	Hourly
14	Personnel			
15		Management	0	0
16		DROP	50,000	50
17		Mechanic	50,000	50
18		Photog	40,000	40
19		Electronics technician	40,000	40
20		Administrative staff	25,000	25
21		Sales and customer service		
22		Human resource		
23				
24	Repair and fabrication tools			
25	Electronics test equipment and spares			
26				
27	Utilities	Total	4,700	
28		Internet access	1,200	
29		Web domain & hosting	500	
30		Cellphone service	1,200	
31		Electricity and gas	1,800	
32	Advertising and tradeshow appearances		10,000	
33	Insurance		4,000	
34	Legal services		5,000	
35	Accounting services		2,000	
36				
37	Callout length in hours	4		
38	Frequency per week	3		
39	Maintenance hrs per callout	1		
40				
41	Price			
42	Per callout	$ 850		
43	Contract price	$ -		
44	Upfront payment	$ -		

Table 11.2 shows the calculations that the spreadsheet software automatically performs:

The formula to calculate monthly revenue is straightforward. The formula for revenue is:

= D50*B42*B38*(52/12)

where

- D50 is the number of customers for that month,
- B42 is the price per callout
- B38 is the frequency of callouts per week
- 52 divided by 12 converts weekly revenue to monthly revenue.

That formula results in the values shown in row 53. The formula is entered in the spreadsheet, but hidden in the illustration.

The spreadsheet must be constructed to handle fixed and variable costs differently.

Insurance premiums are a simple example of an undeniably fixed cost. Almost any aviation liability and hull insurance policy requires the insured to pay an annual premium. Cell *C33* in Table 11.1 shows an assumed annual premium of $4,000. Assuming that the financial projections are constructed on a monthly basis, the annual premium must be divided by 12 to get the monthly premium.

Many of the expense entries simply divide annual fixed costs by 12 to get monthly cost.

The formula for calculating labor cost entered (but hidden) in Row 59 is:

= D50*D16*B37*B38*(52/12)

where

- D50 is the number of customers
- D16 is the DROP hourly compensation
- B37 is the length of a callout in hours
- B38 is the number of callouts per week
- 52 divided by 12 converts weekly DROP compensation to monthly

Certain labor costs, like those for DROPs or mechanics, may be either fixed or variable, depending on the contractual and compensation arrangements for the DROP and mechanic. If the DROP is paid only when he flies, and the mechanic is paid only while he is working on a drone, their labor costs are variable. The spreadsheet assumes hourly costs for both. Thus the entry in the formula for DROP labor costs is hourly rate of pay, as shown in cells *D16* and *D17* of Table 11.1. Thus the cell entries in rows 59 and 57 reflect DROP and

Table 11.2 Business plan profit and loss

	A	B	C	D	E	F	G	H	I	J	K	L	M
49	Month			1	2	3	4	5	6	7	8	9	10
50	Number of customers					1	1	1	2	2	2	3	3
51													
52	Profit and loss												
53	Revenue			$ -	$ -	$ 11,050	$ 11,050	$ 11,050	$ 22,100	$ 22,100	$ 22,100	$ 33,150	$ 33,150
54													
55	Expenses												
56	Vehicle amortization			$ -	$ -	$ 643	$ 643	$ 643	$ 1,285	$ 1,285	$ 1,285	$ 1,928	$ 1,928
57	Maintenance			$ -	$ -	$ 650	$ 650	$ 650	$ 1,300	$ 1,300	$ 1,300	$ 1,950	$ 1,950
58	Insurance			$ 333	$ 333	$ 333	$ 333	$ 333	$ 333	$ 333	$ 333	$ 333	$ 333
59	Labor cost (no photog)			0	0	2600	2600	2600	5200	5200	5200	7800	7800
60	Rent			$ 600	$ 600	$ 600	$ 600	$ 600	$ 600	$ 600	$ 600	$ 600	$ 600
61	Utilities			$ 392	$ 392	$ 392	$ 392	$ 392	$ 392	$ 392	$ 392	$ 392	$ 392
62	Advertising and trade shows			$ 833	$ 833	$ 833	$ 833	$ 833	$ 833	$ 833	$ 833	$ 833	$ 833
63	Legal and accounting services			$ 583	$ 583	$ 583	$ 583	$ 583	$ 583	$ 583	$ 583	$ 583	$ 583
64	Total expenses			$ 2,742	$ 2,742	$ 6,634	$ 6,634	$ 6,634	$ 10,527	$ 10,527	$ 10,527	$ 14,420	$ 14,420
65													
66	Profit/loss			$ (2,742)	$ (2,742)	$ 4,416	$ 4,416	$ 4,416	$ 11,573	$ 11,573	$ 11,573	$ 18,730	$ 18,730

mechanic costs respectively simply by incorporating the assumed hourly rate and multiplying it by hours flown for that month.

The entrepreneur may decide that he can satisfy customer needs better and may actually experience lower costs if he hires DROPs on a salaried basis rather than paying them on an hourly basis only when they fly. Chapter 10 explains the practical difficulties in split shifts and other arrangements that seek to reduce idle time of labor inputs. If DROPs were salaried, the formula in Row 59 would start with the annual figure in *C16* and perform a different calculation to allocate the fixed cost for each DROP over the hours flown.

Alternatively, and more simply, if the enterprise hires DROPs on a salaried basis, they become a fixed cost and their compensation is entered in the spreadsheet just like the insurance premium. Their annual salary is divided by 12 and plugged into each cell on the DROP cost row (row 59) for each month.

One must be careful to include only similar units in the profit and loss section of the spreadsheet. For example, each figure entered in rows 72 through 79 must be in dollars per month.

11.7.3 Determining capital needs

The cash flow analysis reveals the total amount of capital required. In the spreadsheet shown in tables 11.1 to 11.3, both cash flow and profit-and-loss perspectives are available from the same spreadsheet.

The amount of investment capital required by a startup is the cumulative deficit it projects before it begins earning a profit. Table 11.3 shows a portion of the spreadsheet for a drone business that shows cash flow, just after the profit or loss calculation.

Line 66 shows profit and loss. Lines 68–82 present the cash flow analysis. The figures relevant to determining capital needs are shown in line 82. The investment capital need is always the largest negative figure in the cumulative cash flow analysis.

This portion of the spreadsheet shows upfront cash outlays for vehicle purchase (cell *D73*), the upfront insurance premium (cell *D74*), and all of the expenditures for advertising and tradeshows bunched into the first three months (cells *D79*, *E79*, and *F79*).

Customer revenue does not begin to come in until the third month (cell *F69*). The resulting cumulative cash flow is shown in row 82. It peaks at $21,562 in cell *E82*, and that is the amount of capital that must be invested in the business to allow it to get started. Monthly cash flow turns positive in the third month (cell *F81*), and cumulative cash flow turns positive in month 6 (cell I82).

As § 11.7.2 explains, spreadsheet software allows a financial planner to vary assumptions and discover their effect on profitability and cash flow. The

Table 11.3 Business plan cashflow

	A	B	C	D	E	F	G	H	I	J	K	L	M
66	Profit/loss			$ (2,742)	$ (2,742)	$ 4,416	$ 4,416	$ 4,416	$ 11,573	$ 11,573	$ 11,573	$ 18,730	$ 18,730
67													
58	Cash flow												
69	Cash in			$ -	$ -	$ 11,050	$ 11,050	$ 11,050	$ 22,100	$ 22,100	$ 22,100	$ 33,150	$ 33,150
70	Contract payments			$ -		$ 10,000	$ 10,000	$ 10,000	$ 10,000	$ 10,000	$ 10,000	$ 10,000	$ 10,000
71													
72	Cash out												
73	Vehicle purchase			$ 7,712					$ 7,712			$ 7,712	
74	Insurance premium			$ 4,000									
75	Rent			$ 600	$ 600	$ 600	$ 600	$ 600	$ 600	$ 600	$ 600	$ 600	$ 600
76	Compensation			$ -	$ -	$ 3,250	$ 3,250	$ 3,250	$ 6,500	$ 6,500	$ 6,500	$ 9,750	$ 9,750
77	Rent			$ 600	$ 600	$ 600	$ 600	$ 600	$ 600	$ 600	$ 600	$ 600	$ 600
78	Utilities			$ 392	$ 392	$ 392	$ 392	$ 392	$ 392	$ 392	$ 392	$ 392	$ 392
79	Advertising and trade shows			$ 3,333	$ 3,333	$ 3,333							
80													
81	Net cashflow for month			$ (16,637)	$ (4,925)	$ 2,875	$ 6,208	$ 6,208	$ 6,296	$ 14,008	$ 14,008	$ 14,096	$ 21,808
82	Cumulative cashflow			$ (16,637)	$ (21,562)	$ (18,687)	$ (12,479)	$ (6,270)	$ 26	$ 14,034	$ 28,043	$ 42,139	$ 63,947
83													

authors did exactly that in a series of hypothetical scenarios for launching a drone operations business.

For example, line 3 in Table 11.1 assumes that the cost of each drone is $4,000. If the entrepreneur concludes that a drone with acceptable performance will cost $15,000 instead, he simply overwrites the $4,000 figure with $15,000, and the spreadsheet recalculates everything else, including the effect on timing and amount of profitability automatically.

As another example, the entrepreneur may decide that the enterprise requires the full-time attention of a general manager. That could be the entrepreneur himself, in which case the salary would represent what he needs in order to quit his day job. Or, it could be just the right kind of eager drone enthusiast, with good management skills and entrepreneurial energy. An annual salary of $50,000 is modest, but enough to live on prudently. That has the effect of increasing the total investment requirement to $29,145 deferring positive monthly cash flow until month 4.

11.7.4 Determining sources of capital

Cash outflows will exceed cash inflows at the beginning. No business generates revenue before it makes expenditures to put itself in a position to deliver product to its customers, before it actually receives any revenue from the customers. Only the rarest customer is willing to pay so far upfront to finance the early stages of the startup. In fact, a typical new business does not begin to break even until the third year of operation. Therefore the enterprise must have a source of money to cover its deficits before it begins to generate a profit. That money is *startup capital*. Chapter 10 reviews the basic alternative sources of capital and identifies the pros and cons of each.

11.7.5 The business plan narrative

The only way to do a satisfactory job of basic financial planning is to write a business plan. Part of the plan requires constructing a spreadsheet with good estimates of costs and revenues. Sections 11.7.2 through 11.7.4 analyze that part of the plan.

Even before the spreadsheet representing the quantitative aspects of financial planning is complete, narrative portions of the business plan can and should be written. The US Small Business Administration suggests that the basic outline of a business plan includes the following sections:

I. *Table of contents*
II. *Executive Summary*
III. *Company Description*
IV. *Market Analysis*
V. *Organization & Management*
VI. *Service or Product Line*

VII. *Marketing & Sales*
VIII. *Funding Request*
IX. *Financial Projections*

This is a reasonable starting point, but the business plan must also contain a risk-factors section, as discussed in § 11.7.6. Furthermore, it would be more logical to put the Service or Product Line section before the Organization and Management section, and the Financial Projections before the Funding Request. The Market Analysis section must contain subsections on:

- Customers' needs
- Competitive assessment

The Funding Request section should describe:

- How much investment is required
- What form it will take (equity, loans, entrepreneur investment)
- The terms

11.7.6 Pitching it without violating the securities laws

Federal and state securities law significantly constrains raising capital through offers to the general public. A *public offering* is a permissible under the federal securities laws only after a *registration statement* including a *prospectus* is filed with and approved by the US Securities and Exchange Commission (SEC). An enterprise may, however, raise equity capital through various private offerings under certain exceptions to the registration requirement. The SEC regulations providing for these exemptions are known *Reg D*. One option under Reg D permits intrastate offerings to investors located in the same state as the issuer. Another option does not limit investors to one state, but still permits up to 35 investors who are not *accredited*. Accredited investors are those that have certain levels of wealth, income, and sophistication.

The 2013 JOBS Act[2] created another exception to the registration requirement, allowing smaller enterprises to raise capital through *crowdsourcing*, familiar to many people in the form of Kickstarter. Kickstarter, launched in 2009, is an Internet-based service that matches creative projects with contributors willing to help fund them. Designed for contributions rather than investments in order to avoid violating the securities laws, Kickstarter permits enterprises to solicit donations, but not investments. The difference is that an investor expects to receive a return based on the success of an enterprise managed by someone else. Kickstarter contributors expect no financial return at all and are essentially making a gift to the soliciting enterprise. Kickstarter claims to have channeled nearly $2 billion into nearly 100,000 projects. It is used increasingly by profit-seeking enterprises, however, to solicit contributions for which the reward is a product, once it is in production.

The crowdsourcing exemption could not be implemented until the SEC issued final regulations, which it was quite sluggish in doing, finally issuing them in mid-2015.[3] Pending promulgation of SEC regulations, a few states allowed raising investment capital through crowdsourcing. In August 2015, XTI Aircraft announced a crowdfunding campaign under new SEC rules to test the waters for a larger capital campaign to finance its 6-seat Trifan 600, which can take off vertically and cruise at 340 knots. The VTOL aircraft has a range of up to 1,200 miles.

Raising capital through private placement, as contrasted with registered public offerings, involves identifying potential investors and providing them with a *solicitation* and *offering memorandum*. Those that decide to invest enter into a *subscription agreement* with the issuer.

The exemptions, including the one for crowdsourcing, do not exempt the issuer from federal and state statutory prohibitions against fraud. Often, when enterprises are unsuccessful and investors lose their capital, they claim that they were misled, in violation of the statutory anti-fraud provisions. That results in an expensive lawsuit for the issuer. For example, in In re Flight Transportation Corp. Securities Litigation,[4] the SEC and investors alleged that a helicopter air-taxi and tour operator in bankruptcy had violated the securities laws by overstating gross revenues, miles flown, and number of helicopter and charter flights, in representations made to investors.

The likelihood of such a suit and success by the plaintiff investor is greatly reduced by a good *risk factors* section in the offering memorandum. A good risk factor section lists every conceivable thing that might go wrong and explains how it could doom the enterprise and disappoint investors.

It is counterintuitive to write a document whose main purpose is to persuade investors that they can earn a good rate of return based on the success of the issuer, and then to include a section that emphasizes all the reasons that the business may fail. But the reality is that most startup businesses fail. When they do, the issuer and its management will be in a much better position if they have disclosed all of the risks, so that investors make the investment while knowing the potential negative outcomes. Moreover, experienced investors are accustomed to reading good risk factor sections, and are not likely to put off by such sections' pessimism.

11.8 Business entity structures

The entrepreneur must decide what legal form the business should take. The simplest possibility is to organize the business as a sole proprietorship, which means that the entrepreneur, in his individual capacity, is also the business. Sole proprietorship permits the individual to do business under a tradename. Possible trademark infringement occurs if the name is similar to one already in use by someone else or as to which someone owns a registered trademark. Sole proprietors are personally liable for the debts of the business.

Alternatively, the entrepreneur can organize the business as a separate legal entity. A corporation is the oldest form; newer forms, such as limited liability companies (LLCs), resemble corporations but are simpler to organize and operate. Most states permit organizers to establish corporations or LLCs through a relatively simple web-based process offered by the registration agency—usually the secretary of state. Organization of an entity usually requires payment of a fee in the hundreds of dollars.

A separate business entity such as a corporation or LLC offers two advantages: limited liability and allocation of ownership interests according to the bylaws of the corporation or the operating agreement of the LLC. Limited liability means that the organizers are not liable in their individual capacities for the debts, including judgments resulting from lawsuits of the business entity. For limited liability to be effective, however, the entrepreneur, the likely manager of the business enterprise, must observe certain formalities, such as segregating the assets of the business entity from personal assets, usually by setting up separate bank account, and by keeping careful records of transactions between the entrepreneur and the business enterprise. If he does not do this, someone might sue the business entity and "pierce the corporate veil," meaning that the owners—the entrepreneur—will be personally liable.

Imagination, creativity, and negotiated agreement are the only limits on how ownership interests can be allocated. The simplest approach is for each person investing capital, including the entrepreneur, to own shares, (or membership interests in the case of an LLC) in proportion to their capital contributions.

But that is not the only way to do it. Multiple classes of shares or membership interests can exist, giving control rights to some classes but not others, and varying levels of rights to the income of the entity, and to its assets if it is when it is dissolved.

Typically, an entrepreneur establishing a separate business entity wants to retain control and persuade investors to accept economic interests only and not share in control. Some investors may be willing to do this; others will insist on some degree of control. All of this must be spelled out in the bylaws or the operating agreement. Great care and precision is necessary to prevent subsequent embarrassing and costly controversies.

11.9 Starting a new drone line of business within an existing enterprise

As § 11.2 explains, innovation is often easier for startup enterprises than for established ones. As it reports, 80 percent of the first 500 section 333 exemptions for commercials drone services were awarded to startup businesses. But established enterprises also must innovate, or the market will gradually erode their market share as its customers begin to prefer new alternatives. Multiple reasons may motivate an established provider of aerial helicopter support, for example, to establish a new drone line of business.

Most importantly, if it does that, it controls the dialogue with its existing customers. Together, they can decide on the best mix of helicopter and drone support. If the same enterprises sticks to helicopters and yields the drone market to startups, it loses control of the dialogue, and the drone startup has every interest in persuading the customers to cut back as far as possible on their use of helicopters and to replace them with drones.

Existing enterprises regularly consider starting new lines of business. Whether a new line of business is desirable essentially involves a decision to invest enterprise resources in the new line of business. Rational investment decisions, whether they are made within an existing firm or by an independent provider of capital, examine the same factors: do the factors indicate that the investment will produce a rate of return equal to or better than alternatives. Accordingly, senior management of an existing enterprise, whether it is a relatively small commercial helicopter or airplane operator, or whether it is a larger firm providing helicopter or other aviation services or a television network, needs the same kind of analysis suggested earlier in this chapter regarding the market, the competition, a specific product to be offered, financial projections, and a business plan.

So, whether one is in an entrepreneur deciding whether to start a new, independent business controlled by the entrepreneur, or an employee at any level of an established enterprise, the questions and the tasks to be performed are the same. The only difference is that the independent entrepreneur will be making presentations to outside investors, while the internal entrepreneur will be trying to persuade more senior executives who, in turn, may have to persuade the board of directors.

Notes

1 AUVSI, Snapshot of the First 500 Commercial UAS Exemptions (Aug. 2015).
2 Public Law 112-106, 126 Stat. 306, amending Section 3(b) of the Securities Act of 1933.
3 SEC, Amendments for Small and Additional Issues Exemptions Under the Securities Act (Regulation A), 80 Fed.Reg. 21805 (Apr. 20, 2015, effective Jun. 19, 2015).
4 730 F.2d 1128, 1130 n.2 (8th Cir. 1984) (approving in material part settlements of class-action securities fraud claims as part of bankruptcy reorganization).

12 Grasping the future

12.1 Introduction

Predicting the future is neither foolhardy nor avoidable; every entrepreneur has to do it. Sometimes, as in the case of major defense systems, projections have to look forward 20 years or more; in other cases it is sufficient to know what to expect in, say, three years.

In making and evaluating forecasts about aviation's future, one must be mindful that the principles of physics, the laws of thermodynamics, and the teachings of economics are unlikely to change. The core principle of Newtonian physics says that an aircraft with greater mass will always require more thrust to move it; the second law of thermodynamics means that some energy is wasted in any process; economics' law of diminishing returns says that exponential growth will not go on forever.

The future can be different from the present, not because these laws will change; it can be different because of breakthroughs in engineering, such as those highlighted in chapters 2 and 3, which enable new kinds of mission performance within the constraints imposed by theory. Then the regulated market decides which of the breakthroughs will be translated into successful products that expand operational possibilities.

This chapter builds upon the mission profiles in chapter 1, the analysis of technology in chapter 3, the assessment of economics in chapter 10, and the analysis of regulation in chapters 7 and 8 to project likely trends in product development and operations, in aviation generally, and for drones, in particular.

12.2 Determinants of the future

The future of drone use will be determined by a combination of technological innovation, economics, politics, public opinion, and entrepreneurship.

As chapter 3 concludes, breakthrough improvements in battery technology will extend the reach of electric propulsion systems into larger aircraft, dominating the drone field and beginning to appear in smaller manned aircraft as well.

Continued miniaturization and increases in speed and processing power of electronics will allow greater and more reliable autonomous operation of drones and manned aircraft. The regulatory burden to obtain certification for new avionics systems for drones will be eased as products are developed that can be certified for both drones and manned aircraft.

The politics of drones will sharpen as their operations become more pervasive. Significant accidents will occur, and the public will react, generally insisting on regulatory action to control what the public perceives—not always accurately—to be some underappreciated risk.

Litigation will become more frequent as victims of drone accidents try to establish liability by operators and manufacturers. The litigation will result in more hard data on economic exposure to mishaps and thus allow more reliable insurance underwriting decisions and the settling down of the market for liability insurance.

As drones are embedded in a wider variety of economic activities, conflict and pressures on regulators will sharpen. Economic interests favoring wider and more flexible drone use will develop arguments for relaxing regulatory constraints. They will especially challenge constraints borrowed directly from legacy regulation of manned flight that impose large economic burdens, chiefly irrational requirements for DROP training and certification, and detailed, protracted requirements for certification of microdrones. At the same time, more accidents and more intense public reaction will put pressure on regulators to tighten standards, probably with respect to operating rules.

As chapters 1 and 10 suggest, the drivers for a wider deployment of microdrones are different from the drivers for wider use of macrodrones. Microdrones can support activities for which support by manned aircraft is infeasible or undesirable because of safety risks, operating limitations, or cost. Macrodrone deployment will depend more directly on relative cost, compared to manned aircraft. It also will depend on the relative cost compared to ground-based equipment such as cranes for lifting, for example, shipping containers.

It is reasonable to segment the potential for wide use of drones into two clusters, one where their benefits are most obvious and largest in magnitude, while the risks are lowest, and a second cluster in which the benefits are less obvious and the risks greater. The risks are lowest in remote areas where drones are flown entirely or mostly over property owned or controlled by the operator. They are greatest in congested areas where flight occurs over other people's property. The benefits are most obvious for microdrones flying low level missions for close-up surveillance and image capture under conditions that exclude manned aircraft. The benefits for macrodrones are most obvious for macrodrones flying at high altitudes with flight times measured in the dozens of hours. Those flight times exceed those practically available in manned aircraft, because of limitations on aircraft endurance or human flight crew tolerance for long airborne missions.

The condition of labor markets always influences the propagation of new technologies. Factory automation exploded to reduce labor costs. Other economic activities are automated to compensate for labor shortages. The same thing will be true of drones. In some cases, business managers will prefer drones over manned aircraft, because they want to save crew costs. Crew cost savings will be greater when DROPCON technologies materialize that permit a single DROP to operate multiple drones.

The deepening pilot shortage may influence the pace of drone adoption. Data now available on the number of young people undertaking flight training and numbers of older pilots reaching retirement age make it clear that a shortage is developing and will worsen. Whether drones can compete depends on the labor-intensiveness of drone operations and on the relative attractiveness of DROP careers. As chapter 4 explains, DROP training and certification for some vehicles—especially microdrones—will be less intensive than pilot training and certification. For those jobs, the amount of money that a young person considering a career in aviation must invest to become a DROP will be substantially less than that required to become a pilot. Moreover, if the demand for DROPs exceeds the supply, salaries for DROPs will rise, perhaps above the levels for some pilots. If a young person choosing between a career as a DROP and a career as a pilot sees higher salaries and lower education costs, he will be drawn to the field, offset, of course, by the perceived romance of being in a real helicopter or airplane cockpit.

12.3 Social, economic, and environmental forces

Markets always are influenced by law, and law is driven by politics, which, in turn, is driven by public opinion. How market-oriented economies function always depends not only on economics and technology, but also on the social and environmental forces that shape decisions made by market actors, and on regulators' efforts to improve the functioning of markets and to mitigate harmful externalities.

The following subsections identify environmental forces, concerning energy, noise, privacy, automation and robotics, terrorism, and personal liberty that will shape regulation, aviation markets, and the role of drones within those markets.

12.3.1 Energy

Despite the fall in oil prices beginning in 2014, occasioned by technologies for extracting natural gas and petroleum from previously inaccessible reservoirs in the United States, concerns about energy use are sufficiently embedded in the public mind and in the strategic planning of aircraft manufacturers and purchasers that it will continue to shape aircraft design. Not only will the public continue to be concerned about undue dependence on hostile regimes in the Middle East, Russia, Venezuela, and elsewhere, a growing acceptance of the climate-change crisis will further increase pressure to reduce dependence on carbon fuels.

That will reinforce incentives to invest public and private money in battery improvements, which are necessary for all forms of alternative energy, from electrically powered vehicles, to effective use of solar and wind power. It also will continue to encourage aircraft manufacturers and engine manufacturers to improve the performance of aircraft engines, to expand the use of electric propulsion, and to reduce weight by greater use of composites.

12.3.2 Noise

Some of the most ferocious battles about aviation concern siting of airports and heliports in noise-sensitive areas. Excessive noise is a rallying cry for concentrated local interests who seek to restrict existing airports and oppose new ones.

Moreover, noise limitations are sufficiently embedded in existing statutes and regulations that aircraft designers will continue to focus a part of their R&D efforts on reducing noise, particularly aircraft-engine and helicopter-rotor noise. Some of the opposition to wider deployment of drones will undoubtedly be based on noise, but the principal effect of public sensitivity to noise will be a stimulus to develop new technologies that produce less noise. Already a significant share of aircraft-engine and helicopter R&D focuses on designs that reduce noise footprints. The Airbus fenestron on the EC135, resembling a ducted fan, and Blue Edge rotor blades with hockey-stick-shaped rotor tips on other Airbus models, are early examples of the results.

12.3.3 Privacy

Although much of the early opposition to civilian drones was fueled by concerns about invasion of personal privacy, the politics of privacy in United States suggests that privacy-based opposition to drones will not coalesce into an effective political barrier. Despite much alarm raised by privacy commentators, the public tolerates an enormous and growing store of personally identifiable data collected by Internet service providers and brokered to advertisers to enable targeted advertising. Similarly, the disclosure by Edward Snowden of the US National Security Agency's (NSA) routine collection of a substantial fraction of all emails sent and all telephone calls made in the United States engendered surprisingly little uproar. Every indication is that the public will tolerate substantial intrusions into personal privacy, as long as they believe substantial benefits exist—free access to a wide variety of resources on the Web, in the case of targeted advertising, and an edge in the battle against terrorism, in the case of NSA data collection.

As long as there is a credible story about how drones improve economic activity, consumer welfare, and public safety, privacy concerns are likely to stay in the background—or if in the foreground, not to affect policy significantly.

Only if some egregious example of a drone-based privacy intrusion receives wide publicity, is the public likely to react effectively, and even then, it is likely to focus, not on the vehicles themselves, but on particular patterns of misuse or the identity of the intruders.

Privacy, however, can be a code word for not being annoyed. In that respect, the opposition will be more powerful. Even if opponents do not articulate the basis for their opposition accurately, what matters is the level of opposition.

12.3.4 Terrorism

All of the evidence suggests that the threat of terrorism will get worse instead of better. The rise of ISIS, even as Al Qaeda has diminished, the increasing attenuation between terrorism and traditional states, the growing number of failed states, and the onset of domestic terrorists acting alone, will lead to an intensifying sense of vulnerability, without any new ideas on how to mount an effective "war against terrorism."

It's not surprising that the general concern about terrorism, bordering on paranoia, focuses on every new technology, fantasizing alarmingly about how terrorists can use it. Almost immediately after an intoxicated federal employee lost control of his DJI Phantom and it crashed on the White House lawn, the Secret Service and others were talking about how likely it is that terrorists will make use of microdrone technology—even at the Phantom level. Already, domestic security agencies around the world are exploring interdiction measures for hostile drones.

The reality is that drones, especially microdrones, are not particularly attractive tools for terrorists. Not many explosives will fit in a 2.6 pound gross-weight Phantom. In fact, resourceful terrorists can use almost anything as a weapon—a Boeing 767 as in the case of the September 11 attacks; fertilizer, as in the case of the Oklahoma City bombing; pressure cookers, as in the case of the Boston Marathon massacre. Restrictive regulation of all of these things in an effort to reduce the terrorist threat is nonsensical; it would shut down modern life. Still, antiterrorism measures are going to be designed around public fear as much as logic. Credible stories, however fanciful, about terrorism and drones will continue to sit in the political background. Obviously, if a terrorist actually uses a drone to carry out or to attempt an attack, that obstacle to productive drone use will intensify greatly.

12.3.5 Personal liberty and resistance to government

Suspicion of government and concerns about governmental encroachment on personal liberty long have been part of the political equation in the United States—and elsewhere. Jeffersonian and Jacksonian Democrats, Southern Secessionists, William Jennings Bryan Populists, demonizers of the New Deal, and, more recently, the Tea Party Movement, put most of their political energy into resisting expansion of governmental "interference" in private and commercial life.

The influence of these voices waxes and wanes over the decades, but these essentially libertarian concerns will be prominent for the next decade, at least. One effect will be to preserve a broader space for commercial drone operation than would exist in a political environment favoring greater regulatory intervention. An opposing effect may arise from a perception than any drone is mainly a government-sanctioned tool for trespassing on or over private property.

12.4 Cross-pollination of technologies

The civilian drone industry will benefit from technologies developed in other industries. Other industries will benefit from drone technology.

12.4.1 Migration of military technology and configurations to civilian markets

Historically, new aviation technology and concepts often have been pioneered by Defense Department-supported research and development activities. Radar, jet engines, and macrodrones are clear examples. Most large defense contractors also have one or more commercial lines of business. Boeing, Northrop Grumman, Sikorsky (now part of Lockheed-Martin), Pratt & Whitney, and General Electric are clear examples. Airbus inverts the model. It is primarily a commercial aircraft company with defense lines of business as well.

The synergy between defense and commercial aviation R&D originates in the fact that armed forces and intelligence agencies have large budgets to support development of new concepts and to buy new types of aircraft. While budgets have some upper limits, and therefore cost is a consideration, decision-making in the defense context is not driven by the need to project profitability and return on investment. The customer—the air force, army, marines, or navy, does not have to make a profit or attract private capital.

Equally important is the fact that military and naval operations must tolerate a higher level of risk than is acceptable in the civilian context. So they can extend the envelope of experimental technologies and experience from system failures and mishaps that would discourage purely civilian developers. It does not matter if DROPs occasionally lose control of Predators or Reapers as long as they accomplish their missions most of the time. A drone crash in a combat area presents considerations entirely different from those resulting from a crash in a peaceful civilian area.

Moreover, substantial portions of US airspace are reserved or restricted to allow military flight testing and training.

As US involvement in the Iraq and Afghanistan conflicts has wound down, defense R&D and procurement budgets have shrunk significantly, causing defense contractors to shift their strategies to place greater emphasis on commercial products. On the one hand, less money and fewer new projects decrease opportunities to explore innovative technologies, but they also provides stronger incentives for defense contractors to migrate what they have learned and will learn through defense-sponsored projects to the civilian sector.

This is already true in the macrodrone world (and for some larger microdrones): many of the novel products being pitched for civilian sale are adaptations of configurations developed with defense dollars. The K-Max, the WATT, and the Puma-AE are clear examples. The K-Max is an optionally piloted helicopter that proved its capacity to operate unmanned in Afghanistan while it ferried weapons and supplies to troops. Drone Aviation Corporation's WATT is a tethered drone intended for civilian applications, which draws upon its manufacturer's experience in designing and building tethered drones for the armed forces. Aerovironment's Puma-AE is a fixed-wing drone sold both for defense and civilian applications.

Not every such hope for migration is fulfilled, of course. Fifty years ago, Lockheed's efforts to sell commercial versions of the C-5 transport were unsuccessful. Now, the high price of civilian derivatives of some military and intelligence drone systems is startling, compared to products developed entirely with civilian dollars. It remains to be seen whether civilian demand for them will develop at those prices.

12.4.2 Migration of new manned aircraft designs

Manned aircraft engineers are innovating, slowed sometimes by the burden of FAA airworthiness certification for new systems. Sometimes, miniaturized, lightweight, low-power consumption systems developed for drones will interest manned-aircraft system engineers and provide data that can persuade the FAA to certificate them for broader aviation use. In other cases, however, innovation beginning in the manned aircraft sector will influence drone development.

Airplanes have looked pretty much the same since the 1930s, two wings with ailerons to control roll, pitch and yaw controls at the tail; engines in the nose or suspended from the wings in nacelles. Helicopters have looked pretty much the same since the 1950s: one main rotor with controllable pitch blades; one tail rotor to compensate for engine torque.

Microdrones do not usually look like manned aircraft. Miniaturization of electronics and improvements in battery technology and other components of electric propulsion systems have enabled the designs of microdrones that will influence manned aircraft design over the next 20 years.

In the macrodrone category, however, manned aircraft and unmanned aircraft design is quite similar, because manned and unmanned—or optionally piloted—versions of essentially the same aircraft fly similar missions. Accordingly, new manned aircraft configurations are quite relevant to macrodrone design.

12.4.2.1 VTOL

VTOL (*vertical takeoff and landing*) aircraft now in military and commercial service are attracting wider interest and will influence drone configurations. The terminology is misleading. By definition, a helicopter is a VTOL aircraft because it takes off vertically, but in professional usage, VTOL has come to signify aircraft other than helicopters that are capable of taking off vertically. Typically they use shrouded or ducted fans in their wings or tilt their wings or rotors to direct thrust downward as well as backward. The V-22 Osprey is an example of a tilt-rotor design, imitated by the commercial AgustaWestland AW609. The Harrier and the newer F-35B achieve VTOL capability by deflecting jet thrust downward.

Tilt-rotor configurations will receive a boost as the Agusta-Westland AW609 enters service with Bristow, one of the largest global oil and gas platform helicopter operators, and attracts additional customers. This will put tilt-rotor configurations on center stage for customers. The AW609 and imitators will continue to be significantly more expensive than commercial airplane and helicopter designs, and the market will sort out which missions justify the additional cost.

Agusta-Westland promotes the AW609 as offering new flexibility and productivity for oil and gas support, because its airplane capabilities enable it to reduce trip time substantially while its helicopter capabilities enable it to operate from drilling and pumping platforms and existing heliports on land. It promotes it for EMS, because it can not only rescue accident victims from confined and unprepared areas; it also has the range and speed to perform long-distance transfers. It is not yet clear why the same vehicle needs to do both, however, because immediate trauma transport typically is to the nearest trauma hospital, while transfer to another hospital occurs later, after the victim has been stabilized.

Agusta-Westland also promotes the vehicle for executive charter, because of its capability to operate from almost anywhere and fly cross country to the destination almost as fast as executive jets.

XTI Aircraft's TriFan 600 used a crowdsourcing campaign to raise $50 million in investment capital to begin development. The preliminary design features two turboshaft engines driving three ducted fans, two of them tiltable to power forward flight. XTI is promoting the five-passenger aircraft for high-end charters, private ownership, and EMS. It features a maximum cruise speed of 340 knots and a range of 800–1,200 nautical miles, fly by wire control systems, and the ability to take off and land like a helicopter.

The concept is beginning to show up in drones. The Aerovel Flexrotor is a hybrid VTOL/fixed wing configuration weighing 45 pounds, with a wing span of 3 ft and a length of 2 ft. It is capable of flying for up to 40 hours on ocean and land surveillance missions. Its payload accommodates an electro-optical camera. Further engine development may permit it to carry heavier payloads and other kinds of sensors. The vehicle's single prop permits it to take off and land vertically. Once in flight, it deploys an empennage, dives to gain forward speed, and then transitions to horizontal flight, functioning like any fixed-wing airplane. It is powered by a small gasoline engine.

Its vendor[1] aims it at the civilian market by stressing the lack of ground infrastructure and affordability. Its first commercial customer, Precision Integrated, headquartered in Newberg, OR, is a helicopter operator beginning to offer drone support.[2]

Significantly, Amazon's public presentation of a preliminary design for packaged delivery features a hybrid VTOL airplane/helicopter design.

12.4.2.2 Propfan propulsion systems

Propfan propulsion—mostly a new name for turboprops with better propeller blade design and improved coupling to more powerful turboshaft engines—will dilute the longstanding preoccupation with fanjets and generate technology that will expand the options for macrodrone designers.

12.4.2.3 Hybrids

The incrementalism of improvements in battery technology, combined with the attractiveness of electric propulsion, will stimulate interest in hybrid aircraft power.

Because battery technology is such a pervasive limitation on realizing the full advantages of electric propulsion systems, broad experimentation with regard to hybrid propulsion systems is likely. Hybrid systems long have been prominent in some industries; the diesel electric locomotive is a paradigm. It eventually eclipsed pure electric motive power in the railroad industry. It appears that many major automobile manufacturers are concluding that hybrid approaches are superior to all electric vehicles in the automobile and truck markets.

Airbus, which long has built its commercial strategy around technological innovation, is emphasizing hybrid approaches in its development of electric aircraft. Airbus is investing substantial capital in a family of hybrid aircraft, both rotary and fixed wing. Whether or not it stays exactly on schedule, it will introduce some of the smaller versions into the trainer and drone markets in 2016 or 2017. This will attract growing customer attention and generate data about operational reliability, flexibility and costs. Depending on the vehicles' actual fuel consumption, range, and endurance, these configurations may become attractive alternatives to more conventional designs, or they may be sidelined as technological curiosities.

Early concept development of interest to the airlines involves one large turbofan or turboshaft engine that provides only part of the thrust while turning an alternator that recharges batteries that drive electrically powered fans that provide most of the thrust, especially for takeoffs.

12.4.2.4 Enhanced autopilots

The frontier of autopilot capability will continue to advance. Now, category III autopilots, capable of flying an entire flight profile, from departure to landing and ground taxi afterwards, are installed on a substantial part of the air transport fleet and airports. As their cost comes down, they will also be installed on more of the existing inventory of airplanes and airports and become a regular option for new purchases.

Three-axis autopilots for helicopters, initially excluding the takeoff and landing phases, will penetrate more deeply into commercial helicopter fleet, beginning with larger helicopters for oil and gas support, and gradually expanding into single-engine light helicopters. This will have the effect of reducing the accident rate for inadvertent flight into IMC, chiefly in the law enforcement and EMS arenas.

Optionally manned aircraft and greater reliance on automation of manned aircraft will pave the way for eventual broader acceptance of macrodrones. In an August 2015 incident the crew of a Delta Airlines flight used the autopilot to escape a likely disaster. The aircraft was severely damaged by hail to the point that the flight crew could not see through the windshield. The crew produced a safe outcome by relying on the autoland capability of the aircraft. This is the kind of experience that will support arguments that optionally manned aircraft can operate safely.

12.4.2.5 NextGen navigation systems

NextGen is not a prediction; it is a mandate. By 2020, all manned aircraft (with a few relatively unimportant exceptions) will be equipped with ADS-B out, and the utility of much-cheaper ADS-B in will encourage its ubiquity as well. ADS-B is a system that permits aircraft to transmit position, flight speed, and direction data to each other while in flight, so that they can avoid each other automatically.

Prices for systems developed for smaller general aviation aircraft already have fallen into the $2,000–$3,000 range and will fall further by 2020. The expansion of drone fleets and the FAA's quest for systems to allow safe beyond line of sight (BLOS) drone flight will encourage developers to reduce weights, and power consumption so that they can take advantage of the exploding drone market.

ADS-B out-equipped drones will be electronically visible to manned aircraft equipped with ADS-B in. ADS-B in-equipped drones will be electronically aware of the position of manned aircraft in the vicinity. While prices, weights, and power requirements may not fall far enough to make installation of ADS-B in and out systems feasible on smaller microdrones, operators who want to operate larger microdrones BLOS and macrodrones BLOS and at higher altitudes will recognize, along with regulators and the traditional aviation community, that ADS-B out capability on a drone makes detection and avoidance a near certainty. Installation of such systems is a straight-forward way to inoculate against concerns about collision avoidance that otherwise will delay or prevent many useful missions. The cost will become tolerable for operators who seek to perform those missions.

ADS-B in capability is cheaper, lighter, and requires less electrical power than ADS-B out. For relatively short-range control and telemetry links, latency will not interfere with a DROP's improved ability to anticipate the need to yield right of way to manned aircraft in the vicinity.

The end result of integrating drones into NextGen will sweep away many risk-based objections to opening up the airspace more widely to drone flight alongside airplanes and helicopters. NextGen data communications will gradually replace AM voice radio communication for delivering and confirming ATC clearances. This will make it easier to integrate drones.

12.4.3 Migration of drone technology to manned aviation

Weight, power consumption, and costs of increasingly sophisticated navigation, control, and communication systems deter the deployment of new technologies in manned aircraft. Regulatory requirements for FAA certification of such systems increase all of these detrimental attributes, as prescriptive, protracted, and costly regulatory requirements seek to verify the safety of new technologies. The FAA, other policymakers, aircraft designers, pirates, and operators constantly arm wrestle over how to streamline the regulatory requirements.

The consumer electronics revolution has already had an impact on manned aviation. Virtually every pilot carries one or more iPads or other tablet computers to run navigation software such as Foreflight, even though the equipment is not certificated by the FAA. The FAA has been reluctant to allow consumer-products not certificated for aviation use to be installed on the aircraft to meet navigation-system requirements.

The spread of drones may demolish some of these barriers. Almost everyone agrees that it is infeasible to subject microdrone navigation and control systems to traditional airworthiness and type certification; yet thousands of people are flying those systems safely with the aid of miniaturized consumer-grade electronics. Increasingly, helicopter and airplane communities are going to demand better answers as to why they cannot use the latest drone-proven technologies to meet regulatory requirements to operate their aircraft.

The momentum of efforts to allow partially or completely autonomous motor vehicles, considered in § 12.4.5, will accelerate this phenomenon.

12.4.4 Migration of developments in photography

Consumer demand for high-resolution photographs and video in affordable multi-function cameras will continue to press innovation by camera manufacturers forward. Each new model of DSLR and of action cameras like the GoPro will have better resolution, better color fidelity in low-light conditions, better streaming and remote control capability, and weigh less. Remote zooming of lenses is growing in popularity. This pattern of innovation by suppliers already well-established in the markets will expand the quality of what drone-mounted cameras can do.

Professional cinema cameras will decrease in price, weight, and size. The BlackMagic Design Pocket Cinema Camera (BMPCC) is a sub $1,000 camera capable of recording in a RAW codec. Conventional cameras with the same RAW capabilities cost upward of $5,000. BlackMagic Design recently released a compact drone camera, the Micro Cinema Camera. This camera weighs only 11 ounces and has remote control capability, a built in output for video downlink, and the same RAW codec as the BMPCC for less than $1,000.

Present-day limitations on video technology relate more to the infrastructure necessary to deliver imagery already available from leading-edge cameras. Compression and bandwidth constraints frequently decrease resolution of transmitted images before they arrive at the destination. 4K has become the watchword of the high end of the market, but most television distribution systems cannot yet handle it. Only a fraction of homes have 4K receivers and displays. All this will change, but may continue to lag camera innovations.

Prices and weights for IR and laser sensors will continue to fall, encouraged in part by the growing demand for that kind of imagery captured by drones.

Apart from hardware, the impetus to improve collision avoidance software—especially for drones—will lead to substantial improvements in image processing software. For example, further work is necessary to extend the now error-prone and clunky algorithms for scanning a raster image captured by a camera or a laser sensor to decide whether an object is a bird, a plane, a helicopter, a tree, a wind turbine, or a birthday balloon.. These developments will boost efforts to develop reliable and affordable collision avoidance systems, enabling drones safely to operate beyond the line of sight of the DROP.

Onboard processing power and battery life may limit the complexity of image analysis. Faster and more energy efficient processors could solve this issue. Alternatively, drones may rely on cloud computing by uploading the image to a "cloud server" that solves the complex mathematical formula of image analysis in real time and forwards the results to the drone.

An open source Python library, OpenCV, could be adapted for this purpose. OpenCV is a facial detection program that performs real-time image analysis. To avoid the requirement of enormous computational power, this program uses cascading functions that break up the image into larger blocks and move on to the next block if the program does not find general signs of a face instead of performing an in-depth analysis on each block. OpenCV learns by analyzing images with faces and without faces. By this principle, the software could learn what a drone is by feeding the algorithm images of drones and images without drones.

12.4.5 Migration of autonomous ground vehicle technologies to aviation

The momentum of motor-vehicle automation already is considerable. Major automobile manufacturers—and Google—are beginning to demonstrate the safe operation of fully automated automobiles and trucks, albeit mostly with safety drivers. NHTSA is encouraging the process.

Significant investment in autonomous driving capabilities for passenger automobiles and trucks, already visible in piecemeal implementations on newer, high-end models, will result in greater reliance on the necessary subsystems and a more hospitable legal and regulatory environment for them. Self-driving cars and trucks are already emerging from laboratories and test tracks to operate on regular streets and highways. California, Nevada, and the District of Columbia have legalized them.

The authors test drove a Tesla Model S equipped with an autopilot on 18 December 2015. The autopilot did a good job of speed and directional control as long as it could see lines marking lanes, but had difficulty with exit ramps and left turn lanes, where the lines were interrupted. It similarly offered precise control of the following distance in heavy surface-street traffic, bringing the Tesla to a complete stop when the car ahead stopped at traffic lights and starting up again when traffic began to move. It could not, however, navigate a preplanned trip. Tesla is collecting live data from all of its vehicles on the road and expects to expand autopilot capabilities as its geospatial database becomes more complete.

As the systems prove themselves, automobile manufacturer advertising budgets and interest by the general press will keep them prominent in the public mind. The advances in sensors, navigation algorithms, and collision avoidance and stationary object identification, will facilitate miniaturization and certification of motor vehicle systems in aircraft. Proponents of such systems for aviation will be able to call on test data generated by certification of highways systems.

Equally important, as autonomous driving systems prove their potential to improve safety, public perception of automated transportation systems will change. As a result, both the general public and regulators will react more favorably to claims that a drone is as safe as—or safer than—a manned aircraft because it has automation even better than that available on the family automobile.

The large sales base of passenger automobiles supports a level of research and development and reliability testing far beyond what the aviation community, including its drone component, can afford. As fully autonomous automobiles become acceptable, the technology employed to make that possible will migrate to the aviation world, both reducing the cost of the components of autonomous safety systems and increasing their acceptability.

12.4.6 Migration of maritime technologies

For a variety of reasons, automated cargo ships may sail before the full range of micro-and macrodrones are integrated with the National Airspace System and before genuinely self-driving cars crowd the highways. The technology for autonomous ship navigation already exists, and is it used for specialized ships that carry cargo to offshore oil platforms. The legal barriers to open-ocean operation of automated cargo ships are few, because the oceans, beyond national territorial limits of 3 to 20 miles, are not subject to regulation by any nation state. The treaty framework for open-ocean maritime commerce focuses on specific matters, such as piracy, and does not cover general operations. Beyond that, public opposition to automated maritime commerce is likely to be muted, as long as only manned ships enter territorial waters and harbors.

12.5 Consumer markets

The central focus of this book is commercial use of drones. But one should not forget that the commercialization of microdrones began with recreational users and a combination of model airplane hobbyists, amateur photographers, and people interested in the latest technological toys. Together, they represented a market big enough to buy hundreds of thousands of drones within a year after Jeff Bezos' famous 60 Minutes interview. This demand produced enough revenue for multiple competitors to make startling advances in microdrone technology within that year. Automatic hover, automatic takeoff and landing, automatic flight plan execution, and much improved gimbals and video downlinking technology now are common in entry-level vehicles.

The amateur photography market is huge, as sales of DSLRs and the pace of innovation in that marketplace attests. The reality is that microdrones are a valuable addition to any photographer's toolkit, enabling her to get points of view that simply are not available from the ground. Microdrones equipped with good cameras are not much harder to operate than high-end DSLR cameras, and cost about the same as a good DSLR or a good zoom lens.

The demand for microdrones will continue to be strong for this market segment, which simply wants to take good pictures and video. Some of them will sell their work product, but that will be opportunistic and infrequent. Essentially these microdrones will keep the Vimeo, YouTube, Instagram, and Flickr pipelines full. At the end of 2014, 300 hours of new YouTube videos were being posted every minute.

This phenomenon is essentially noncommercial. In journalism, moviemaking, and music, the low barriers to entry enabled by the Internet have made it clear that millions of talented people are willing to put substantial effort into creating things and making them available for free. Much of the open-source software movement, and thousands of "how to" videos on YouTube offer proof of this phenomenon.

These users will fuel continued competitive pressure to improve microdrone technology—and provide plenty of money to fund the necessary development.

But the most rapid advances will be in camera, gimbal, downlink, control, and navigation system development rather than in the drone vehicles themselves. Zoomable cameras lenses will become commonplace, as well as reliable systems to keep the camera pointed at its subject regardless of drone movement, with "follow me" functionality. Flight plan execution systems will become far more reliable, and integrate GPS and IMU data so variations in GPS signal do not disorient the navigation system.

And, of course, the pressure will be on everyone to improve battery technology, although incremental advances are more likely for the foreseeable future then dramatic breakthroughs.

12.6 Airspace management

The numbers of drones and drone flights will overwhelm the air traffic control system designed for smaller numbers of manned aircraft. Something like the NAMID concept, described in chapter 3, similar in its material respects to what Amazon has proposed, will be implemented. This will allow microdrones to operate below 400 ft AGL in a matrix of Internet-linked command and control and geospatial databases, which direct the flights of drones that navigate without outside aid—except for GPS information—and maintain traffic separation by onboard peer-to-peer data communications.

One of the major development challenges for long-distance macrodrone operations is to figure out how to manage latency in a BLOS control link. As the 2015 ICAO report on RPAS says, "The distinction between RLOS [radio line of sight] and BRLOS [beyond radio line of sight] concerns whether any part of the communications link introduces appreciable or variable delay into the communications."[3]

12.7 Choosing between helicopters and drones

The arena of competition mostly involves a choice between helicopters and drones. Only a few fixed-wing applications will require endurance of the dozens of hours where macrodrones excel such as the German/Qatari Q01. There also may be a handful of missions where aircrew fatigue or boredom would make macrodrones attractive. Otherwise, fixed wing aircraft will continue to perform the same missions they do today, because their acquisition and operation costs will be lower.

In the rotary wing arena, two realities will become manifest: First, microdrones will perform many new aerial support missions not performed at all now because the cost and capability of microdrones shift the supply curve so far back on the demand curve. To some extent, the lower-cost will cause customers to substitute microdrone support for specific types of helicopter support, where endurance and transit time and speed are not major barriers to the attractiveness of unmanned aircraft.

Otherwise, the competitive battle will be waged between helicopters and macrodrones. In this arena, it's hard to imagine that macrodrones will ever cost less than manned helicopters because the vehicles and their control systems must be more complex to deliver the requisite reliability and adaptability, given the unavailability of an onboard pilot to intervene if something goes wrong. And operators questions whether they should pay more for a macrodrone. "The bottom line, will the government and business community be willing to pay for the new technology?" one operator said. "Why would I make the significant investment to be unable to compete at a competitive price?" said another.[4]

But some aircraft missions put the lives of their crew in jeopardy. Tower maintenance, hazardous spills monitoring, and delivery of relief supplies are examples. These are activities where operators will be willing to pay a higher price for macrodrones in order to eliminate the risks to the crew, either for moral reasons, or because of differences in insurance costs, as underwriting for drones matures.

As is appropriate in a market economy, the choice between drones and manned aircraft for missions they both can perform will be dominated by economics.

For the most part, as technology advances and the regulatory environment crystallizes, almost any mission that a manned aircraft can fly a drone will also be able to fly. In some cases, the danger to aircrews will tilt the choice toward drones as it has in the military and naval contexts. But even here, cost matters—a lot. Some missions exist where a microdrone costing less than 1 percent of a helicopter's cost will enjoy enormous economic advantages, and customers will be tempted to work around endurance, range, and altitude limitations.

In other cases, likely with macrodrones, the development costs associated with airworthiness and type certification and troubleshooting and maintenance costs associated with far more complex navigation and control subsystems for macrodrones may drive their price up to the point that most customers may prefer simply to use manned helicopters. A well-trained human pilot can do lots of things by instinct. He has the innate capability to adapt to the unexpected—a quality that would have to be programmed painstakingly into automated subsystems at enormous cost. Even then, some questions always would remain about reliability and unpredictability of automated system performance in unprogrammed emergencies. Customers are going to be less excited than engineers and aviation journalists when a drone shows that it—mostly, maybe—can do as good a job as a pilot. Customers mainly will be more interested in how much it costs.

HAI, the main helicopter trade association, argues that helicopter operators will add drones to their fleets, just as many of them now offer a mix of helicopters and fixed wing aircraft.[5] "The most valuable rotorcraft pilot in the future might be one who can climb out of the cockpit, pick up a UAS controller, and start another mission."[6]

12.8 Market structure

One set of questions relates to industry and labor-market structure. Whether commercial drone support is offered by specialized firms, or is integrated with helicopter and airplane services offered by existing firms is a subject explored in chapters 10 and 11.

12.8.1 Vehicle production and certification

Even if the FAA takes a restrictive approach limiting the market for commercial drone operations of the United States, the international market will be robust. Major vendors will continue to improve their products and sell them aggressively, without caring too much whether the purchasers intend to use them for hobbyist purposes or for purposes the relevant regulators consider to be commercial. There is not much the US can do to dry up the supply, because so many of the vendors are overseas.

The economics of microdrones will be worked out with vehicles not having airworthiness and type certification, as other, more sophisticated models begin the trek toward certification. By 2017 or early 2018, type-certificated vehicles will enter the market and they will have a significant regulatory advantage, although their higher cost will mitigate the advantage to an uncertain extent.

12.8.2 Labor markets

It is unclear how much the pool of commercial DROPs will overlap with the pool of helicopter and airplane pilots. Some experienced helicopter pilots will embrace the idea of learning a new flying skill and expanding their range of job opportunities. Others will, for reasons of inertia, or because they are attached to being in the cockpit, eschew the possibilities.

The result may be different for people just beginning their aviation careers. For them, the possibility of becoming a DROP represents an alternative to becoming a pilot. In some cases, they may persist in obtaining pilot qualification but seek DROP qualification as an additional rating, much as some commercial helicopter pilots also seek airplane ratings.

A pilot and mechanic shortage is developing for airplanes and helicopters, expected to grow more acute over the next 15–20 years. A big reason for the diminution in supply is that young people seeking exciting technology-focused careers, not touched by the increasingly distant romance of the early aviation pioneers and deterred by difficulty and cost of learning to fly, are going to Silicon Valley instead of to flight schools.

The growing pilot shortage will intensify over the next 10 years, and the result will be more opportunities for lower time pilots, and improved compensation. That may lessen the pressure on pilot aspirants to hedge their bets by obtaining DROP qualification as well. It is too soon to tell how strong the relative demand will be for DROPs. It may rival or exceed the demand for pilots. It may be that the demand for this new type of airmen will also bid up compensation. In any event the traditional commercial airplane and helicopter requirement for more than 1,000 hours total time and experience in particular types of aircraft will not restrict the demand for DROPs.

Pilot preferences also will be influenced by idiosyncratic factors. Some pilots are drawn by the physical and instinctive aspects of flying-flying an airplane or helicopter by feel and muscle memory, with only as much reference to instruments and automated systems as the regulations or the profile of a particular flight require. Others are drawn by the technology and get more gratification from mastering and precisely operating complex technological flight control and navigation systems than from hand flying the aircraft. The former group will be repelled by drones; the latter group will be drawn to them.

The wider deployment of drones may help relieve the pilot shortage for two reasons. As § 12.10 explains, wider use of drones will diminish the demand for certain uses of helicopters and may also diminish the demand for certain specialized airplane operations. More significantly, new entrants to aviation community who are not interested in aviation per se, but rather in photography or information technology, will face lower barriers to entry to aviation through drones. These people, who grow up naturally using the airspace for a particular purpose, will step into other aspects.

As young people motivated more by electronics, computing, photography, and image processing, than by the image of the barnstorming aviator with his white scarf blowing the wind are drawn to flying drones, this new segment of the aviation community may become an unanticipated supply of pilots and mechanics for airplanes and helicopters. Their proximity to the traditional aviation community will expose them to the technology challenges in crafting the future of the airplane and helicopter industry, shaped by electronics, computer, photography and image processing technology as much as drones and increasingly borrowing those technologies from the drone industry. Some of them will be drawn to working on manned aircraft, as well—as pilots, engineers, managers, and entrepreneurs. They will start with drones and move to more sophisticated and therefore interesting helicopters and airplanes.

Established flight schools, now offering DROP as well as pilot training, and newer specialized DROP training programs will proliferate. This opens up new opportunities for pilots and aspiring pilots with DROP experience to give DROP instruction.

But all of this may prove unnecessary. Just as radio amateurs and model aircraft flyers learn from one another and from practical experience, so will microdrone DROPs, whether or not they have formal DROP training in a flight school or university environment is unnecessary to fly a microdrone well and safely.

12.9 US regulatory environment

The FAA's final general rule for microdrones, largely reflecting the content of its NPRM, simplifies the regulatory requirements considerably. No longer must commercial operators contemplating routine microdrone missions shoulder the burden of specific petitions for exemption.

Eventually, the FAA will respond to pressure to relax LOS and daytime-only limitations. It also will permit certain operations over people. Depending on accident experience and on the configuration of political pressures, the agency will experiment with relaxing some of the limitations, for particular classes of operation.

Then the question is: what will the level of noncompliance be? Hundreds of people will continue to buy microdrones and fly them, in ignorance or defiance of FAA requirements. The FAA will commence enforcement actions against a few, but only a small fraction will be called to account.

Noncompliance will predominate among individuals—freelance photographers, for example—and very small businesses. As the size of an operator increases, so does its risk-averseness because it has more at stake. Larger operators are less likely to be deterred by airman certificate, visual observer, and detailed operating rule and handbook requirements, so more of them will go the legal route.

Even under the FAA's final rule for microdrones and initial proposals for macrodrones, the legal and regulatory environment will remain murky, well into the 2020s. As the use of drones grows, grass-roots pressure will grow commensurately for state legislatures and municipal legislative bodies to restrict their use. In some cases, this will be driven by public reaction to actual accidents; in others it will be driven by exaggerated reports of drone interference with manned aircraft and by vague concerns about invasions of privacy. In this context, privacy will be broadly defined to include peace and quiet in public spaces like parks and beaches.

At least some state and local legislative bodies will respond, and often their responses will not be carefully crafted. As chapter 6 explains, concentrated interests trump diffuse interests in the political process, and anti-drone interests will be concentrated at the local level, while pro-drone interests will be diffuse, simply because no drone interest group will ever have the resources to cover every one of the 40,000 municipalities in the United States.

The pro-drone interests do have the resources to litigate the permissible scope of every state and local regulation, and they will pick test cases and file lawsuits challenging more extreme state and local measures. Depending on the outcome of these cases, and the degree to which the challenged state and local regulation interferes with commercial microdrone uses that are, on their face, safe, momentum may develop for the FAA and the Congress to redefine the boundaries between federal and permissible state and local regulations to make them clearer. The course of legislative initiatives is even more unpredictable than the course of litigation, and drone operators will be forced to operate in an uncertain legal environment for the next 5–10 years, even after the FAA begins to put final rules in place.

12.10 Effect on other sectors of the aviation industry

As many other sections of this book argue, microdrones will mostly make aviation support available in parts of the economy that cannot be served by helicopters and airplanes because of cost or unacceptable risks to aircrews. Macrodrones will compete head-to-head with manned aircraft based on price and performance.

Beyond the role of drones in complementing manned aircraft for the uses highlighted in chapter 1, other, more ambitious applications may lie in the future. Logistics is an example. As airspace management systems discussed in § 12.6 are proven, small-package delivery in certain parts of the country will become more common. This is not a function presently performed by helicopters, so the effect on the market for helicopter services will be modest. But unmanned cargo aircraft are making their debut in the military context, and they will spread to the civilian market. One can imagine Federal Express, United Parcel Service, and DHL beginning to use unmanned freighters on some of their routes.

Amazon's proposal for segregated drone airspace is already stirring up a firestorm of controversy, because it suggests that helicopters and airplanes would be excluded from this low-level airspace. Nevertheless, although compromises in the proposal are certain, some form of segregated airspace with peer-to-peer collision avoidance, as Amazon proposes—essentially the same NAMID concept set forth in chapter 3—is certain to be developed in the next 5–10 years.

The result will be a requirement for manned aircraft to carry equipment that cooperates with NAMID.

The only sectors relatively immune from drone competition will be those carrying passengers: rides and tours, off-shore oil gas support, EMS, and high-income charter.

12.11 Long-distance cargo transport

Unmanned air cargo flights will occur earlier than widespread microdrone package delivery. Air cargo aircraft can operate in the existing system for controlling IFR flights when autopilot technology is adapted. ATC communications with the DROP can be handled relatively easily. The necessary development relates mostly to certificating reliable control links. Over-ocean flights, particularly, present few safety risks.

ICAO agrees. Its 2011 report on drones says, "Larger and more complex RPA—able to undertake more challenging tasks—will most likely begin to operate in controlled airspace where all traffic is known and where ATC is able to provide separation from other traffic. This could conceivably lead to routine unmanned commercial cargo flights."[7]

Microdrone package delivery, on the other hand, requires development and deployment of a complete new system such as NAMID for low-level airspace management and sense-and-avoid. That will take some time. NAMID is explored in chapter 3.

12.12 People onboard?

Most observers of drone development, including the authors of this book, believe that drones will not carry passengers anytime in the near future.

Nevertheless, the public does accept autonomous transport of people in some limited circumstances. No one objects to riding in an automatic elevator; few remember the time when almost all elevators had human operators. Airline passengers do not balk at riding on airport trams which have no operators. But these vehicles do not move in three dimensions; they have fixed guideways—vertical for elevators; horizontal for airport trams.

Innovation is driven by imagination, and there will be a growing number of engineers that imagine the day when drones will carry people, especially as drones become commonplace performing other functions.

Concrete exploration of the possibility will begin in the recreational context. There is demand for adventure and high risk sports, like bungee-jumping, some types of extreme rock climbing, hot air balloons, and skydiving. It is, therefore, conceivable that someone will offer drone rides in carefully limited circumstances, maybe just up and around the traffic pattern at an airport or theme park. As that occurs in a few places, confidence will grow in the safety of passenger transport—at least if the experience is relatively accident free. Then a handful of operators will deploy drones for passenger transport, most likely in specialized, large budget markets such as oil and gas support, where the passengers can be assigned to drone flights as a part of their job. Early experimentation and demonstration also is more likely to occur in optionally piloted aircraft, where data will be available from the pilot as to how much she has to intervene. Depending on cost and safety records, it could spread from there, likely not reaching general passenger transport until the middle part of the twenty-first century.

Except for life-threatening emergencies, such as some rescue operations, the likelihood is low that anytime within the next decade, the public or customers will be willing to accept widespread drone transportation of people. Even though pilot error figures prominently an accident causes, automation failures figure prominently in military drone mishaps. In any event, the widespread perception that one is safer with a pilot on will be hard to dislodge, no matter what the objective data suggest.

At the very least, statistics for drone accidents will have to be better than statistics for manned aircraft accidents before any argument exists that drones carrying people should be tried out.

12.13 False starts

Several parts of the book make the point that microdrone deployment already is common, while most macrodrone deployment lies in the future. But that observation refers to the eventual steady state; it doesn't consider how it develops over time. The pattern is clear under the section 333 exemptions. It is working just like the spread of any new technology. Vendors and entrepreneurs promote their products, trying to make sure that the message reaches those most likely to become customers. A few bold potential users agree to try it out, and then they and their vendors work hard to gain publicity for their activities. Print and online trade magazines write about the early adopters—they already are—and word spreads in each industry. Skeptics exist in every industry, and even the most attractive new technology spreads only as the skepticism is gradually overcome.

Not every new idea works, and what may seem like a great application may turn out to cost too much, be too much trouble to implement, or produce substandard results. If any of those things happen, the technology will stop spreading, and users will gradually abandon it.

This will surely be true for some drone uses. If anyone knew which ones and if there were a consensus, the false start would not occur in the first place. Experience is the best teacher, and market economies provide for considerable experience.

12.14 Optionally piloted aircraft as a transitional path

Optionally piloted aircraft represent an attractive transitional pathway from manned aircraft to unmanned aircraft, especially for missions for which major questions exist about whether the risks of unmanned aircraft are acceptable. Optionally piloted aircraft would help operators build confidence in the systems designed to fly the aircraft without human intervention. A low frequency of pilot intervention to correct for automated system errors would reduce concerns about flying the vehicle without a pilot on board. Conversely, if experience shows that regular intervention is necessary, the aircraft can remain in service with a pilot on board for most missions. In effect, it would represent a point in technological evolution where the aircraft has an autopilot that is more automatic then presently is typical.

The disadvantage is that engineers would have to design an aircraft that meets the requirements both of a drone and of a manned aircraft, which might result in greater weight than for either a drone or a manned aircraft, and one that has to satisfy safety standards focused on the risks of undermanned operation as well as standards focused on the risks of manned operation.

But this is the most plausible transition for unmanned technology designed to carry passengers, and may also be necessary to achieve popular acceptance of large-capacity cargo operations flying congested routes.

12.15 Engines of change

Disruptive technologies often have effects far beyond what their early adopters anticipated. Railroads turned out to be useful, not primarily to link one steamboat port to another, but to weave a transportation web across the whole country, making steamboats obsolete except as tourist curiosities. The telephone proved popular, not as a mechanism for receiving music, as its inventor, Alexander Graham Bell, first anticipated, but as a ubiquitous mechanism for person-to-person communication, eventually making telegrams obsolete. Transport aircraft turned out not only to be vehicles for handfuls of the rich, able to fly mostly when the weather was good, but to be the mainstay of modern all-weather transportation for everyone. The jet aircraft engine turned out not only to be a boon for high-performance fighter aircraft, but to represent the propulsion system of choice for most airplanes and helicopters.

Now come the drones. Their ability to carry sophisticated imaging equipment aloft in less than 5-pound vehicles, to hover in one position despite the wind, to fly simple missions, and to take off and land automatically, all with little more than operator oversight, at prices a close to $1,000, qualifies them as a disruptive technology by anyone's definition. The capabilities of the microdrones, now in operation by the hundreds of thousands, mark a path for the evolution of their technologies into larger configurations, which already have captured the imagination of military commanders and planners.

That their availability will make aviation support available where it has never penetrated below is indisputable. How that will play out for industries using them to improve their business models can be sketched, but only experience can fill-in the outlines. How they will transform manned aviation can only be guessed at now, but the possibilities for optionally piloted, single pilot, and completely unmanned aircraft to carry packages and larger items and then, eventually, to carry passengers, is plausible.

Their novelty already is stirring up a political firestorm in many quarters between those who are ecstatic about drones' potential and those who are horrified about the threats they post to life and limb and to privacy. It is too soon to tell how operational reality will balance risk and utility. It is certain that political conflicts over their use and efforts to craft a sensible regulatory framework will challenge the creativity of advocates, lawyers, regulators, government officials, entrepreneurs and engineers.

Profound developments around the world have been made possible by aviation's first 120 years. A walk through an aviation museum a century from now will be just as fascinating as it is today, and the drone exhibit will be a substantial part of the museum.

Notes

1 www.aerovelco.com/flexrotor/.
2 www.flyprecision.com/integrated.
3 *Manual on Remotely Piloted Aircraft Systems* at p. 2–2, www.wyvernltd.com/wp-content/uploads/2015/05/ICAO-10019-RPAS.pdf.
4 Jen Boyer, *Unmanned K-Max Completes Firefighting Demo*, Vertical, Dec. 2015/Jan. 2016 at pp. 30, 31–32 (quoting aerial firefighting contractors).
5 Douglas Nelms, *Unmanned Aircraft: The Next Big Thing for Your Business*, Rotor, Fall, 2015 at 22, 23.
6 Ibid. at 27.
7 ICAO, *Unmanned Aircraft Systems*, Cir 328, AN/190, § 3.15, at p. 8 (2011), www.icao.int/Meetings/UAS/Documents/Circular%20328_en.pdf.

Glossary

3-D three-dimensional.

AADI American Association of Drone Instructors—a non-profit organization offering training and certification of DROPs.

Accelerometer a device that measures acceleration along each of three axes.

Accredited an entity that has received direct or indirect governmental approval to engage in its activities, especially pertaining to educational and training institutions.

Accrual a principle of financial accounting that spreads expenditures or revenues over the periods in which the costs are incurred or the income earned.

Adjudication a formal legal process intended to apply the law to the facts associated with a particular case.

ADM Aeronautical decision making—application of principles of judgment to factors influencing safety of flight.

Administrative Appeals Tribunal an administrative agency in some countries responsible for appellate review of initial decisions by administrative agencies.

Administrative law a body of law that limits the authority and activities of governmental administrative bodies.

Administrative Procedure Law a generic term for the body of law regulating the authority and procedures of administrative agencies.

ADS-B Automatic dependent surveillance-broadcast—a collision avoidance system that relies on peer-to-peer transmission of GPS-determined position from one aircraft to another.

ADS-B in an ADS-B receiver.

ADS-B out an ADS-B transmitter.

Aeromodeling a hobby involving construction and flying of facsimiles of aircraft, usually conducted in a club context.

Aerovel Flexrotor long-range robotic aircraft with more than 40 hour endurance, manufactured by Aerovel.

Aircraft a vehicle capable of navigating through the air.

Airfoil an object shaped to generate lift as it moves through the air—airplane wings, helicopter rotor blades, and drone rotor blades.

Airframe that part of an aircraft excluding its propulsion system.

AirRobot a leading UK microdrone manufacturer.

Airworthiness Directive A mandatory order from the FAA, requiring modification of a specific aircraft system or component to ensure safe continued operation.

Al Qaeda a militant Islamic group responsible for terrorist attacks on civilians, including the September 11, 2001 attack on the World Trade Center in New York.

ALJ Administrative law judge—a quasi-judicial public employee assigned to an administrative agency, who holds adjudicator hearings and makes preliminary or final decisions based on the record.

ALPA Air Line Pilots Association—the principal US trade union representing airline pilots.

Amazon the leading online ecommerce vendor.

Amplifier a device for increasing the strength of radio signals.

ANAC Agência Nacional de Aviação Civil, in Portuguese.

APA Administrative Procedure Act—the US statute that prescribes procedures for administrative-agency rulemaking and adjudication and sets the standards for judicial review of agency decisions.

Approach control a part of the air traffic control system responsible for traffic separation and aircraft guidance in the vicinity of airports.

ATC air traffic control—an infrastructure responsible for separating aircraft, usually by means of radar surveillance and voice radio communications with pilots.

Autopilot a system that flies an aircraft without pilot intervention except to set basic parameters such as heading, altitude, speed, and geographic waypoints.

Autorotation an aerodynamic phenomenon that permits a helicopter to glide without power as its descent provides airflow to spin its rotors.

AUVSI Association for Unmanned Vehicle Systems International—the leading drone trade association.

AW169 a helicopter model produced by Agusta-Westland that employs tilt-wing technology to permit it to land and takeoff like a helicopter and to fly like an airplane in cruise configuration.

AWACS Airborne warning and control system—an aircraft equipped to provide air traffic control and radar-based threat warnings to combat aircraft.

Balance sheet a financial accounting statement that presents a snapshot of the financial condition of an enterprise at one point in time.

Barometric altimeter an instrument that shows altitude by measuring atmospheric pressure.

Biplane a fixed wing aircraft with two wings stacked on top of each other.

Bit depth the number of different shades that can be represented by a single photocell or pixel on a video sensor.

BLOS beyond line of sight, referring to a drone further away than its operator can see.

BNSF Burlington Northern Santa Fe railroad.

Bureau of Labor Statistics a unit within the United States Department of Labor responsible for gathering and analyzing labor market statistics.

B-VLOS beyond visual line of sight. See BLOS.

C-5 transport one of the largest cargo transports in the world, manufactured by Lockheed Martin.

CAA Civil Aeronautics Authority—the predecessor of FAA, responsible for aviation regulation, accident investigation, and developing air commerce in the early days of US aviation.

CAAC Civil Aviation Administration of China.

Callout a summons from a flight services customer for a contractor to fly an aircraft or drone for a particular mission.

Camber the curvature of an airfoil, front to back.

CAOSC Congested areas operations safety case template—UK microdrone regulatory regime for operations over congested areas.

Carefree mode a microdrone flight mode in which the drone responds to forward, back, left, and right commands with respect to the DROP's perspective, rather than from the perspective of the vehicle.

CASA Civil Aviation Safety Authority—Australian administrative agency responsible for aviation regulation.

Cell a component of a battery generating a voltage specific to a particular battery chemistry; a unit of a spreadsheet, representing the intersection of a row and a column.

CFI Certified Flight Instructor—a pilot licensed to give flight instruction.

Chattel mortgage a legal encumbrance or lien on an item of personal property intended to secure repayment of a loan.

Checkride a practical test of a pilot's flying ability.

Chord the front-to-back dimension of an airfoil.

Class B airspace controlled airspace requiring certain equipment, pilot qualifications, and an ATC clearance to enter; usually associated with the busiest airports.

Class D airspace controlled airspace requiring ATC clearances for takeoffs and landings and radio communications for transit; usually associated with the smallest airports that have control towers.

Clocking the process for synchronizing the operations of computer hardware components by means of a central clock which periodically pulses them.

Cloud server an Internet-connected computer intended to provide storage of files belonging to multiple users who subscribe to a file-storage service.

CNN Cable News Network—a major cable television and Internet news network.

COA Certificate of authority—an FAA authorization to operate an aircraft outside the strict limits of the FARs; Certificate of authority—a document accompanying section 333 exemptions specifying additional limitations.

Codec coder/decoder—an integrated circuit chip or a software application that encodes and decodes digital signals.

Columns the areas of a spreadsheet running vertically, comprising multiple cells.

Commerce clause Article I, § 8, clause 3 of the United States Constitution, giving Congress the power to regulate interstate and foreign commerce and interpreted to limit state authority to interfere with interstate or foreign commerce.

Common law a body of law developed by courts rather than being prescribed by legislatures.

Commutator a device for switching the polarity of a current applied to an electric motor armature.

Competitive assessment analysis of the competitive position of a business enterprise, relative to its competitors.

Conservatism principle a financial accounting principle that requires booking expenses as early as possible, and revenues as late as possible.

Console a device from which a drone can be operated.

Control forces the forces applied by control tabs such as elevators, ailerons, and rudders on aircraft to adjust pitch, roll, and yaw.

Control link one or more radio-frequency channels that send control signals from a DROP to a drone.

Conventional law the body of international law originating in treaties.

COWA Certificate of waiver and authority—US Public Safety Agency's exemption from common civil regulations.

CPSC Consumer product safety commission—a US government agency with authority to regulate the safety of consumer products.

CPU Central processing unit—a collection of digital logic circuits on a computer that perform manipulations on data.

Cropdusting application of chemicals or fertilizers to agricultural crops by air.

Crowdsourcing soliciting financial investments or contributions from the general public through the Internet.

CTAF Common traffic advisory frequency—designated VHF frequencies intended for aircraft to exchange information about their positions.

Customary international law a body of international law originating, not in treaties, but in the actual behavior of states.

Cyberwarfare attacks on computers, storage devices, digital networks, and automated systems intended to disrupt their operations.

DARs Designated airworthiness representatives—a private individual designated by the FAA to perform inspections necessary for airworthiness certification of an aircraft.

De novo on a clean slate; mainly referring to judicial review of facts or law without deferring to the findings of a subordinate body.

Debt a debtor's legal obligation to pay a creditor.

Decoding transforming an encoded digital signal into an uncompressed or analog version of the information represented by the signal.

Dejero a leading manufacturer of cellular bonding devices, which allow transmission of high-bandwidth audio or video signals through only a multiplicity of cell telephone circuits.

Depletion an accounting concept referring to the progressive exhaustion of limited physical resources such as timber, coal, or oil.

Depreciation an accounting concept that allocates the progressive exhaustion of economic value of an asset to time periods in which the asset is used.

Designated pilot examiner a private certified flight instructor authorized by the FAA to conduct pilot flight tests.

DGCA Indian aviation regulatory agency.

Dissymmetry of lift an aerodynamic phenomenon affecting helicopters in forward flight in which the larger velocity of the advancing rotor blade, compared with the smaller velocity of the retreating blade creates a roll moment on the helicopter.

DJI S1000 a popular octocopter.

DoC Declaration of conformity—legal certification of compliance with legal requirements by the manufacturer or vendor of a device; common in regulation of ground vehicles and consumer electronics.

Downlink a radio channel used to transfer information, particularly video, from an aircraft to the ground.

Drone operator certificate an airman license proposed by the FAA in its February 2015 NPRM.

DROP a drone operator.

DROPCON the console used by a DROP to control a drone.

DROSOP a drone system operator.

DROTOG a drone photographer.

DSLR a digital single lens reflex camera.

Dualist a theory in international law treating international as separate from national law.

EAA Experimental aircraft association—the principal US assocation for designers and builders of experimental aircraft.

EASA European Aviation Safety Agency.

ECJ European Court of Justice.

EMS Emergency medical service—commonly used to refer to helicopters that rescue accident victims in the field and transport them to hospitals.

ENAC Ecole Nationale de l'Aviation Civile—French national aviation academy; Ente Navionale Per l'Aviazione Civile—Italian civil aviation authority.

Encoding a process that transforms digital information into other forms, including compression.

ENG electronic news gathering—collecting news imagery through digital means, on the ground or in the air.

ENG microdrones a microdrone used to collect news electronically.

Equity an investor's ownership interest in a business enterprise.

ESC Electronic speed control—a hardware device on electronic propulsion systems that integrates electrical power with digital instructions from a controller to generate pulsed current to drive AC motors.

ETL Effective translational lift—an aerodynamic phenomenon that causes a helicopter rotor to generate more lift at the same RPM and pitch when forward flight causes it to move out of its downwash.

E-VLOS Extended visual line of sight—control of a drone beyond unaided visual perception by the DROP, aided by visual observers or electronic aids such as FPV, q.v.

Exciter the component of a radio transmitter responsible for generating a carrier signal at the proper frequency.

F-35B a high-performance fighter aircraft manufactured by Lockheed-Martin.

FAA inspector an FAA employee authorized to enforce compliance with the FARs, q.v.

FAA Federal Aviation Administration—a component of the US Department of Transportation broadly authorized by statute to establish and enforce regulations for air commerce.

Facebook a popular social media application on the Internet.

FADECs Fully automatic digital engine control systems

FARs Federal Aviation Regulations—the body of FAA regulations concerning airman and aircraft certification and flight rules.

FCC Federal Communications Commission—an independent US regulatory agency authorized by statute to regulate radio frequency devices and operations.

FCS Flight control system—the logic embedded in hardware and software mounted on a drone that converts DROP control signals into commands for motor torque.

FDA Food and Drug Administration—an independent US administrative agency authorized by statute to establish and enforce rules to ensure the safety of food and drugs.

Fixed costs an accounting concept referring to costs that are incurred periodically without regard to the level of operations.

Flickr an Internet site that allows subscribers to post and view photographs and videos.

Flight test an examination administered by the FAA or its designees to ensure pilot or DROP proficiency in flying.

Flip-flop an electronic circuit capable of two states and transitioning between the two; a fundamental component of digital logic.

Fly-by-wire control systems a means for controlling an aircraft that employs electrical connections between pilot controls and control surfaces instead of relying on purely mechanical linkages.

Formal rulemaking a legal process used infrequently by administrative agencies to make rules based on evidence formally introduced at live hearings; contrasted with *informal rulemaking*, q.v.

FPV First person view—A means for providing DROPs with video images captured by a drone in real time.

Frequency hopping an RF modulation scheme that splits the signal into very small pieces and transmits each on a separate frequency.

Fully depreciated an accounting concept that refers to an asset whose entire value has been charged as an expense during preceding periods of time.

GAAP Generally accepted accounting principles—rules applied by certified public accountants to assure the integrity of financial accounts.

GAO Government Accountability Office (formerly Government Accounting Office)—an independent agency within the legislative branch charged with conducting investigations of government operations.

GCS group communication system—computer software.

Generalists professionals, especially lawyers, who do not specialize narrowly, but instead apply broad knowledge and experience.

Geo-fencing a software-enabled technique that excludes drones from certain areas.

GoPro a popular portable camera intended to capture action video.

GPS Geographic positioning system—a system enabling ground-based vehicles or aircraft to determine their position in space by triangulating signals received from specialized satellites.

Ground resonance a dynamic physical phenomenon in which helicopters with more than two main rotor blades experience a dynamically unstable bouncing from one side to another, eventually destroying the aircraft.

Gs multiples of gravitational forces resulting from the acceleration of aircraft in flight.

Gyrocopter an air vehicle that relies on forward flight rather than engine torque to spin its rotor and generate lift.

HAI Helicopter Association International—a prominent helicopter trade association.

Harrier a VTOL combat jet manufactured by Hawker Siddeley and Boeing; in service with the Royal Navy and the United States Marine Corps.

Heavy lift a common term used in the utility helicopter industry for sling load operations.

HEMS Helicopter emergency medical services.

Hexacopter a drone with five motors and rotors.

Hi-rail vehicle a truck configured with flanged wheels that can be raised and lowered to allow it to ride on rails.

Hot fuel the practice of refueling an aircraft—particularly a helicopter—while its engines and rotors are turning.

Hunters US Army Short Range UAV system for division and corps commanders.

ICAO International Civil Aviation Organization—the principal treaty-based international aviation body, part of the UN System.

IMC Instrument meteorological conditions—weather conditions too poor for visual flight.

IMU Inertial measurement unit—a system comprising multiple accelerometers and logic that permits calculating position based on acceleration along its three axes.

Income statement a financial accounting report that discloses income, expenses, and profit for a specific time period.

Informal rulemaking a legal process typically used by administrative agencies to make rules based on agency expertise, research and on written comments submitted by the public; contrasted with *formal rulemaking*, q.v.

Inspire a DJI microdrone.

IR infrared.

IRIS+ a 3DRobotics microdrone.

ISIS a militant Islamic group.

IT information technology.

JARUS Joint Authorities for Rulemaking on Unmanned Systems—a non-governmental international coordinating body of national aviation regulators.

Jet engines heat engines that develop thrust by accelerating airflow through the engine rather than by generating mechanical torque.

JOBS Act a US statute that allows business enterprises to sell securities over the Internet through a process known as crowdsourcing.

Judicial review action by a court to determine if an administrative agency action is authorized and adopted through the requisite procedure.

Kickstarter an Internet site that allows projects of any size to solicit and receive donations.

Kitty Hawk a place in North Carolina where the Wright Brothers conducted most of their flight tests.

K-Max an optionally manned helicopter that flew unmanned cargo missions in Afghanistan for the United States Marine Corps.

Kosovo Conflict an armed conflict in the Balkans involving a NATO bombing campaign in 1998 against Serbian forces, the establishment of a UN-run political trusteeship, followed by independence and statehood in 2008.

Leading-edge slats lift augmentation devices that drop down from the leading edge of an airfoil to increase camber and introduce higher pressure air to defer boundary layer separation.

Lien a legal claim to a piece of property that becomes possessory if the debt which it secures is not paid.

LiveU a leading manufacturer of cellular bonding devices, which allow transmission of high-bandwidth audio or video signals through only a multiplicity of cell telephone circuits.

LLC Limited liability company—a form of business organization that insulates its owners from liability; similar to a corporation, but with simpler structure and requirements.

Longline a helicopter operation that involves suspending a human or cargo load from a line beneath the helicopter; a part of the public switched telephone network handling long distance traffic.

LOS Line of Sight—visual contact between DROP and drone; required by most microdrone regulations.

Lumpiness a quality related to the minimum practical size of economic assets; a resource whose minimum size is greater is said to be lumpier than one with smaller minimum size.

Macrodrone an unmanned aircraft system weighing more than 55 pounds.

Magnetometer a device that uses the Earth's magnetic field to determine direction.

Market research analysis of the structure of a market and of the relative advantages of competitors of the entity conducting the analysis.

Market structure the number, size, and relationships among buyers and sellers of a particular good or service.

Matching principle a financial accounting rule that requires revenue to be recognized in the same period in which expenses pertaining to it are recognized.

MB/s megabyte per second—a measure of the speed of digital communications streams.

MBA Master of Business Administration—a professional degree in management, usually earned after a bachelor's degree.

Merits review a decision by a court based on legal and factual arguments pertinent to the main claims; distinguished from jurisdiction or procedural review.

MHz megahertz—a measure of frequency of a radio signal in millions of cycles per second.

Microcomputer revolution the replacement of mainframe and minicomputers by smaller pieces of integrated, general purpose computer power.

Microdrone an unmanned aircraft system weighing less than 55 pounds.

Microeconomics the study of economic behavior at the industry or individual seller or buyer level.

Microwave a radio signal with a frequency between 1 GHz and 300 GHz.

Microwave transmitter a device for sending microwave radio signals through space.

Mididrone an intermediate size unmanned aircraft system.

MIT Massachusetts Institute of Technology—a prestigious technological university located in Cambridge, MA.

MLIT Ministry of land, transport and tourism of the Japanese government.

Modulation transformation of an electronic signal by combining a baseband signal containing formation with a carrier signal capable of propagation through the intended transmission medium.

Monist a theory in international law that incorporates international norms into national law without the necessity of explicit legislation.

Monopoly a market in which only one seller exists.

Monopsony a market in which only one buyer exists.

mp3 a compression standard for audio information.

mp4 a compression standard for video information.

MPEG-2 a compression standard for audio and video information.

MPEG-4 a compression standard for video information; same as mp4.

MSL above mean sea level; a reference point for determining aircraft altitude.

Multicopter a drone with more than one rotor.

NAMID Neighborhood access for microdrones—a theoretical airspace management system for allowing microdrones to share low-level airspace.

NAS National airspace system—an integrated system of navigation, air traffic control infrastructure, aircraft, airman certification, and operating rules in the United States.

NASA National Aeronautics and Space Administration—an agency of the US government charged with conducting research and development into aeronautics and astronautics and with the US civilian space flight program.

NATO North Atlantic Treaty Organization—the principal military defense treaty linking the United States with European states.

NCS Navigation control system—the subsystem on a drone responsible for receiving DROP commands and sensor data and computing the necessary currents to motors for flight control to fly the path directed by the DROP.

Newtonian mechanics a branch of physics concerning the behavior of physical objects subjected to force.

NextGen a major transformation of the US airspace management system that relies predominantly on GPS navigation rather than ground-based radar and navigational radio; mandatory for all US aircraft by 2020.

NHTSA National Highway Traffic Safety Administration—an agency of the US government charged with safety regulation of truck and automobiles.

NLRA National Labor Relations Act—a federal statute granting private-sector employees outside the railroad and airline industries the right to engage in collective bargaining.

Northrop-Grumman a major US aerospace manufacturer; builder of Global Hawk, Fire Scout, and Bat Unmanned Aircraft System, a fixed wing internal-combustion-powered macrodrone.

NPRM Notice of proposed rulemaking—the initial solicitation of public comment on administrative agency proposed rules; more specifically, the notice of proposed microdrone rules published by the FAA in Feb. 2015.

NSA National Security Agency—an agency of the US Defense Department responsible for communications security and communications intelligence.

NTSB National Transportation Safety Board—an agency of the US government authorized to investigate transportation accidents and to adjudicate civil penalties imposed on airmen by the FAA.

Octocopter a drone with eight motors and rotors.

Offering memorandum a document published by a business enterprise setting out the terms on which it solicits capital investment.

OIRA Office of Information and Regulatory Affairs—an agency within the US Office of Management and Budget (OMB), q.v., charged with reviewing administrative agency regulations and regulatory proposals.

Oligopoly an industry structure with only a few sellers, having the power to influence prices.

OMB Office of Management and Budget—an agency within the Executive Office of the President responsible for ensuring that administrative agencies implement policies determined by the President.

OpenCV open source computer vision—originally intended to develop CPU code optimization.

Oscillator an electronic circuit capable of generating radio-frequency (RF) signals; central feature of any radio transmitter.

OSHA Occupational Safety and Health Administration—a US administrative agency charged with regulating workplace health and safety.

Parabolic antenna a dish-shaped directional antenna; common for microwave communications.

Parliament an elected legislative body, in which the majority party chooses a government, which performs executive functions.

PCB power control board—an integrated circuit board that accepts inputs from the navigation control system to meter electrical power to multicopter motors.

Perfect competition a market structure in which no seller or buyer is big enough to influence price.

Phantom a model of microdrone sold by DJI—the most popular microdrone in the United States.

PHOTOCON photographer's console—a ground-based unit that displays drone-captured video and allows the operator to control gimbal and camera settings.

Pixelation a type of video distortion in which the parts of an image captured by individual photosensors are set apart from each other rather than being blended into a holistic photographic frame.

PNG portable network graphics—a common standard for lossless image compression.

Polymer a synthetic inorganic material commonly used for structural components and tubing.

Predator a long-range US military macrodrone used for surveillance and warfare.

Preemption the displacement of state or local law by superior federal law.

Price elasticity of demand the tendency of demand for a product to vary with price.

Price elasticity of supply the tendency of supply of a product to vary with price.

Product a good or service delivered or performed by a supplier in a market; a good or service demanded by buyers in a market.

Production and operations management that branch of business administration concerned with organizing and supervising human and physical assets to meet business goals.

Production function an equation showing how factors of production such as land, capital, labor, and technology are combined to produce one or more products or services.

Production management see Production and operations management.

Profit and loss statement an accounting statement that summarizes income and expenses for a particular time period, on an accrual basis.

Pro-forma a shorthand expression for pro-formal financial statements—projected income statements, balance sheets, and sources-and-uses of funds; a common component of a business plan.

Propagation the means by which an electromagnet signal moves from one place to another.

Prospectus a statement of plans and risk factors for a new enterprise; part of the disclosure to potential investors required by securities law.

PTO permit to operate—an FAA regulatory concept that would allow only those drone operators with a permit to fly commercially.

Public aircraft An aircraft owned or operated by a unit of federal, state, or local government, except from certain FARs.

Public law a statute adopted by the United States Congress; that aspect of law pertaining to relations between governments or between governments and their people, distinguished from *private law*.

Public offering the offer by a business enterprise to see securities to the general public.

Puma-AE a model of fixed-wing microdrone developed for the US armed services, civilian models of which are now available.

Quadcopter a drone with four motors and rotors.

Radar Radio detection and ranging—a system for using microwave radio pulses for determining the position and range of objects that reflect the signals.

RAF Royal Air Force—the air force of the United Kingdom.

RAW codec a variety of standardized image format used by digital camera containing the unprocessed data from the sensor.

RC radio-controlled—a term commonly used to refer to model aircraft.

Reaper a large macrodrone flown by the US armed forces and intelligence agencies.

Reciprocating engine an engine in which combustion causes back-and-forth motion of one or more pistons and in which lateral motion is converted to rotary motion by means of crankshafts.

RED camera a sophisticated, small form-factor professional camera with a single sensor.

Registration statement a set of documents including a prospectus, which a company must file with the US Securities and Exchange Commission before it proceeds with a public offering.

Return to home common function of unmanned vehicles in which the vehicle is triggered and returns to its point of departure autonomously.

RF radio frequency—electromagnetic wave frequencies that lie in the range extending from around 3 kHz to 300 Ghz.

Risk factors that part of a business plan that analyzes specific risks jeopardizing success; common aviation term used to discuss a flight's level of risk.

RLA Railway Labor Act, an American law governing labor relations in the railway and airline industries.

Rotary engine An early type of internal combustion engine usually with an odd number of cylinders in radial configuration; an internal combustion engine, chiefly the Wankel design, which uses rotary motion rather than reciprocating pistons to convert the heat of combustion into torque.

RPA remotely piloted aircraft.

RPAS remotely piloted aircraft systems.

RPVs remotely piloted vehicles.

RTDNA Radio and Television Digital News Association—major organization for television and news directors.

Rudder primary control surface used to steer aircraft and counter adverse yaw and p-factor.

SARP technical specifications adopted by the Council of ICAO.

Scan Eagle small long endurance UAV built by Boeing Insitu, primarily used by militaries.

SDK software development kit offered by some drone vendors.

SEC Securities and Exchange Commission—an independent US administrative agency responsible for regulation of investment securities markets and enforcing of securities laws.

Section 333 exemption FAA authorization for drone operations by a specific operator under the authority of section 333 of the 2012 FAA Modernization and Reform Act.

Security threat assessment TSA review of individuals employed in aviation to mitigate security threats.

SFOC Special flight operating certificate—a Canadian authorization for drone operations by specific operators.

Shannon's law the relationship between bandwidth and the maximum rate at which error-free digits can be transmitted over a channel in the presence of noise.

Signal propagation behavior of radio waves when they are transmitted or propagated from one point on Earth to another in various parts of the atmosphere.

Solar cell a device converting solar radiation into electricity.

Sources and uses of funds an accounting statement that summarizes actual cash flowing in and cash flowing out of an enterprise, as contrasted with in income statement, which reports accrued income and expenses.

Spoiler a flap on an aircraft wing that can be extended to create drag and reduce speed.

Spread-spectrum modulation a means for dividing an RF signal and transmitting it on a series of closely spaced frequencies; used in the Wi-Fi 802.11 standard.

Stall an aerodynamic phenomenon in which the airflow over the top of an airfoil separates when the angle of attack compared to the relative wind exceeds its critical value.

Statement of cashflow same as *sources and uses of funds*.

SUA small unmanned aircraft—FAA's proposed "micro" category for a small vehicle weighing 4.4 pounds or less.

sUAS small unmanned aerial system.

SUSA small unmanned surveillance aircraft—with reference to the Civil Aviation Authority, an aircraft of 20kg or less.

Subscription agreement an agreement between an enterprise and an investor regarding investment on specific terms.

Supercomputer a computer with a high level of computational capacity compared to a general purpose unit.

Supremacy clause a provision in Article Six, Clause 2 of the US Constitution that establishes the US Constitution federal statutes and treats it as the "supreme law of the land."

Swash plate device used to translate an input to a rotating unit; a part of most helicopter rotor hubs.

SYSOP systems operator.

TBO time between overhaul—a measure of an aircraft's component overhaul.

TCAS traffic collision avoidance system of traffic alert and collision avoidance system used in aerial navigation to help reduce collisions between aircraft.

Tethered drone a microdrone attached to the ground by a line that secures it and usually carries electrical power to it.

TFO tactical flight officer—term commonly used in airborne law enforcement support.

Three-axis stabilized gimbal a mount using gyros for stabilization around all three axes.

TIS traffic information system for aerial navigation; a ground-based system for retransmitting ADS-B out signals and integrating them with ground-based radar returns.

TOC transmission operations center—the name of the individual who runs transmission operations control at television news stations.

TRACON service provided by ground-based controllers who direct aircraft on the ground and through controlled airspace.

TRACON terminal radar control in the US National Airspace System—ATC facility that provides radar-based guidance to aircraft arriving at and departing from busy tower-controlled airports.

Transport Canada Canadian government department which is responsible for developing regulations, policies and services of transportation in Canada.

TSA Transportation Security Agency—a US administrative agency charged with protecting aviation security.

Turboprop aircraft turbine engine that drives a propeller.

Twitter online social networking service through which users send short messages.

Type certificate issued to signify the airworthiness of an aircraft manufacturing design.

Type rating FAA certification of a particular model of aircraft and of an airman qualified to fly that aircraft.

U-2 a US reconnaissance aircraft.

UAS America Fund a major US drone public affairs organization.

UAS unmanned aerial system.

UAV Coalition a public affairs-oriented group comprising major drone interests.

UAV unmanned aerial vehicle.

UPS United Parcel Service—an international shipping company.

US Census Bureau a component of the United States Department of Commerce responsible for collecting, analyzing, and reporting on data about the population and economic units.

US Department of Labor a cabinet-level department of the US executive branch responsible for labor statistics and a variety of regulatory programs focused on workplaces.

US Small Business Administration an independent agency of the US government responsible for promoting and assisting small business.

USAF United States Air Force—aerial warfare service branch of the US armed forces.

UV ultraviolet light in the electromagnetic radiation.

V-22 Osprey VTOL aircraft, multi-mission tilt rotor.

Variable costs costs that change in proportion to the good or service that a business produces.

Vimeo a video sharing website.

VLF very low frequency—portion at the bottom of the radio spectrum.

VLOS visual line of sight—controlling a drone by means of visual contact by the DROP.

VMC visual metrological conditions—an aviation flight category in which visual flight rules flight is permitted.

VOR VHF omnidirectional range, a radio navigation aid used in aviation.

VORTAC navigational aid for aircraft pilots utilizing a VHF omnidirectional range (VOR) beacon and a tactical air navigation system (TACAN), providing azimuth information to pilots.

VTOL (vertical takeoff and landing) an aircraft that can hover, take off, and land vertically.

WAAS wide area augmentation system—an air navigation aid developed by the FAA that augments GPS for improved accuracy, integrity and availability.

Wankel a type of internal combustion engine using eccentric rotary design to convert pressure into rotating motion, commonly called a rotary engine.

Watt derived unit of power in the International System of Units; used to express the rate of energy conversion or transfer with respect to time

WATT a commercial tethered drone.

Weight and balance a term commonly used in aviation referring to where the center of gravity of an aircraft is located. Limits are established by the manufacturer; operating the aircraft within those limits ensures control limits under normal operation won't be exceeded.

Wi-Fi local area wireless computer networking utilizing the 2.4 GHz UHF and 5 GHz SHF ISM radio bands.

Writ of certiorari an order from a court, especially the Supreme Court of the United States compelling a lower body to send up its official record of a case for review by the higher court.

Yagi antenna directional antenna consisting of multiple parallel dipole elements in a line usually made of metal rods.

Yamaha RMAX Japanese-built unmanned helicopter developed by the Yamaha Motor Company in the 1990s; two-bladed, gasoline-powered, remote-controlled by a line of sight user. Popular for agricultural work.

YouTube a video sharing website.

Index